MARTENS AND FISHERS *(MARTES)* IN HUMAN-ALTERED ENVIRONMENTS:
An International Perspective

Springer

An American marten pursuing its most common prey, the red-backed vole. Drawing by Mark McCollough.

MARTENS AND FISHERS *(MARTES)* IN HUMAN-ALTERED ENVIRONMENTS:
An International Perspective

edited by

Daniel J. Harrison
Department of Wildlife Ecology
The University of Maine
Orono, Maine, USA

Angela K. Fuller
Department of Wildlife Ecology
The University of Maine
Orono, Maine, USA

Gilbert Proulx
Alpha Wildlife Research & Management Ltd.
Sherwood Park, Alberta, Canada

Cover Illustrations:
The endangered Newfoundland marten (*Martes Americana atrata*), a genetically unique subspecies of American martens. Only 300-500 individuals survive on the island of Newfoundland, Canada. Photo rights of Barrett and MacKay Photography, Inc.

Extensive foresting harvesting, road building and habitat fragmentation are potential threats to habitat for martens and fishers in many parts of their geographic ranges. Photo by Daniel Harrison.

Library of Congress Cataloging-in-Publication Data

A C.I.P. Catalogue record for this book is available
from the Library of Congress.

ISBN 0-387-22580-3 e-ISBN 0-387-22915 Printed on acid-free paper.

Printed in the United States of America.

9 8 7 6 5 4 3 2 1 SPIN 11054672

springeronline.com

This book is dedicated

to

John "Jack" McPhee

1937–2003

Long-Time Telemetry Pilot, Naturalist, and Friend

Contents

List of Figures

List of Tables

CONTRIBUTORS

Keith B. Aubry
USDA Forest Service
Pacific Northwest Research Station
Olympia, Washington 98512-9193, USA
Phone: 360-753-7685; E-mail: kaubry@fs.fed.us

Johnny D.S. Birks
The Vincent Wildlife Trust
3&4 Bronsil Courtyard
Eastnor, Ledbury, Herefordshire
HR8 1EP, UK
Phone: 44 01531 636441; E-mail: johnnybirks@vwt.org.uk

Tony C. Braithwaite
Nant-y-Llyn, Ffarmers, Llanwrda
Carmarthenshire SA19 8PX, UK

Laûrier Breton
Société de la faune et des parcs du Québec
Direction du développement de la faune
675 boul. René-Lévesque Est
11e étage, boîte 92
Québec G1R 5V7, Canada

Rachael C. Brookes
Institute of Genetics, Q.M.C.
University of Nottingham
Nottingham NG7 2UH, UK

Steven W. Buskirk
Department of Zoology and Physiology
Box 3166
University of Wyoming
Laramie, WY 82071, USA

Michel Cantin
Société de la faune et des parcs du Québec
Direction de l'aménagement de la faune de la Capitale Nationale
9530 de la faune, Charlesbourg
Québec G1G 5H9, Canada

Fraser B. Corbould
Peace/Williston Fish and Wildlife Compensation Program
1011 Fourth Avenue, 3rd Floor
Prince George, British Columbia
V2L 3H9, Canada

Angus Davison
Institute of Genetics
Q.M.C., University of Nottingham
Nottingham NG7 2UH, UK

Steven H. Ferguson
Fisheries and Oceans Canada
501 University Crescent
Winnipeg, Manitoba R3T 2N6, Canada
Phone: 204-983-5057; E-mail: fergusonsh@dfo-mpo.gc.ca

Names in **Bold** = senior authors

Clément Fortin
1320 Jacques-Cartier Sud
Tewkesbury, Québec
G0A 4P0 Canada
Phone: 418-848-3627; E-mail:
crfortin@mediom.qc.ca

Herbert C. Frost
Great Basin Cooperative Ecosystem
Studies Unit
University of Nevada, Reno
1000 Valley Road/186
Reno, Nevada 89512, USA
Phone: 775-784-4616; E-mail:
bert_frost@nps.gov

Alton S. Harestad
Department of Biological Sciences,
Simon Fraser University,
Burnaby, British Columbia
V5A 1S6, Canada

Daniel J. Harrison
Department of Wildlife Ecology
5755 Nutting Hall, Rm. 210
University of Maine
Orono, Maine 04469-5755, USA

Christopher L. Hoving
Department of Wildlife Ecology and
Maine Cooperative Fish and
Wildlife Research Unit
5755 Nutting Hall, Rm. 210
University of Maine
Orono, Maine, 04469-5755, USA
(Present address: Michigan DNR,
Wildlife Division
621 N. 10th St.
Plainwell, MI 49080, USA)

Maria João Santos
Universidade de Lisboa
Centro de de Biologia Ambiental
Faculdade de Ciências
Campo Grande
Bloco C_2 - 3º Piso, 1749-016
Lisboa, Portugal

William B. Krohn
Maine Cooperative Fish and
Wildlife Research Unit
USGS Biological Resources
Division
5755 Nutting Hall, Room 210
University of Maine
Orono, Maine 04469-5755, USA
Phone: 207-581-2870; E-mail:
wkrohn@umenfa.maine.edu

Serge Larivière
Delta Waterfowl Foundation
R. R. #1, Box 1
Portage La Prairie
Manitoba R1N 3A1, Canada

Sofia Lourenço
Rua Gonçalves Zarco nº 5 - 12º Esq.
2685-211 Portela – Loures, Portugal

Lem Mayo
Department of Environment and
Conservation
Parks and Natural Areas Division
33 Reid's Lane
Deer Lake, Newfoundland
A8A 2A3, Canada

Names in **Bold** = senior authors

John E. Messenger
The Vincent Wildlife Trust
3&4 Bronsil Courtyard
Eastnor, Ledbury, Herefordshire
HR8 1EP, UK

Vladimir Monakhov
Institute of Plant and Animal Ecology
Ekaterinburg "8 Marta" Str
202 620144 Russian Federation

Robert Patenaude
Jardin Zoologique du Québec
Ministère de l'Environnement
9530 rue de la Faune
Charlesbourg, Québec G1G 5H9, Canada

David C. Payer
U S Fish and Wildlife Service
Arctic National Wildlife Refuge
101 12th Avenue, Room 236, Box 20
Fairbanks, Alaska 99701, USA
Phone: 907-455-1830; E-mail: david_payer@fws.gov

Iris Pereira
Universidade de Lisboa
Centro de de Biologia Ambiental
Faculdade de Ciências
Campo Grande
Bloco C_2 - 3° Piso, 1749-016
Lisboa, Portugal

David M. Phillips
Department of Wildlife Ecology
5755 Nutting Hall, Rm. 210
University of Maine
Orono, Maine 04469-5755, USA
(Present address: Holderness School Box 1879, Plymouth, New Hampshire 03264, USA).

Bruno Pinto
Rua Paulo Falcão nº99
2775 Parede, Portugal

François Potvin
Société de la faune et des parcs du Québec
675 boul. René-Lévesque est (11e), Boite 92
Québec, Québec G1R 5V7, Canada
Phone: 418 521-3955 ext. 4491; E-mail: francos.potvin@fapaq.gouv.qc.ca

Roger Powell
Department of Zoology and Forestry
North Carolina State University
Raleigh, North Carolina 27695-7617, USA
Phone: 919-315-4561; E-mail: newf@ncsu.edu

Catherine M. Raley
Pacific Northwest Research Station
U.S. Forest Service
3625 93rd Ave. SW
Olympia, Washington 98512, USA

Names in **Bold** = senior authors

Gilbert Proulx
Alpha Wildlife Research & Management Ltd.
229 Lilac Terrace
Sherwood Park, Alberta T8H 1W3, Canada
Phone: 780-464-5228 E-mail: gproulx@alphawildlife.ca

Midori Saeki
6-22, 2-chrome Minamikasugaoka
Ibaraki-city, Osaka
567-0046 Japan

Margarida Santos-Reis
Universidade de Lisboa
Centro de Biologia Ambiental
Faculdade de Ciências
Campo Grande, Bloco C_2-3° Piso
1749-016 Lisboa, Portugal
Phone: 00 351 21 7500000; E-mail: msr@fc.ul.pt

Chris Strachan
The Vincent Wildlife Trust
3&4 Bronsil Courtyard
Eastnor, Ledbury, Herefordshire
HR8 1EP, UK

João Tiago Marques
Rua Central da Quinta da Asseca
nº 14, 2950-426 Palmela
Portugal

Richard D. Weir
Artemis Wildlife Consultants
4515 Hullcar Road
Armstrong, British Columbia
V0E 1B4, Canada
Phone: 250-546-0531; E-mail: rweir@artemiswildlife.com

Samantha M. Wisely
Molecular Genetics Laboratory
Smithsonian Institution
Washington, DC 20008, USA

Andrzej Zalewski
Mammal Research Institute
Polish Academy of Sciences
17-230 Białowieża, Poland
E-mail:
zalewski@bison.zbs.bialowieza.pl

William Zielinski
USDA Forest Service
Pacific Southwest Research Station
Redwood Science Laboratory
1700 Bayview Drive
Arcata, California 95521, USA

Names in **Bold** = senior authors

PREFACE

The genus *Martes* represents 7 species in the family Mustelidae, including 6 species of martens and the fisher (*M. pennanti*), who are phylogenetically and ecologically distinct from other weasels, minks, otters, and badgers. Other members of the genus include the pine marten (*M. martes*) and the stone marten (*M. foina*) of Europe and Asia, the sable (*M. zibellina*) of northern Asia, the Korean peninsula, and some islands of the Japanese archipelago, the indigenous Japanese marten (*M. melampus*) of Japan and the Korean peninsula, the American marten (*M. americana*) of the northern United States and Canada, and the little studied yellow-throated marten (*M. flavigula*) of Asia. As the taxonomic relationship between the yellow-throated marten of southern and southeastern Asia and the Nilgiri marten (*M. gwatkinsi*) of the Indian subcontinent remains questionable, we have taken a conservative taxonomic approach and consider them here as the same species.

All *Martes* have been documented to use forested habitats and 6 species (excluding the stone marten) are generally considered to require complex mid- to late-successional forests throughout much of their geographic ranges. All species in the genus require complex horizontal and vertical structure to provide escape cover, protection from predators, habitat for their prey, access to food resources, and protection from the elements. Martens and the fisher have high metabolic rates, have large spatial requirements, have high surface area to volume ratios for animals that often inhabit high latitudes, and often require among the largest home range areas per unit body weight of any group of mammals. Resulting from these unique life history characteristics, this genus is particularly sensitive to human influences on their habitats, including habitat loss, stand-scale simplification of forest structure via some forms of logging, and landscape-scale effects of habitat fragmentation. Given their strong associations with structural complexity in forests, martens and the fisher are often considered as useful barometers of forest health and have been used as ecological indicators, flagship, and umbrella species in different parts of the world, particularly in the United States, Canada, and Scandinavia. Thus, efforts to successfully conserve and manage martens and fishers are associated with the ecological fates of other forest dependent species and can greatly influence ecosystem integrity within forests that are increasingly shared among wildlife and humans.

Human populations continue to increase exponentially at the global scale and less than 7% of the world's land area is protected. Further, many protected areas within the range of the world's *Martes* experience managed and unmanaged forms of direct exploitation of these species and their habitats. Martens and the fisher often live in landscapes where harvesting of wood and extraction of minerals and energy resources provide the most significant economic returns. Further, these species live in complex and ever-changing ecological communities where their interspecific interactions, food resources, and habitat structure are affected by global processes such as international wood fiber markets and climate change. If viable populations of these species are to exist outside of the scarce inviolate parks and reserves scattered throughout the globe, then humans are challenged to understand the functional effects of their activities at the level of the individual and population and at multiple spatial scales ranging from the microhabitat, patch, landscape, and the metapopulation.

Historically, martens and the fisher (with the possible exclusion of the stone marten who has adapted to take advantage of the unnatural structural complexity, cover, and food resources that are enhanced in some human-dominated landscapes) have been associated with forested areas with low human populations. This has contributed to a general perception that these species are intolerant to humans and cannot adapt to human alterations of their habitat. Indeed, recent research has indicated that these species, which are often considered valuable furbearers, are vulnerable to over-exploitation and changes in population structure associated with overharvesting, increased access for humans via forest roads and trails, and indiscriminate killing. The American marten and the fisher were extirpated throughout many remote areas of North America during the late 1800s and early 1900s as a result of unregulated trapping and shooting for their furs, despite that other habitat conditions remained favorable. These species have been subsequently restored to many areas of their former range despite increasing human populations and access; many of these populations again support sustainable, regulated harvests in habitats significantly altered by humans. Thus, one of our primary challenges is to understand the resiliency and limits of *Martes* populations to sustain human-caused forms of mortality.

The historical (pre-1985) literature also focused on the stand-scale associations of martens and the fisher with mature and over-mature forests and of the relationship of these species with pristine forests. Recent studies in both North America and Europe have indicated that the relationships of Martes with humans may be more complex than previously understood. Martens and the fisher have been documented to use a range of forest types and seral stages throughout their geographic ranges; however, unifying principles supporting the requirement for complex horizontal and vertical structure are emerging. Recent studies have reported *Martes* successfully co-existing in some areas with human activities such as logging; these examples provide promising evidence that our increasing knowledge may be used (in some places) to mitigate human influences on habitat, and to provide opportunities for these species to co-exist in some landscapes altered by humans. New knowledge also suggests that broad-scale processes such as fragmentation of habitat across landscapes increasingly threaten the world's *Martes*, and that processes such as climate change may threaten the integrity of the natural communities where these species interact with a multitude of the world's flora and fauna. Again, our challenge is to understand the conditions where humans and martens are compatible and incompatible, and to promote land use practices that allow *Martes* to be representatively distributed and viable.

The 14 chapters of this book address I) the status, distribution, and life history of martens (7 species) throughout the world; II) the habitat and interspecific relationships relationships (3 species) at multiple spatial scales in North America and Europe; and III) new management and research approaches for evaluating and studying martens, the fisher, and their habitats. All of these papers provide tools and insights for better understanding *Martes* in landscapes that are significantly altered by humans. Monumental gaps continue to exist that hinder our understanding of the relationships of humans with some species, most notably the Japanese marten and yellow-throated marten. In the past 2 decades we have made great strides in our fundamental understanding of how animals with these unique life history traits perceive and utilize habitats, respond to habitat change, and how their populations function and perform under different forms of human management and mismanagement. Hopefully this knowledge will enhance our basic understanding of all species of *Martes* and will help us to achieve the goal of conserving viable populations and representative distributions of the world's *Martes*, their habitats, and associated ecological communities in our new millennium.

Daniel J. Harrison
Angela K. Fuller
Gilbert Proulx

ACKNOWLEDGMENTS

We thank the 44 reviewers who shared their scientific expertise and knowledge of *Martes* while reviewing one or more chapters of this book: W. A. Adair, Utah State University; S. M. Arthur, Alaska Department of Fish and Game; K. B. Aubry, United States Department of Agriculture - Forest Service; J. D. S. Birks, The Vincent Wildlife Trust; J. A. Bissonette, United States Geological Survey, Utah Cooperative Fish and Wildlife Research Unit; L. Boitani, UniversitB di Roma "La Sapienza"; J. Bowman, Carleton University; H. N. Bryant, Royal Saskatchewan Museum; M. Brown, New York State Department of Environmental Conservation; S. W. Buskirk, University of Wyoming; T. G. Chapin, Ecology and Environment, Inc.; H. C. Frost, University of Nevada; S. H. Ferguson, Lakehead University; C. Fortin, Société de la faune et des parcs du Québec (retired); J. M. Fryxell, University of Guelph; T. K. Fuller, University of Massachusetts; J. W. Gosse, Terra Nova National Park; H. I. Griffiths, University of Hull; C. D. Hargis, United States Department of Agriculture - Forest Service; H. J. Harlow, University of Wyoming; A. S. Harestad, Simon Fraser University; M. J. Henault, Société de la faune et des parcs du Québec; W. J. Jakubas, Maine Department of Inland Fisheries and Wildlife; D. D. Katnik, Maine Department of Inland Fisheries and Wildlife; W. B. Krohn, United States Geological Survey - Maine Cooperative Fish and Wildlife Research Unit; T. E. Kucera, University of California, Berkeley; J. Messenger, The Vincent Wildlife Trust; E. C. O'Doherty, United States Department of Agriculture-Forest Service; T. F. Paragi, Alaska Department of Fish and Game; D. C. Payer, Arctic National Wildlife Refuge; K. G. Poole, Timberland Consultants; F. Potvin, Société de la faune et des parcs du Québec; R. A. Powell, North Carolina State University; M. G. Raphael, United States Department of Agriculture-Forest Service; J. M. Rhymer, University of Maine; J. F. Robitaille, Laurentian University; M. Santos-Reis, Lisbon University; T. L. Serfass, Frostburg State University; J. D. Steventon, Ministry of Forests, British Columbia; I. D. Thompson, Canadian Forest Service; R. L. Truex, United States Department of Agriculture Forest Service; R. D. Weir, Artemis Wildlife Consultants; E. C. York, Santa Monica National Recreation Area; and W. J. Zielinski, United States Department of Agriculture-Forest Service.

We also extend our thanks to Theresa Libby who contributed her word processing skills while spending countless hours incorporating revisions from the editors. Mark McCollough, wildlife biologist and artist, graciously shared his drawing of an American marten pursuing a red-backed vole.

The Maine Agricultural and Forest Experiment Station, the Department of Wildlife Ecology at The University of Maine, Natural Resources Canada–Canadian Forest Service, Alpha Wildlife Research & Management Ltd., Corner Brook Pulp and Paper Ltd., and the Newfoundland-Labrador Inland Fish and Wildlife Division provided funding and logistical assistance in support of this collaborative effort.

Barbara Harrity, Maine Agricultural and Forest Experiment Station, The University of Maine, served as layout, design, and copy editor. Her expertise and efficiency greatly assisted the authors during the publication and printing stages of this project.

Part I

Status, Distribution, and Life History

Chapter 1

IS MUSTELID LIFE HISTORY DIFFERENT?

Steven Ferguson and Serge Larivière

Abstract: The relationship between life-history variation and population processes may form a foundation for developing conservation strategies. Researchers have argued that mustelids require special conservation practices due to their unique habitat requirements and K-selected life-history strategy. We used the comparative method to test whether life-history and behavioral traits of mustelids differed from those of other carnivores. Controlling for phylogeny, we documented that mustelids are characterized by shorter gestation ($P = 0.09$) relative to other terrestrial carnivores. Moreover, mustelids have a longer period of estrus, and are more sexually dimorphic, live at lower densities, and occupy larger home ranges. The amount of energy (evapotranspiration) did not differ between the environments of mustelids and other carnivores, but mustelids lived with greater variation in energy (seasonality). We argue that mustelids have evolved "bet-hedging" life-history adaptations to unpredictable environments that include a trade-off between adult survival and reproductive effort. Thus, conservation measures to promote persistence of mustelid populations should consider environmental unpredictability, and ensure low trapping rates of adults.

1. INTRODUCTION

Environmental complexity (Gittleman 1986) and high seasonality (King 1980) may characterize the environment in which mustelidae (hereafter referred to as mustelids) evolved, and hence may help explain differences in life histories relative to other carnivores. Terrestrial mustelids (excludes mink *Mustela vison,* and otter *Lontra* and *Enhydra* species) are adapted to forested habitats, where spatio-temporal variation is greater than grasslands or savannahs (Eisenberg 1981). Characteristics of their environment likely relate to life history adaptations that promote fitness for that environment. For example, Oftedal (1984) argued that forest-dependent species live in an environment that is nutritionally limiting relative to open environments, and therefore carnivore species have evolved later sexual maturity as part of slower growth. Similarly, specific life history adaptations will correlate with management considerations. For example, mustelid populations are predicted to support trapping of juveniles but not adults (Ferguson and Larivière 2002).

Recent results suggest that many populations of carnivore species, including mustelids, are over-exploited by humans and living in habitats considerably altered by human activities (Ruggiero et al. 1994, Fuller and Kittredge 1996, Mech 1996). The result is the extinction of subspecies (Kucera et al. 1995) and the isolation of populations (e.g., Snyder and Bissonette 1987, Gibilisco 1994, Zielinski et al. 2001). In contrast, some populations of North American carnivores, including mustelids, can withstand high trapping pressure (Hodgman et al. 1994, Oehler and Litvaitis 1996, Larivière et al. 2000). For fisheries, evidence suggests a relationship between life histories and tolerance to exploitation (Trippel 1995, Jennings et al. 1998). The role of life histories in determining conservation methods, such as done for birds (Saether et al. 1996) and for carnivores (Ferguson and Larivière 2002), remains largely unexplored for mustelids.

Our goal is to provide a method for predicting vulnerability to overexploitation of harvested populations based on particular life histories (e.g., Sutherland and Reynolds 1998). For example, species that invest less maternal energy in progeny may tolerate the trapping of juveniles without significantly affecting population density. Conversely, these same species may not abide trapping of adults, which are more valuable to maintaining successful population demography. Also, species with life history adaptations to unpredictable climatic conditions or a heterogeneous distribution of energy across time and space may require the conservation of these environmental conditions to provide the demographic advantages over competitors that have life histories adapted to predictable environments.

We describe differences in life-history strategies between mustelids and other North American carnivores to explore whether mustelids warrant special conservation strategies. We used the comparative approach to control for non-independence of species data (Harvey and Pagel 1991). Previously, Ferguson and Larivière (2002) grouped some mustelid species with bears (*Ursus*) into a group called "bet-hedgers" that, relative to other carnivores, lived in unpredictable low energy environments and are characterized by low maternal investment in reproduction while extending the chronology of reproductive events. Specific predictions include later age at sexual maturity, longer interbirth interval, greater longevity, shorter gestation length, smaller neonate mass, and shorter duration of weaning relative to non-mustelid carnivores. As well, we predict that relative to other carnivores, mustelids inhabit highly seasonal environments, live at lower population densities, have larger home ranges, have longer estrus periods, have a greater likelihood of using multi-male mating systems (versus monogamy or polygyny), and have greater sexual dimorphism.

2. METHODS

2.1 Phylogeny and Data

Extant members of Mustelidae are diagnosed as a monophyletic group on the basis of the carnassial notch on the upper fourth premolar, the loss of the upper second molar, as well as enlarged scent glands (Martin 1989, Wozencraft 1989, Bryant et al. 1993). We used the phylogenetic tree proposed by Bininda-Emonds et al. (1999) and the taxonomy of Wozencraft (1993), except that we considered skunks as a separate family, Mephitidae (Dragoo and Honeycutt 1997, see Ferguson and Larivière 2002). The data consisted of 6 families, 21 genera, and 38 species of North American terrestrial carnivores of which 10 were mustelids. We did not use information for marine carnivores (i.e., pinnipeds and sea otter *Enhydra lutis*), as this group possesses unique life-history traits distinct from terrestrial carnivores (Ferguson et al. 1996).

We obtained data on life-history and behavioral traits from published sources (e.g., *Mammalian Species* articles). See Ferguson and Larivière (2002) for the complete data set. Where more than one value was available, we used the mean and if a range was reported we used the midpoint. All data were $\log_{10}$ transformed before analysis to meet assumptions of normality (Harvey and Pagel 1991). Gestation length refers to the time from implantation to parturition and, therefore, does not include the period of delayed implantation.

We estimated productivity and variation in productivity within the historical geographic range (Novak et al. 1987, Nowak 1991) of each carnivore species in North America (Ferguson et al. 1996). We estimated site-specific actual evapotranspiration (mm m-2 y-1) for a set (n = 112) of weather stations located across North America that provided greater than 30 years of continuous weather information (Zeveloff and Boyce 1988). Tables and equations of Thornthwaite and Mather (1957) and climate data were used to calculate energy and seasonality as the total and the coefficient of variation (CV) of monthly (n = 12) values of actual evapotranspiration respectively. Actual evapotranspiration represents the amount of rainfall returned to the atmosphere and is calculated from a site's latitude, soil and vegetation type, and mean monthly temperature and rainfall. Actual evapotranspiration generally increases with a site's solar input, precipitation, and soil capacity and is highly correlated with primary productivity (Rosenzweig 1968). Hence, actual evapotranspiration is used as a productivity surrogate in a variety of studies (e.g., Currie 1991, Ferguson and McLoughlin 2000, Kaspari et al. 2000). We used Lieth's (1976) algorithm to correlate actual evapotranspiration to total net primary productivity. Large primary productivity values indicate greater energy within a species' geographic range. Similarly, large CV values indicate large seasonality within the range of a species.

Mating systems are often coded as categorical data, although the information can also be interpreted as a continuous variable (Garland et al. 1993). We grouped mating systems as polygynous (one male mating >3 females in one area over a relatively short breeding season), multi-male mating (one male mating 1–3 females over a large area and over a relatively long breeding season), and monogamous (one male generally breeds with one female) using the following three category-ordered variables: 3 = polygyny, 2 = multi-male, 1 = monogamy. Multi-male mating occurs in populations where males increase their range during the mating season to encompass a number of female ranges and females are often mated by a number of males (Schenk and Kovacs 1995, Schenk et al. 1999). Mating system was compared using analysis of covariance with female body mass as the covariate. Although mating system was treated as a continuous variable, only one species (*Mephitis mephitis*) was considered polygynous and, therefore, the results are comparable to treating the data as categorical.

2.2 Statistical Analyses

We tested whether mustelids have predictable differences in life-history and behavioral traits compared to other carnivores (see introduction). Phylogenetic corrections are necessary when variation in the observed data set results from phylogenetic structure, creating non-independence of data points (Harvey and Pagel 1991). We tested for the hierarchical pattern of variation in life-history and behavioral traits using nested analysis of variance at three taxonomic levels (species, genus, family). Nested ANOVA provides a suggestion of the taxonomic level that should be used for analysis (Harvey and Pagel 1991). We assume that most variation occurring at the family level indicates the need for phylogenetic correction methods. Conversely, if most variation occurred at the species level then phylogenetic corrections may not be necessary. This selection criterion is somewhat arbitrary and therefore we provide both phylogenetically corrected and conventional statistical results.

We used Monte Carlo algorithms to incorporate phylogenetic structure (i.e., phylogenetic tree) from 38 species (2 polytomies) to estimate statistical parameters for phylogenetic analysis of covariance (ANCOVA) (Garland et al. 1993). Initial limits corresponding to life-history and behavioral traits were obtained from the average of all species values. We performed simulations according to the gradual model of speciation that assumes variance changes are proportional to branch lengths. For each simulated dataset (n = 1,000), we calculated phylogenetically corrected estimates of ANCOVA parameters using general linear models. Conventional ANCOVA statistics were calculated from the observed sample data and compared to the distribution of simulated test statistics. ANCOVA adjusts for differences associated with body mass between groups and enables the assessment of differences in traits due to groups alone. Least-squared means of adjusted trait values represent the predicted mean value for traits after regressing traits on body mass for each group. The ANCOVA model used Type III sum of squares to determine the statistical difference between the least-squared (adjusted) means associated with each group. The phylogenetically corrected critical value of differences due to group (mustelids and others) was set at alpha = 0.10 from the 90th percentile of the simulated distribution. Significant differences are reported in least-squared means that control for body size variation.

3. RESULTS

We found considerable differences among traits as to what phylogenetic level most variation occurred (Table 1.1). Most variation in species traits was attributable to differences within family (median = 42.3, range = 0.9–88.8) and within species (median = 53.0, range = 0.0–94.8), but relatively little variance was explained at the level of genera (median = 9.5, range = 2.0–37.1). The greatest variance in traits occurred at the family level relative to genera or species level for mating system, weaning duration, gestation length, neonate mass, age at maturity, litter size, and interbirth interval. These results indicate that phylogenetic correction methods are necessary for statistical comparisons of these life-history traits. Once we corrected for phylogeny, only gestation length differed between mustelids and other carnivores (P = 0.09; Table 1.2). Mustelids had shorter gestation length (Fig. 1.1) relative to other terrestrial carnivores. Although not significant, the general trend was for mustelids to have smaller neonates, smaller litter size, later age at maturity, longer interbirth interval, and longer life relative to other carnivores (Table 1.2).

All mustelids have multi-male mating systems. In comparison, other terrestrial carnivores adopt monogamous (32%), multi-male (64%) and polygynous (4%) mating systems. Despite these apparent differences, mating systems did not differ between the two groups once we corrected for phylogenetic effects (Table 1.2).

Table 1.1. Comparison of relative percent of variance attributable at the species- (n = 38), genera- (n = 21) and family- (n = 6) level for 8 life-history and 7 behavior traits for species of North American carnivores using a nested analysis of variance for each variable. Bold indicates the level with greatest percent of variation explained.

Life-history and Behavior Traits	Percent of Total Variance		
	Family	Genera	Species
Mating system	**88.8**	11.2	0.0
Weaning duration	**64.5**	2.0	33.5
Gestation length	**61.3**	11.6	27.1
Neonate mass	**52.2**	37.1	10.6
Age at maturity	**51.2**	5.3	43.5
Litter size	**48.1**	32.3	19.5
Interbirth interval	**47.9**	6.0	46.1
Longevity	42.3	4.7	**53.0**
Male home range	16.3	19.9	**63.8**
Seasonality	29.8	2.4	**68.2**
Duration of estrus	7.9	22.5	**69.6**
Female home range	3.8	25.8	**70.4**
Sexual dimorphism	21.3	4.4	**74.3**
Population density	1.0	9.5	**89.5**
Primary productivity	0.9	4.3	**94.8**

Longevity, male home range size, seasonality, duration of estrus, female home range size, sexual dimorphism, population density, and energy (primary productivity) had the greatest variance attributable to the species level (Table 1.1). This pattern of variation suggests that phylogenetic correction may not be necessary for these variables. Using conventional ANCOVA statistics, we found that mustelids differed from other carnivores in sexual dimorphism, population density, male home range size, and length of estrus. Relative to other carnivores, mustelids had greater sexual dimorphism ($P = 0.05$; Fig. 1.2), lower population density ($P = 0.09$; Fig. 1.3), larger male home range size ($P = 0.04$; Fig. 1.4), and longer estrus periods ($P = 0.02$; Fig. 1.5). A significant interaction effect occurred in sexual dimorphism between mustelids and other carnivores indicating a difference in slope: larger mustelids were less dimorphic, whereas larger carnivores were more dimorphic (Fig. 1.2). Comparing environmental variables, mustelids lived in more seasonal environments ($P = 0.01$; Fig. 1.6) but energy (primary productivity) in these environments did not differ from other terrestrial carnivores ($P = 0.33$; Table 1.2).

Figure 1.1. Relationship between gestation length (days) and female body mass (g) for mustelids ($n = 10$) and other terrestrial carnivores in North America ($n = 28$)

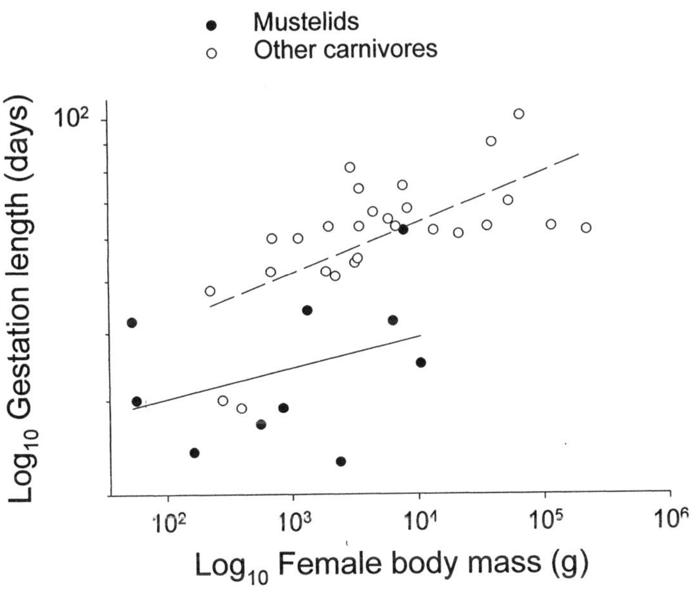

Figure 1.2. Relationship between sexual dimorphism (male/female mass) and female body mass (g) for mustelids ($n = 11$) and other terrestrial carnivores in North America ($n = 27$)

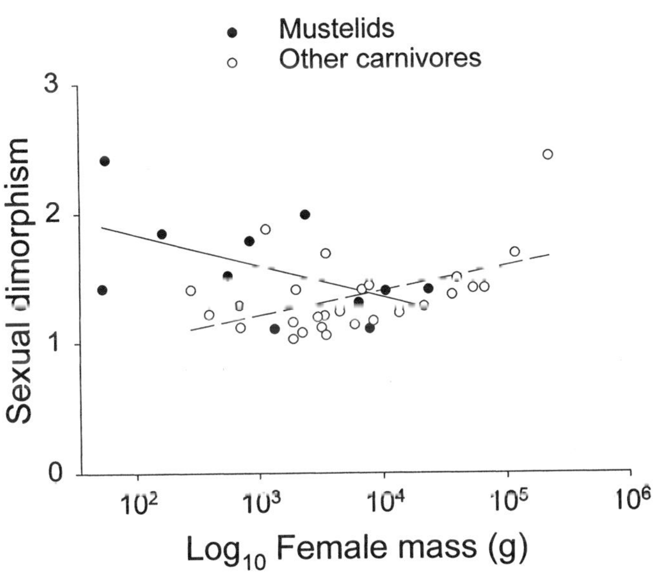

Table 1.2. Difference between mustelids ($n = 10$) and other terrestrial North American carnivores ($n = 28$) for 8 life-history and 7 behavior traits using analysis of covariance tests. The covariate is female body mass except in the case of male home range size when male body mass is the covariate. Models of evolutionary change were no phylogenetic change (standard ANCOVA) or gradual evolutionary change (proportional to branch lengths). Bold represents significant differences using conventional or phylogenetic tests based on results from Table 1.

Life-history and behavior traits	Conventional ANCOVA				Least-square means		Phylogenetic correction	
	n	Observed *F*	Critical *F*	*P*	Mustelids	Other	Critical *F*	*P*
Weaning duration	33	0.11	4.16	0.74[a]	70.80	76.70	22.8	0.88
Gestation length	38	20.50	4.10	<0.01	37.90	57.80	23.3	**0.09**
Neonate mass	34	7.68	4.15	0.01	47.00	86.50	37.2	0.49
Age at maturity	37	0.41	4.11	0.53	542.00	481.00	36.8	0.85
Litter size	38	0.27	4.10	0.61	3.20	3.50	23.7	0.71
Interbirth interval	37	1.69	4.11	0.20	430.00	379.00	26.4	0.59
Longevity	36	0.06	4.12	0.81	4,656.00	4,550.00	36.2	0.99
Male home range	31	4.72	4.17	**0.04**	28.90	8.80	28.3	0.32
Seasonality	38	8.64	4.10	**0.01**	1.14	0.87	22.7	0.20
Duration of estrus	23	6.49	4.34	**0.02**	19.30	5.10	56.3	0.35
Female home range	31	2.59	4.17	0.12	21.30	6.80	29.9	0.44
Sexual dimorphism	38	4.18	4.10	**0.05**[a]	1.58	1.32	23.8	0.33
Population density	30	3.02	4.18	**0.09**	0.31	1.07	26.0	0.46
Primary productivity	38	0.97	4.10	0.33	987.00	1,082.00	27.3	0.71
Mating system	38	2.02	4.10	0.16[a]	1.99	1.72	25.3	0.78

[a]Slopes differ significantly between groups.

Figure 1.3. Relationship between population density (individuals/km²) and female body mass (g) for mustelids ($n = 10$) and other terrestrial carnivores in North America ($n = 20$).

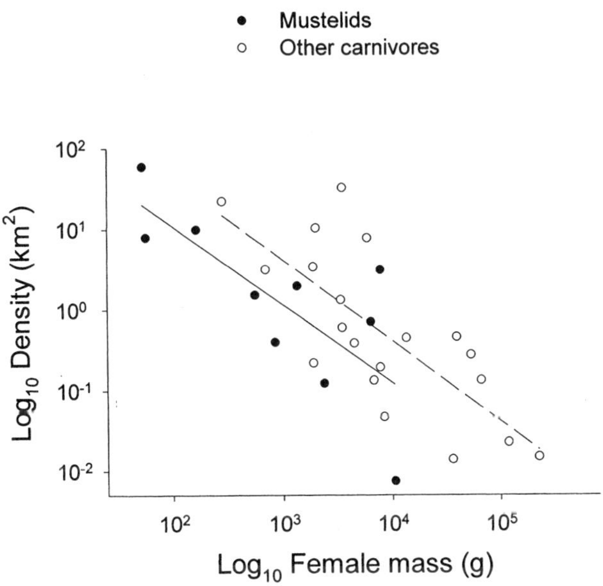

Figure 1.4. Relationship between male home range size (km²) and male body mass (g) for mustelids ($n = 10$) and other terrestrial carnivores in North America ($n = 21$).

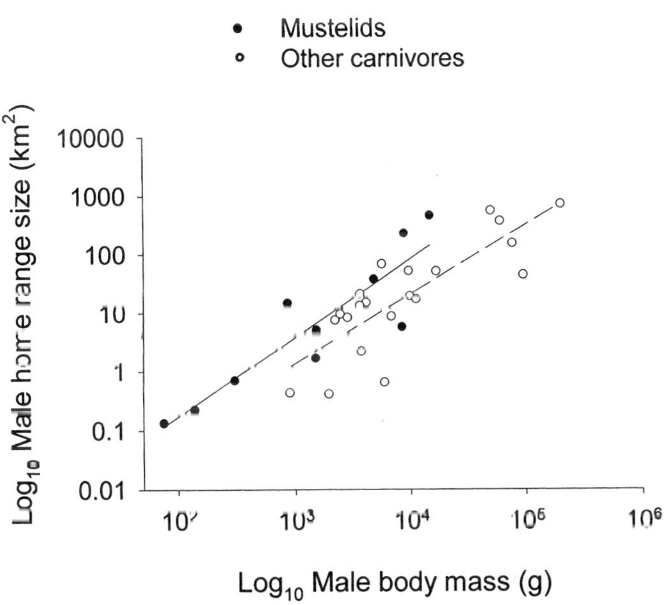

Figure 1.5. Relationship between duration of estrus (days) and female body mass (g) for mustelids ($n = 6$) and other terrestrial carnivores in North America ($n = 17$).

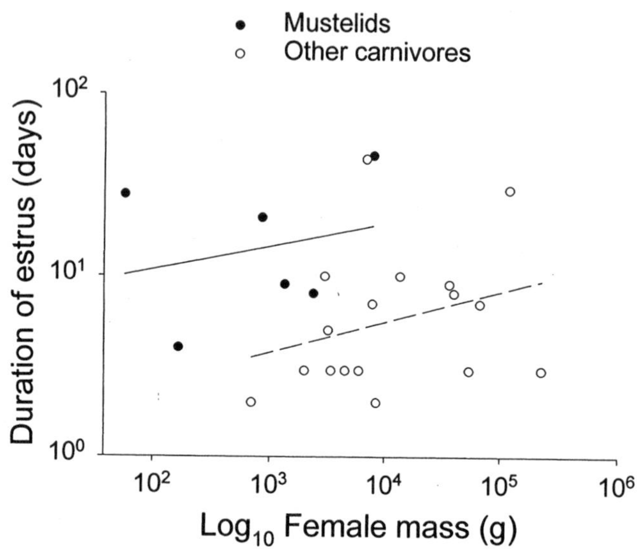

Figure 1.6. Relationship between seasonality (coefficient of variation) and female body mass (g) for mustelids ($n = 10$) and other terrestrial carnivores in North America ($n = 28$).

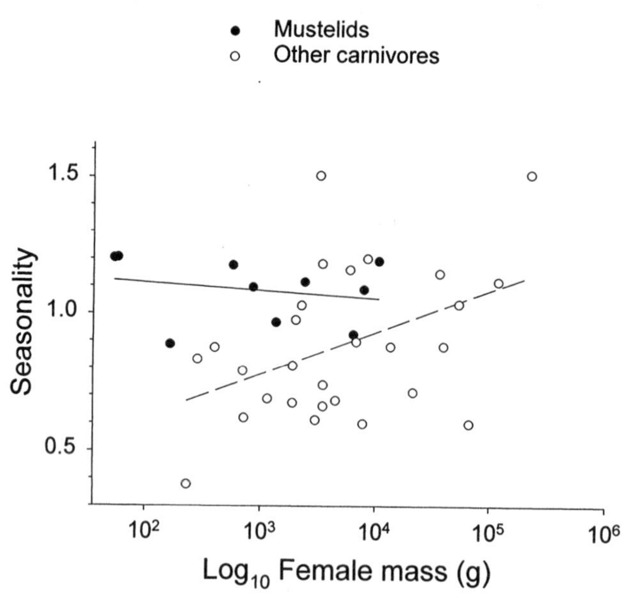

4. DISCUSSION

Previous research has found that mustelid life history is relatively different from many other carnivore groups in social system (Johnson et al. 2000), home range size (Lindstedt et al. 1986, Buskirk and McDonald 1989), delayed implantation (Sandell 1990, Ferguson et al. 1996), induced ovulation (Amstislavsky and Ternovskaya 2000, Larivière and Ferguson 2003), sexual dimorphism (Erlinge 1979), and baculum size (Larivière and Ferguson 2002). Mustelids did not exhibit the K-selected life history strategies of high longevity, slow growth, and low fecundity, but rather were characterized by the life-history adaptations referred to as bet-hedgers (Ferguson and Larivière 2002). Our results provide the first statistical evidence using the comparative approach to identify a suite of interacting life history and behavioral traits of mustelids that differ from other terrestrial carnivores. These interacting traits may have relevance to management and conservation by suggesting that mustelids require different conservation strategies.

Mustelids inhabit highly seasonal environments, have larger home ranges, and lower population densities compared to other terrestrial carnivores. Low densities imply small populations (Gaston 1996), and small populations are predisposed to stochastic and genetic changes that lead to extinction (Gilpin and Soulé 1986). For example, wolverines (*Gulo gulo*) occur at extremely low densities (0.006–0.02/km^2, Pasitschniak-Arts and Larivière 1995) and range widely, which predisposes them to impacts from humans (Finch 1992). Furthermore, low densities and large home ranges suggest that mustelids require larger areas for their conservation, and that they may be more sensitive to trapping than other carnivores (Kyle and Strobeck 2002). Large home ranges also suggest that mustelids are more likely to be affected by human activities (Wilson et al. 2000). Already, some mustelid species have undergone distributional losses (reviewed by Ruggiero et al. 1994) that have been attributed to humans. Undoubtedly, maintenance or preservation of large patches of suitable habitat will remain one of the priorities for conservation of mustelids, especially relatively large-bodied species that inhabit boreal forests (e.g., *Martes*, Helldin 2000, Potvin et al. 2000, Rondinini and Boitani 2002)

Mustelids likely evolved in temperate areas (King 1986) characterized by high seasonality. Adaptations for life in seasonal high-latitude environments that are more unpredictable (Ferguson and Messier 1996) include the evolution of delayed implantation in mustelids (Sandell 1980, Ferguson et al. 1996). Life history comparisons suggest that mustelids have generally evolved a "bet-hedging" life-history strategy that maximizes reproduction in unpredictable seasonal environments occurring at high latitude/altitude. Previously, we identified a group called "bet-hedgers" that consisted of ursids (black *Ursus americanus*, brown *U. arctos*, and polar bears *U. maritimus*) and forest-dwelling mustelids (martens *Martes americana*, fishers *M. pennanti*, and wolverines) that were characterized by short gestation, small neonate mass, large litters, late maturation and long life (Ferguson and Larivière 2002). If juvenile survival responds more strongly to environmental conditions than adult survival, then the best option for a parent is to keep its own survival probability high and reproductive effort low (Both et al. 1999, Lindstrom 1999). The pattern of low maternal investment in offspring for mustelids relative to other carnivores was indicated by short gestation length and small neonates, although the latter was not significant with phylogenetic corrections. The unpredictability of high-latitude seasonal environments is intensified by the time delay between reproductive decisions made by the parents and the environmental conditions that the offspring face at birth.

Environmental unpredictability is a key component of mustelid environments, and one component that managers often fail to address. Most forest animals, including many mustelids, are adapted to the natural disturbance regimes of fires, windfalls, and disease (Ruggiero et al. 1994). For example, martens, fishers and in southern parts of their range, wolverines are generally thought to require large areas of old-growth forest (Hornocker and Hash1981, Powell 1993, Buskirk and Powell 1994) rather than the mixed landscapes of different-aged stands created by disturbance such as fire or logging. Thus, forest management guidelines (e.g., Watt et al. 1996) specify the legal requirement of maintaining old growth forest for mustelids. Nevertheless, studies have found mustelids surviving and reproducing in younger forests (e.g., Banci 1987, Arthur and Krohn 1991, Chapin et al. 1997, Potvin et al. 2000), suggesting that mustelid population dynamics are adapted to highly dynamic environments, such as occurs with fire-cycles in boreal forests or spruce budworm (*Choristoneura fumiferana*) cycles in Acadian forests (Attiwill 1994). In fact, mustelids may depend on unpredictability to 'out-compete' more generalist carnivores, which are characterized by greater fecundity and higher recruitment (Ferguson and Larivière 2002).

The American marten provides a well-studied example of mustelid life history. The marten was historically distributed throughout the northern boreal, mixed Acadian forests, and northeastern Appalachian forests of North America (Gibilisco 1994). Martens have low reproductive potential and hence require protection from loss of habitat (Snyder 1986, Forsey et al. 1995). Relative to other mammals, martens display a prolonged time to sexual maturity, litter size is as expected on the basis of body size, interbirth interval may be shorter than allometric predictions, yearly reproductive output of pregnant female martens is low, and longevity is high (Buskirk and Ruggiero 1994). Trapping has contributed to the loss of martens in some areas, including the north-central states and eastern Canada (Buskirk and Ruggiero 1994). A successful method of restoring mustelid populations in Minnesota, U.S.A., was to close the trapping season to conserve martens and fishers (Mech 1996). In addition to trapping, marten populations can fluctuate in response to resource conditions that result from cyclic changes in prey density and loss of physical structure of the forest, such as timber harvesting (Fryxell et al. 1999, Helldin 2000).

The life histories of mustelids have management and conservation implications in an increasingly fragmented habitat because of anthropogenic causes. Mustelids exhibit multi-male mating systems, long estrus periods, delayed implantation, induced ovulation, and large sexual size dimorphism relative to other terrestrial carnivores. As well, mustelids live in seasonal environments characterized by snowfall in winter and demanding energetic conditions (Wilbert et al. 2000), and they occur at low densities and range over large areas. These reproductive and behavioral traits relate to a multi-male mating system adapted to the environmental conditions that make it difficult for male and female mustelids to get together. The multi-male mating system promotes sexual selection (Rowe and Arnqvist 2002) and increases genetic variation (Petrie et al. 1998). A conservation consequence of the multi-male mating system and associated genetic variation may be increased local population extinctions. Thus, there is a need to retain gene flow via linked populations among fragmented habitat to sustain populations that are sensitive to inbreeding (Schwartz et al. 2002).

Increasing concern for the conservation status of many mustelids (Fuller and Kittredge 1996) makes assessments of their vulnerability to over-trapping and habitat loss more important (Ruggerio et al. 1994). Our analyses suggest that mustelids show life history adaptations to high latitude environments characterized by variability. The conservation outcome of these adaptations includes the need to maintain genetic linkages among populations and the need to maintain environmental variability across time and space. Environmental variability preserves the advantage afforded by mustelid life histories over their carnivore competitors. The approach of comparing life histories should have general applicability to other taxa, as conservation biologists search for general resource and spatial requirements that can be used to identify minimum conditions necessary for long-term population persistence (Smallwood 1999). Forested landscapes are rapidly being converted to intensive human uses (Turner 1987, Chapin et al. 1998) and traditional forest management results in fragmented habitats, thereby leading to loss of biological diversity (Wallin et al. 1994, Hargis et al. 1999). We argue that a broader understanding of the relationship between life-history patterns and population processes may facilitate the development of general principles to help managers understand the impact of forest disturbance and trapping on mustelids.

5. ACKNOWLEDGMENTS

Bowater Pulp & Paper Inc. provided funding to the senior author for this research. The Institute for Wetlands and Waterfowl Research, Ducks Unlimited Inc., provided the second author time to pursue this research. T. Garland, Jr. provided critical advice on comparative analysis.

6. LITERATURE CITED

Amstislavsky, S., and Y. Ternovskaya. 2000. Reproduction in mustelids. Animal Reproductive Science 60/61:571–581.

Arthur, S. M., and W. B. Krohn. 1991. Activity patterns, movements, and reproductive ecology of fishers in southcentral Maine. Journal of Mammalogy 72:379–385.

Attiwill, P. M. 1994. The disturbance of forest ecosystems: the ecological basis for conservation management. Forest Ecology and Management 63:247–300.

Banci, V. 1987. Ecology and behavior of wolverine in Yukon. Dissertation, Simon Fraser University, Burnaby, British Columbia, Canada.

Bininda-Emonds, O. R. P., J. L. Gittleman, and A. Purvis. 1999. Building large trees by combining phylogenetic information: a complete phylogeny of the extant Carnivora (Mammalia). Biological Review 74:143–175.

Both, C., M. E. Visser, and N. Verboven. 1999. Density-dependent recruitment rates in great tits: the importance of being heavier. Proceedings Royal Society of London, Series B 266:465–469.

Bryant, H. N., A. P. Russell, and W. D. Fitch. 1993. Phylogenetic relationships within the extant Mustelidae (Carnivora): appraisal of the cladistic status of the Simpsonian subfamilies. Zoological Journal of the Linnean Society 108:301–334.

Buskirk, S. W., and L. L. McDonald. 1989. Analysis of variability in home-range size of the American marten. Journal of Wildlife Management 53:997–1004.

_____, and R. A. Powell. 1994. Habitat ecology of fishers and American martens. Pages 283–296 *in* S. W. Buskirk, A. S. Harestad, and M. G. Raphael, editors. Martens, sables, and fishers: biology and conservation. Ithaca, New York, Cornell University Press, USA.

_____, and L. F. Ruggiero. 1994. American Marten. Pages 7–37 *in* L. F. Ruggiero, K. B. Aubry, S. W. Buskirk, L. J. Lyon, and W. J. Zielinski, editors. The scientific basis for conserving forest carnivores: American marten, fisher, lynx, and wolverine in the western United States. USDA Forest Service General Technical Report RM-254, Fort Collins, Colorado, USA.

Chapin, T. G., D. J. Harrison, and D. M. Phillips. 1997. Seasonal habitat selection by marten in an untrapped forest preserve. Journal of Wildlife Management 61:707–717.

_____, D. J. Harrison, and D. D. Katnik. 1998. Influence of landscape pattern on habitat use by American martens in an industrial forest. Conservation Biology 12:1327–1337.

Currie, D. 1991. Energy and large-scale patterns of animal- and plant-species richness. American Naturalist 98:399–414.

Dragoo, J. W., and R. L. Honeycutt. 1997. Systematics of mustelid-like carnivores. Journal of Mammalogy 78:426–443.

Eisenberg, J. F. 1981. The mammalian radiations: and analysis of trends in evolution, adaptations and behavior. University of Chicago Press, Chicago, USA.

Erlinge, S. 1979. Adaptive significance of sexual dimorphism in weasels. Oikos 33:233–245.

Ferguson, S. H., and S. Larivière. 2002. Can comparing life histories help conserve carnivores? Animal Conservation 5:1–12.

_____, and P. D. McLoughlin. 2000. Effect of energy availability, seasonality, and geographic range on brown bear life history. Ecography 23:193–200.

_____, and F. Messier. 1996. Ecological implications of a latitudinal gradient in inter-annual climatic variability: a test using fractal and chaos theories. Ecography 19:382–392.

_____, J. A. Virgl, and S. Larivière. 1996. Evolution of delayed implantation and associated grade shifts in life history traits of North American carnivores. Ecoscience 3:7–17.

Finch, D. M. 1992. Threatened, endangered, and vulnerable species of terrestrial vertebrates in the Rocky Mountain region. U.S. Department of Agriculture Forest Service General Technical Report RM-215.

Forsey, O., J. Bissonette, J. Brazil, K. Curnew, J. Lemon, L. Mayo, I. Thompson, L. Bateman, and L. O'Driscoll. 1995. National recovery plan for the Newfoundland marten. Report No. 14, Recovery of Nationally Endangered Wildlife Committee, Ottawa.

Fryxell, J., J. B. Falls, E. A. Falls, R. J. Brooks, L. Dix, and M. Strickland. 1999. Density dependence, prey dependence, and population dynamics of martens in Ontario. Ecology 80:1311–1321.

Fuller, T. K., and D. B. Kittredge, Jr. 1996. Conservation of large forest carnivores. Pages 137–164 *in* R. M. DeGraaf, and R. I. Miller, editors. Conservation of faunal diversity in forested landscapes. Chapman and Hall, London, UK.

Garland, T., Jr., A. W. Dickerman, C. M. Janis, and J. A. Jones. 1993. Phylogenetic analysis of covariance by computer simulation. Systematic Biology 42:265–292.

Gaston, K. J. 1996. Species-range-size distributions: patterns, mechanisms and implications. Trends in Ecology and Evolution 11:197–201.

Gibilisco, C. J. 1994. Distributional dynamics of martens and fishers in North America. Pages 59–71 *in* S. W. Buskirk, A. S. Harestad, M. G. Raphael, editors. Martens, sables, and fishers: biology and conservation. Cornell University Press, Ithaca, New York, USA.

Gilpin, M. E., and M. E. Soulé. 1986. Minimum viable populations: processes of species extinction. Pages 19–34 *in* M. E. Soulé, editor. Conservation biology: the science of scarcity and diversity. Sinauer Associates, Sunderland, Massachusetts, USA.

Gittleman, J. L. 1986. Carnivore life history patterns: allometric, phylogenetic, and ecological associations. American Naturalist 127:744–771.

Hargis, C. D., J. A. Bissonette, and D. L. Turner. 1999. The influence of forest fragmentation and landscape pattern on American martens. Journal of Applied Ecology 36:157–172.

Harvey, P. H., and M. D. Pagel. 1991. The comparative method in evolutionary biology. Oxford University Press, Oxford.

Helldin, J.-O. 2000. Population trends in harvest management of pine marten *Martes martes* in Scandinavia. Wildlife Biology 6:111–120.

Hodgman, T. P., D. J. Harrison, D. D. Katnik, and K. D. Elowe. 1994. Survival in an intensively trapped marten population in Maine. Journal of Wildlife Management 58:593–600.

Hornocker, M. G., and H. S. Hash. 1981. Ecology of the wolverine in northwestern Montana. Canadian Journal of Zoology 59:1286–1301.

Jennings, S., J. D. Reynolds, and S. C. Mills. 1998. Life history correlates of responses to fisheries exploitation. Proceedings Royal Society of London, Series B 265:333–339.

Johnson, D. D. P., D. W. Macdonald, and A. J. Dickman. 2000. An analysis and review of

models of the sociobiology of the Mustelidae. Mammal Review 30:171–196.
Kaspari, M., S. O'Donnell, and J. R. Kercher. 2000. Energy, density, and constraints to species richness: ant assemblages along a productivity gradient. American Naturalist 155:280–293.
King, C. M. 1980. Population biology of the weasel *Mustela nivalis* on British game estates. Holarctic Ecology 3:160–168.
King, J. E. 1986. Seals of the world. Oxford University Press, Oxford, UK.
Kucera, T. E., W. J. Zielinski, and R. H. Barrett. 1995. The current distribution of American marten, *Martes americana*, in California. California Fish and Game 81:96–103.
Kyle, C. J., and C. Strobeck. 2002. Connectivity of peripheral and core populations of North American wolverines. Journal of Mammalogy 83:1141–1150.
Larivière, S., and S. H. Ferguson. 2002. On the evolution of the mammalian baculum: vaginal friction, prolonged intromission or induced ovulation? Mammal Review 32:283–294.
_____, and _____. 2003. Evolution of induced ovulation in North American carnivores. Journal of Mammalogy 84:937–947.
_____, H. Jolicoeur, and M. Crête. 2000. Status and conservation of the gray wolf (*Canis lupus*) in wildlife reserves of Québec. Biological Conservation 94:143–151.
Lieth, H. 1976. The use of correlation models to predict primary productivity from precipitation or evapotranspiration. Pages 392–406 *in* O. L. Lange, L. Kappern, and E. D. Schulze, editors. Water and plant life. Springer-Verlag, Berlin, Germany.
Lindstedt, S. L., B. J. Miller, and S. W. Buskirk. 1986. Home range, time and body size in mammals. Ecology 67:413–418.
Lindstrom, J. 1999. Early development and fitness in birds and mammals. Trends in Ecology and Evolution 14:343–348.
Martin, L. D. 1989. Fossil history of the terrestrial Carnivora. Pages 536–568 *in* J. L. Gittleman, editor. Carnivore behavior, ecology, and evolution. Cornell University Press, Ithaca, New York, USA.
Mech, L. D. 1996. A new era for carnivore conservation. Wildlife Society Bulletin 24:397–401.
Novak, M., J. A. Baker, M. E. Obbard, and B. Malloch. 1987. Wild furbearer Management and Conservation in North America, Ontario Ministry of Natural Resources, Ontario Trappers Association, North Bay, Canada.
Nowak, R. M. 1991. Walker's Mammals of the World. Fifth edition, Vol. II. The John Hopkins University Press, Baltimore, Maryland, USA.
Oftedal, O. T. 1984. Milk composition, milk yield and energy output at peak lactation: a comparative review. Symposia of the Zoological Society of London 51:33–85.
Oehler, J. D., and J. A. Litvaitis. 1996. The role of spatial scale in understanding responses of medium-sized carnivores to forest fragmentation. Canadian Journal of Zoology 74:2070–2079.
Pasitschniak-Arts, M., and S. Larivière. 1995. *Gulo gulo*. Mammalian Species 499:1–10.
Petrie, M., C. Doums, and A. P. Møller. 1998. The degree of extra-pair paternity increases with genetic variability. Proceedings of the National Academy of Science, USA 95:9390–9395.
Potvin, F., L. Bélanger, and K. Lowell. 2000. Marten habitat selection in a clearcut boreal landscape. Conservation Biology 14:844–857.
Powell, R. A. 1993. The fisher: life history, ecology, and behavior. Second edition. University of Minnesota Press, Minneapolis, Minnesota, USA.
Rondinini, C., and L. Boitani. 2002. Habitat use by beech martens in a fragmented landscape. Ecography 25:257–264.
Rosenzweig, M. L. 1968. Net primary productivity of terrestrial communities: Prediction from climatological data. American Naturalist 102:67–74.
Rowe, L., and G. Arnqvist. 2002. Sexually antagonistic coevolution in a mating system: combining experimental and comparative approaches to address evolutionary processes. Evolution

56:754–767.

Ruggiero, L. F., K. B. Aubry, S. W. Buskirk, L. J. Lyon, and W. J. Zielinski. 1994. The scientific basis for conserving forest carnivores: American marten, fisher, lynx, and wolverine in the Western United States. U. S. Department of Agriculture Forest Service General Technical Report RM-254, Fort Collins, Colorado, USA.

Saether, B. -E., T. H. Ringsby, and E. Roskaft. 1996. Life-history variation, population processes and priorities in species conservation: towards a reunion of research paradigms. Oikos 77:217–226.

Sandell, M. 1990. The evolution of seasonal delayed implantation. Quarterly Review of Biology 65:23–42.

Schenk, A., and K. M. Kovacs. 1995. Multiple mating between black bears revealed by DNA fingerprinting. Animal Behavior 50:1483–1490.

_____, M. E. Obbard, and K. M. Kovacs. 1999. Genetic relatedness and home-range overlap among female black bears (*Ursus americanus*) in northern Ontario, Canada. Canadian Journal of Zoology 76:1511–1519.

Schwartz, M. K., L. S. Mills, K. S. McKelvey, L. F. Ruggiero, and F. W. Allendorf. 2002. DNA reveals high dispersal synchronizing the population dynamics of Canada lynx. Nature 415:520–522.

Smallwood, K. S. 1999. Scale domains of abundance amongst species of mammalian Carnivora. Environmental Conservation 26:102–111.

Snyder, J. E. 1986. Updated status report on the marten (Newfoundland population) *Martes americana atrata*. Committee on the Status of Endangered Wildlife in Canada, Ottawa, Canada.

_____, and J. A. Bissonette. 1987. Marten use of clearcuttings and residual forest in western Newfoundland. Canadian Journal of Zoology 65:169–174.

Sutherland, W. J., and J. D. Reynolds. 1998. Sustainable and unsustainable exploitation. Pages 129–141 *in* W. J. Sutherland, editor. Conservation science and action. Blackwell Science, Oxford, UK.

Thornthwaite, C. W., and J. R. Mather. 1957. Instructions and tables for computing potential evapotranspiration and the water balance. Publications in Climatology 10:185–311.

Trippel, E. 1995. Age at maturity as a stress indicator in fishes. BioScience 45:759–771.

Turner, M. G. 1987. Landscape heterogeneity and disturbance. Springer-Verlag, New York, New York, USA.

Wallin, D. O., F. J. Swanson, and B. Marks. 1994. Landscape pattern response to changes in pattern generation rules: land-use legacies in forestry. Ecological Applications 4:569–580.

Watt, W. R., J. A. Baker, D. M. Hogg, J. G. McNicol, and B. J. Naylor. 1996. Forest management guidelines for the provision of marten habitat. Ontario Ministry of Natural Resources Technical Series, Sault St. Marie, ON, Canada.

Wilbert, C. J., S. W. Buskirk, and K. G. Gerow. 2000. Effects of weather and snow on habitat selection by American marten (*Martes americana*). Canadian Journal of Zoology 78:1691–1696.

Wilson, G. M., A. van der Busche, P. K. Kennedy, A. Gunn, and K. Poole. 2000. Genetic variability of wolverines (*Gulu gulo*) from the Northwestern Territories, Canada: conservation implications. Journal of Mammalogy 81:186–196.

Wozencraft, W. C. 1989. The phylogeny of the recent carnivora. Pages 495–535 *in* J. L. Gittleman, editor. Carnivore behavior, ecology, and evolution. Cornell University Press, Ithaca, New York, USA.

_____. 1993. Order carnivora. Pages 279–348 *in* D. E. Wilson and D. M. Reeder, editors. Mammal species of the world: a taxonomic and geographic reference. Second edition. Smithsonian Institution Press, Washington, DC, USA.

Zielinski, W. J., K. M. Slauson, C. R. Carroll, C. J. Kent, and D. G. Kudrna. 2001. Status of American Martens in coastal forests of the Pacific States. Journal of Mammalogy 82:478–490.

Zeveloff, S. E., and M. S. Boyce. 1988. Body size patterns in North American mammal faunas. Pages 123–146 *in* M. S. Boyce, editor. Evolution of life histories of mammals. Yale University Press, New Haven, Connecticut, USA.

Chapter 2

WORLD DISTRIBUTION AND STATUS OF THE GENUS *MARTES* IN 2000

Gilbert Proulx, Keith Aubry, Johnny Birks, Steven Buskirk, Clément Fortin, Herbert Frost, William Krohn, Lem Mayo, Vladimir Monakhov, David Payer, Midori Saeki, Margarida Santos-Reis, Richard Weir, and William Zielinski

***Abstract*:** The genus *Martes* is comprised of 7 species of martens, sables and fishers, most of them forest-dwelling animals with valuable fur, distributed throughout North America, Europe and Asia. The pine marten (*Martes martes*) is indigenous over most of Europe, from Mediterranean biotopes to Fennoscandian taiga, and to western Siberia and Iran. It is found in insular wooded areas, shrublands, and coniferous forests. The stone marten (*M. foina*) occurs from Mongolia and the northern Himalayas to most of Europe. It frequents forests, woodlands and pastures, and is expanding in suburban and urban areas. The sable (*M. zibellina*) occurs in Russia, Mongolia, China, North Korea, and Japan. Over most of its distribution, the sable inhabits coniferous taiga forests with late seral attributes. The yellow-throated marten (*M. flavigula*; including the Nilgiri marten, *M. gwatkinsi*) occurs in sub-tropical and tropical forests from the Himalayas to eastern Russia, south to the Malay Peninsula and Sunda Shelf to Taiwan. The Japanese marten (*M. melampus*) occurs in forests of the main Japanese archipelago and the Korean peninsula. The American marten (*M. americana*) occurs in large contiguous populations in forested habitats of North America north of 35° latitude. It is associated with mesic coniferous and mixed forests with overhead cover and structural complexity near the ground. The fisher (*M. pennanti*) occurs in large contiguous areas across Canada, and in disjunct areas within the United States, north of 35° latitude. Whereas the distribution of *Martes* significantly expanded in many parts of the world over the last 20 years, largely due to several reintroduction programs, many populations are threatened by habitat loss and alteration. There is a need to develop cost-effective survey methods, monitor populations and fur-harvest activities, and assess the effects of natural and anthropogenic disturbance agents on habitat use by *Martes* species.

1. INTRODUCTION

The genus *Martes* occurs in tropical, temperate, and boreal forest zones of the Old and New Worlds. It is comprised of 7 species of martens, sables and fishers (Buskirk 1994), most of them forest-dwelling animals with valuable fur. Their distribution and abundance are strongly influenced by habitat change resulting from forestry and agricultural practices (Brainerd et al. 1994, Kryštufek 2000, Messenger and Birks 2000, Proulx 2000), and the resiliency of populations to trapping and hunting pressure (Banci and Proulx 1999). It is, therefore, important to regularly monitor the distribution of *Martes* species across their range to assess the effects of human activities on population changes, recognize information gaps, develop effective research programs, and implement sound management programs that will ensure the future of these species.

This paper reviews the distribution of the Eurasian pine marten (*Martes martes*), stone marten (*M. foina*), sable (*M. zibellina*), yellow-throated marten (*M. flavigula* including the Nilgiri marten, *M. gwatkinsi*), Japanese marten (*M. melampus*), American marten (*M. americana*), and fisher (*M. pennanti*). It focuses on the conservation status and geographic distribution of extant populations during the last 20 years, discusses factors explaining population trends, and identifies present and future management and research activities addressing these species within their current geographic distributions.

2. DATA COLLECTION AND ACKNOWLEDGMENTS

Basic information on the status and distribution of *Martes* species was obtained from scientific literature and technical reports from various government agencies and conservation organizations. This information was updated with a questionnaire sent to wildlife researchers and agencies in countries where *Martes* species are or might be present. Questionnaires requested information on: 1) conservation status, i.e. endangered, threatened, special concern, furbearer, or other; 2) harvest status, with mean length of trapping/hunting seasons, harvest limits, and characteristics of harvested populations; 3) geographic distribution and variation in abundance from 1980 to 2000; 4) habitat loss or expansion during the last 20 years; 5) factors associated with population changes; and 6) management (e.g., reintroduction programs) or research activities affecting the distribution of species.

There was a marked variation in the quantity and quality of information provided by respondents. The information was first used to define the contemporary distribution of each *Martes* species. Because of taxonomic uncertainties or lack of precise data, changes in geographic distribution and variations in abundance usually did not include sub-species. Information on habitat loss or expansion in various ecosystems was largely subjective and was used only to identify major trends at the country level. Questionnaires were used to differentiate harvested and protected populations, and to identify population trends.

This review could not have been completed without the contribution of many individuals and their agencies. We sincerely thank them all for taking the time to respond to our long and detailed questionnaire. We also thank Pauline Feldstein, Daniel Harrison, Angela Fuller, and two anonymous referees for their comments on an earlier manuscript.

3. SPECIES ACCOUNTS

3.1 The Pine Marten (*Martes martes*)

3.1.1 Distribution

The pine marten is indigenous over most of Europe, from Mediterranean biotopes to Fennoscandian taiga, and to western Siberia and Iran (Clevenger 1994, Helldin 1998, De Marinis et al. 2000) (Fig. 2.1).

Formerly widespread in Britain, the pine marten declined due to habitat loss and persecution and is now mainly confined to northern Scotland, with small, relict populations surviving in parts of England and Wales. Since 1980, the species range has been slowly expanding in Scotland. Martens were re-introduced in 1980–1981 in the southwest portion of the country. Elsewhere in Britain, populations remain isolated, vulnerable, and difficult to monitor (Messenger and Birks 2000). The situation is complicated by the recent confirmation of the presence of *M. americana* (believed to have escaped from fur farms) and evidence of possible introgression with *M. martes* in areas of the latter's relict distribution in northern England (Kyle et al. 2003). In Ireland, the distribution is patchy (Mitchell-Jones et al. 1999), but expanding due to increased coniferous forest and legal protection (P. Sleeman, Department of Zoology and Animal Ecology, National University of Ireland, Cork, Ireland, personal communication).

The occurrence of the pine marten in continental Portugal was unknown until the late 1980s. In her review of the status and distribution of the Portuguese mustelids, Santos-Reis (1983) did not include the pine marten as a resident species. The first mention of the pine marten in Portugal occurred in the Red Data Book for Terrestrial Vertebrates on the basis of carcass analyses (Serviço Nacional de Parques Reservas e Conservação da Natureza 1990). It appears that, because of its scarcity and morphological similarities with the much more abundant stone marten, the inclusion of the pine marten in the Portugal mammalian fauna was delayed. The species is now considered indigenous to Portugal (Santos-Reis and Petrucci-Fonseca 1999). Validated records of the species and responses to questionnaires sent to municipalities (H. Matos and M. Santos-Reis, Faculdade de Ciêcias, Lisbon University, Portugal, unpublished data) confirmed the scarcity of the pine marten in Portugal and suggest the species is scattered in the north and interior portions of the country. It is absent from the Atlantic islands (Azores and Madeira archipelagos).

Still in the Mediterranean region, the pine marten occurs in northern Spain (Clevenger 1993), particularly in the Pyrenean mountains, the Cordillera Cantabrica, and the Atlantic areas (J. Ruiz-Olmo, Servei de Protecció I Gestió de la Fauna, Direcció del Medi Natural, Barcelona, Spain, and J.M. Lopez-Martin, Department of Animal Biology, Barcelona University, Spain, personal communication) (Fig. 2.1). Insular populations occur in the Balearic Islands of Minorca and Majorca (Clevenger 1993). In France, the pine marten mostly occurs in the Pyrenees, Limousin, and the eastern portion of the country, except Provence and Côte d'Azur (Bouchardy and Labrid 1986). It is rare in southwest France and the Mediterranean area but occurs in Corsica (T. Lode, Laboratoire d'Écologie Animale, UFR Sciences, Université d'Angers, France, personal communication). In Italy, the species is present in the forested areas of the peninsula, with a distribution that appears to be very fragmented; insular populations also occur in Sardinia, Sicily and Elba (De Marinis and Masseti 1993, De Marinis et al. 2000, Fornasari et al. 2000; P. Genovesi, National Wildlife Institute, Italy, personal communication).

In Switzerland, the pine marten is believed to be widespread. However, since 1980, most observations have occurred in the western and southern regions (S. Capt, Centre Suisse de Cartographie de la Faune, Neuchâtel, Switzerland, personal communication). In Belgium, the pine marten is restricted to southern regions (Libois 1983). It is present throughout Luxembourg (A. Baghli, National History Museum, Luxembourg, and L. Schley, Service de la Conservation de la Nature, Direction des Eaux et Forêts, Luxembourg, personal communication). The distribution of marten in The Netherlands is patchy (S. Broekhuizen, Wageningen, The Netherlands, personal communication; Muskens et al. 2000). In Denmark, the pine marten is a rare species occurring mainly in the southern forests of the peninsula of Jutland; small populations also occur in the islands of Fyn, Lolland-Falster, and Zealand (T. Asferg, National Environmental Research Institute, Department of Landscape Ecology, Rønde, Denmark, personal communication). Martens are present throughout the forested regions of Germany (M. Stubbe, personal communication). The species is widespread in Austria (A. Kranz, Hunting Association of Styria, Graz, Austria, personal communication) and Hungary (M. T. Apathy, Department of Biology, Eotvos Lorand University, Budapest, Hungary, personal communication) (Fig. 2.1).

In Finland, the pine marten is present in Lapland, at the northern limit of its range (Pulliainen 1984), but its populations reach higher densities in the

Figure 2.1. General distribution of *Martes martes* throughout Europe and western Asia (after King 1977, O'Sullivan 1983, Fayard 1984, Velander 1983, 1991, Balharry et al. 1996, Strachan et al. 1996, Messenger et al. 1997, De Marinis et al. 2000; Muskens et al. 2000; T. Asferg, National Environmental Research Institute, Department of Landscape Ecology, Rønde, Denmark, personal communication; S. Capt, Centre Suisse de Cartographie de la Faune, Neuchâtel, Switzerland, personal communication; M. Dumitru, "Grigore Antipa" National Museum of Natural History, Bucharest, Romania, personal communication; A. Legakis, Zoological Museum, Department of Biology, University of Athens, Greece, personal communication; C. Prigioni, Department of Animal Biology, University of Pavia, Italy, personal communication; P. Sleeman, Department of Zoology and Animal Ecology, National University of Ireland, Cork, Ireland, personal communication; F. Spitzenberger, Museum of Natural History, Vienna, Austria, personal communication).

more forested eastern and southern regions of the country (Helle et al. 1996, Kurki et al. 1997, Kauhala and Helle 2000). It is also more abundant in the central and southern forests of Norway (Helldin 2000, Ryvarden 2001). The pine marten is present throughout Sweden (except Gotland Island; J. O. Helldin, Grimsö Wildlife Research Station, Swedish University of Agricultural Sciences, Riddarhyttan, Sweden, personal communication), Lithuania (Mickevicius and Baranauskas 1992; L. Balciauskas, Institute of Ecology, Vilnius, Lithuania, personal communication), Latvia (Ozolins and Pilats 1995, Ž. Andersone, Kemeri National Park, Latvia, personal communication), the Czech Republic (Andera and Hanzal 1996; M. Andera, Department of Zoology, National Museum, Praha, Czech Republic, personal communication) and Poland (A. Zalewski, Mammal Research Institute, Polish Academy of Science, Poland, personal communication), with no apparent change in distribution over the last 20 years (Fig. 2.1). The species is common in the Carpathian Mountains (Bakeyev 1994), which lie mostly in Romania and the Czech Republic. In Romania, the species occurs in the central region of the country, and along the Hungarian and Ukrainian borders (M. Dumitru, "Grigore Antipa" National Museum of Natural History, Bucharest, Romania, personal communication). The pine marten is present in Slovenia, Macedonia, Bosnia-Herzegovina, and European Turkey, but the limits of its range are poorly defined (Stubbe 1993, Kryštufek 2000). In Bulgaria, it inhabits mountainous forests, preferably over 1,500 m above sea level (ASL) (Grigorov 1986). Between the 1940s and the 1960s, the species was considered in danger of extinction. Since then, it has recovered even though it is still considered as threatened (Spriridonov and Spassov 1998; N. Spassov, National Museum of Natural History, Sofia, Bulgaria, personal communication). The pine marten is widely distributed in Serbia and Montenegro (Mitchell-Jones et al. 1999, M. Paunovic, Zoological Department for Vertebrata, Natural History Museum, Belgrade, Yugoslavia, personal communication). It is also recorded in all the continental parts of Croatia (N. Tvrtkovic, Croatian Natural History Museum, Zagreb, Croatia, personal communication), in eastern Albania (C. Prigioni, Department of Animal Biology, University of Pavia, Italy, personal communication), and northern Greece (A. Legakis, Zoological Museum, Department of Biology, University of Athens, Greece, personal communication) (Fig. 2.1).

In the Siberian taiga, the pine marten is replaced by the closely related *M. zibellina*; some overlap occurs around the Ural Mountains in central Russia, and hybridization between the two species is not uncommon (Helldin 1998). The resulting offspring is called "kidus"; it is not believed to be fertile (Grakov 1994).

3.1.2 Habitat Relations

The pine marten is found in a variety of habitat types including insular wooded areas and shrublands (Clevenger 1993, De Marinis and Masseti 1993), alpine shrublands with coniferous and broad-leaved stands (Fornasari et al. 2000), lowland deciduous forests (Marchesi 1989), mesic pine stands (Fedyk et al. 1984, Jędrzejewski et al. 1993), and spruce-dominated forests (Pulliainen 1984, Brainerd et al. 1994).

Although many respondents were unable to describe marten-habitat relationships, it appears that forested areas continue to be the main strongholds of this species. In Britain and Ireland, small marten populations occur in young and old forests, and riparian woodland. In these heavily deforested countries, pine martens also use alternative three-dimensional habitats provided by rocky mountains and cliffs. It is suggested that these habitats provided refuges for pine martens when forest cover fell to as low as 4%; today rock crevices still provide secure natal denning sites in place of tree cavities that are scarce in modern forests in the British Isles (Birks et al. 2003). In Portugal, the species may be associated with forested hills. In France, Switzerland, Austria, Hungary, Bulgaria,Yugoslavia, Italy, Sweden, Poland, Lithuana, Albania, and Croatia, marten populations reach higher densities in mature or old coniferous, deciduous or mixed forests. While Hayden and Harrington (2000) consider pine marten to be extremely adaptable and opportunistic, respondents reported that martens are usually scarce or absent in agricultural lands, urban developments, and in areas without trees.

The presence of martens in forested areas and, concurrently, their absence in treeless areas, raise concerns about the effects of forestry development in several countries. For example, respondents reported a decrease in mature and old-growth forests, and an increase in <20-year-old stands in Sweden and Latvia during the last 5 years. Because clearcuts (barren or with trees <1 m in height) and fragmentation of mature forest types have been documented to exert negative effects on pine martens (Brainerd et al. 1994, Kurki et al. 1998), recent landscape changes resulting from forestry practices could have long-term effects on the distribution of the species. The scarcity of tree cavities suitable as natal den sites in managed forests may be a limiting factor to pine marten populations (Brainerd et al. 1995, Zalewski 1997, Birks et al. 2003). Also, respondents from Spain, France, Italy, Austria, Switzerland, Albania, Croatia, Bulgaria, Greece and Turkey have identified habitat loss resulting from forestry practices as a major concern for pine marten conservation.

3.1.3 Population Status and Trends

Hunting or trapping of pine martens is permitted in 13 of 25 countries (Table 2.1). In most of these countries, forested regions still cover large areas, and most marten populations are considered stable. In Latvia, the marten population is increasing, likely because of decreased hunting pressure due to a significant reduction in fur prices (Ozolinš and Pilats 1995). In Scandinavia, marten population densities may be limited by red fox (*Vulpes vulpes*) predation and competition (J.-O. Helldin, Grimsö Wildlife Research Station, Swedish University of Agricultural Sciences, Riddarhyttan, Sweden, personal communication; see Storch et al. 1990, Lindström et al. 1995), possibly in combination with modern forestry practices (Brainerd 1997). Fox predation may limit marten populations in other countries, especially where tree cover is low (Birks et al. 2003). In Austria, where martens may be captured as furbearers or pests, animals are trapped year-round, and little is known about population trends (Table 2.1). In France, the pine marten was removed from the national list of potential pest species in 2002 (Moutou 2003).

In most countries where the pine marten is protected, population trends are either increasing due to habitat improvement, or are unknown (Table 2.1). In the latter case, the pine marten is so rare that respondents did not want to risk an assessment. Pine marten populations may be decreasing in Albania due to habitat loss (C. Prigioni, Department of Animal Biology, University of Pavia, Italy, personal communication), and in Portugal, because of forest replacement by *Eucalyptus* plantations, which support fewer prey and resting and denning sites (Santos-Reis, Faculdade de Ciêcias, Lisbon University, Portugal, unpublished data). Respondents from Britain, Germany, Ireland, Luxembourg, and The Netherlands reported that pine martens are threatened by habitat fragmentation, loss of connectivity between populations, increased urbanization and roadkills, increased predation by foxes, illegal or widespread use of toxicants (particularly herbicides and rodenticides), and illegal trapping or shooting by gamekeepers (Strachan et al. 1996). Habitat loss and overharvesting or poaching threatens marten populations in France (T. Lode, Laboratoire d'Écologie Animale, UFR Sciences, Université d'Angers, France, personal communication), Romania (M. Dumitru, "Grigore Antipa" National Museum of Natural History, Bucharest, Romania, personal communication), and Turkey (Ö. E. Can, Turkish Society for the Conservation of Nature, Ankara, Turkey, personal communication).

3.1.4 Research and Management Needs

While information about pine marten populations is limited, there are a few research and management programs that are evaluating population monitoring techniques (Britain), distribution (Hungary), reintroduction (Ireland), reproduction, mortality and dispersal (France and The Netherlands), and general ecology (Poland and Spain). Harvest records are maintained in many countries (Table 2.1) and are the primary data used to monitor populations.

Table 2.1. Responses to questionnaires on the status of pine marten populations since 1995.

	Marten Population Status			Marten Harvest				
Country	No Status	Furbearer/Pest	Special Status	Length of trapping/hunting season (days)	Mean Annual Harvest	Data Collected	Population Trend[a]	Reasons for Population Change
Albania			Protected (Vulnerable or endangered)[b]				D	Habitat loss
Austria		x		365	700–1000	Number and location of captures	U	
Britain			Protected				I	Habitat increase
Bulgaria			Protected (Vulnerable)				I	Protection from hunting
Croatia		x		90	Unknown	Location of captures	U	
Czech Rep.		x		120	4,000–5,000	Number and location of captures	S-I	Unknown
Denmark			Protected (Threatened)				S-I	Habitat increase
France		x		365	10,000	Questionnaires to trappers	U	
Germany		x		≤140	3,000–7,000	Number of captures	S	
Greece	x						S	
Hungary			Protected (Endangered)				S	
Ireland			Protected				I	Habitat increase & legal protection
Italy			Protected				U	
Latvia		x		166	800	Number and location of captures	I	Decreased trapping

Table 2.1. Continued.

Country	Marten Population Status			Marten Harvest				
	No Status	Furbearer/ Pest	Special Status	Length of trapping/hunting season (days)	Mean Annual Harvest	Data Collected	Population Trend[a]	Reasons for Population Change
Lithuania		x		275	200–300	Number of captures[c]	S	
Luxembourg	x						U	
Poland		x		181	900-1,200	Number of captures	S	
Portugal			Protected (Undetermined)				U[d]	
Romania		x		196	1,500	Quotas	S	
Spain			Protected (Threatened)				S-I	No harvest, and habitat recovery in the mountains
Switzerland		x		168	170		S	
The Netherlands			Protected (Vulnerable)				U	
Turkey		x		77	500–1,000		S	
Sweden		x		165	10,000	Number and location of captures	D	Predation by fox
Yugoslavia		x[e]		120	> 1,250		S-I	Habitat conservation

[a]S = stable; I = increasing; D = decreasing; U = unknown.
[b]Hunting is allowed by special permission of the Ministry of Agriculture.
[c]Total number of *Martes martes* and *M. foina*.
[d]Under the International Union for Nature Conservation, the "undetermined" category implies that the species is threatened to an unknown extent. Martens are scarce and its populations are probably in decline due to habitat loss.
[e]Protected in the northern province of Vojvodina.

There is a need to develop cost-effective detection, survey, and monitoring methods for pine marten populations, particularly those at low densities. This should be done in parallel with the development of recovery programs for sparse populations inhabiting fragmented landscapes. The restoration and linkage of woodlands should be promoted to maximize the viability of populations. Management should also include measures to increase the availability of arboreal cavities suitable as natal dens.

3.2 The Stone Marten (*Martes foina*)

3.2.1 Distribution

The stone marten occurs from Mongolia and the northern Himalayas to most of Europe (Fig. 2.2). It is absent from most of the Mediterranean islands except Crete, and from Great Britain and Ireland. The northern limit of its range is Denmark (Lachat 1991). The distribution of the stone marten has increased in many European countries (e.g., Swiss Jura, Denmark, Germany, and Poland) (Godin and Vivier 1995; A. Zalewski, Mammal Research Institute, Polish Academy of Science, Poland, personal communication). In The Netherlands, the stone marten was found along the border with Germany in 1980 (Broekhuizen and Müskens 1984). The species range has now expanded to include central portions of the country, both in the south and the north, (S. Broekhuizen, Wageningen, The Netherlands, personal communication). The stone marten is widespread in Portugal (but absent in the Atlantic islands, and Madeira and Azores; Santos-Reis 1983), France except Corsica (Bouchardy and Libois 1986), Luxembourg (A. Baghli, National History Museum, Luxembourg, and L. Schley, Service de la Conservation de la Nature, Direction des Eaux et Forêts, Luxembourg, personal communication), Switzerland (S. Capt, Centre Suisse de Cartographie de la Faune, Neuchâtel, Switzerland, personal communication), Denmark (T. Asferg, National Environmental Research Institute, Department of Landscape Ecology, Rønde, Denmark, personal communication), Germany (M. Stubbe, Institut für Zoologie, Martin-Luther-Universität, Halle, Germany, personal communication), Austria (A. Kranz, Hunting Association of Styria, Graz, Austria, personal communication), Hun gary (M. T. Apathy, Department of Biology, Eotvos Lorand University, Budapest, Hungary, personal communication), Bulgaria (N. Spassov, National Museum of Natural History, Sofia, Bulgaria, personal communication), Serbia and Montenegro (Milenkovic 1985, Mitchell-Jones et al. 1999, M. Paunovic, Zoological Department for Vertebrata, Natural History Museum, Belgrade, Yugoslavia, personal communication), Greece (A. Legakis, Zoological Museum, Department of Biology, University of Athens, Greece, personal commu-

Figure 2.2. General distribution of *Martes foina* in Europe (after Broekhuizen and Müskens 1984; S. Broekhuisen, Wageningen, The Netherlands, personal communication; M. Dumitru, "Grigore Antipa" National Museum of Natural History, Bucharest, Romania, personal communication; A. Legakis, Zoological Museum, Department of Biology, University of Athens, Greece, personal communication; C. Prigioni, Department of Animal Biology, University of Pavia, Italy, personal communication; F. Spitzenberger, Museum of Natural History, Vienna, Austria, personal communication).

nication), Italy (Serafini and Lovari 1993, Fornasari et al. 2000, De Marinis et al. 2000, P. Genovesi, National Wildlife Institute, Italy, personal communication), the Czech Republic (Andera and Hanzal 1996; M. Andera, Department of Zoology, National Museum, Praha, Czech Republic, personal communication), Albania (except in the Alps; C. Priogioni, Department of Animal Biology, University of Pavia, Italy, personal communication) and Croatia (N. Tvrtkovic, Croatian Natural History Museum, Zagreb, Croatia, personal communication) (Fig. 2.2). In Spain, the stone marten is widespread but absent from coastal environments and areas intensively farmed for cereal crops (E. Virgós, Instituto de Investigación en Recursos Cinegéticos, Spain, personal communication). In Romania, distribution is patchy and partly overlaps that of the pine marten. The stone marten occurs in the northwest between the cities of Cluj and Hunedoara, and in the south, near the cities of Craiova, Brasov, and Galati (M. Dumitru, "Grigore Antipa" National Museum of Natural History, Bucharest, Romania, personal communication).

In Lithuania, the stone marten is not as common as the pine marten, and its distribution is patchy, with greater densities in the south (L. Balciauskas, Institute of Ecology, Vilnius, Lithuania, personal communication) (Fig. 2.2). In Latvia, the species is rare, and considered to be at the periphery of its distribution (Ozolinš and Pilats 1995). A few martens occur in Estonia (Timm 1991). The species is also present in the forests of the Carpathians (Bakeyev 1994). It occurs throughout the Balkans, but its distributional dynamics are poorly documented (Kryštufek 2000).

The stone marten is present in the Ukraine and Russia, with higher populations in areas where hunting is prohibited, such as in Chernobyl near the site of the nuclear catastrophe (Bakeyev 1994). In Russia, the stone marten occurs in the Caucasus and the Crimea, as far east as the Volga River (Fig. 2.2). Whereas the ranges of stone martens and pine martens overlap extensively, population sizes of the 2 species on a site have been reported to be inversely related. Pine martens are more common in extensive forests; stone martens in areas with less forest and more openings (Bakeyev 1994).

The stone marten is best adapted to warm climates and lacks morphological adaptations (i.e., its fur is less dense and its feet are hairless) to survive severe winters with deep snow (Lachat Feller 1993, Bakeyev 1994). However, with increasing populations, the stone marten inhabits mountain forests almost to the subalpine zone (Bakeyev 1994). It occurs to 2,400 m altitude in the Alps, and 2,000 m in the Pyrenees (Saint-Girons 1973). In India, Prater (1971) reports the presence of stone martens in Kashmir and the Himalayas (between 1500 and 3600 m ASL, Pocock 1999) (Fig. 2.3). Choudhury (1997a) reported the stone marten in the middle and higher ranges of the Eastern Himalaya and

Figure 2.3. General distribution of *Martes foina* in Asia (after Chotolchu et al. 1980, Bakeyev 1994, Heilin et al. 1999).

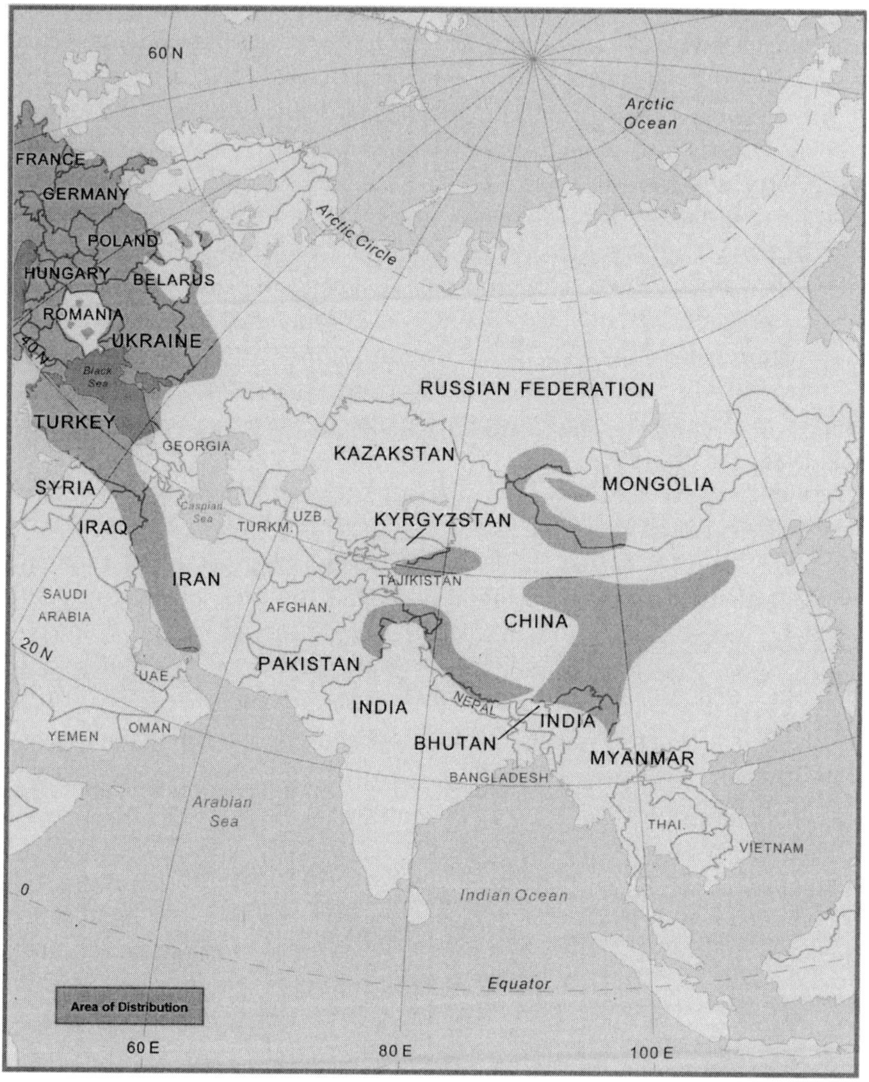

Mishmi Hills, where it coexists with the yellow-throated marten. Mallon (1991) recorded stone martens in northern India, near the border of Pakistan and the People's Republic of China. He believed that they were widely distributed at low densities in mountainous areas. The stone marten also occurs in the Annapurna Mountain Range of Nepal (Oli 1994).

In Iraq, Hatt (1959) believed that the stone marten was probably confined to hilly forestlands (Fig. 2.3). However, recent records on the distribution of this species in Iraq, Iran, and Syria are lacking. The stone marten occurs in central and northwest China where its distribution overlaps with the sable (Helin et al. 1999). The stone marten also occurs in southwest Mongolia, mostly along the Chinese border (Chotolchu et al. 1980) (Fig. 2.3).

3.2.2 Habitat Relations

The stone marten frequents forests (Amores 1980, Mickevicius and Baranauskas 1992), cork oak (*Quercus suber*) woodlands (Santos-Reis et al. 2003), rocky areas (Waechter 1975, Mallon 1991), fields, pastures, gardens and wooded farmlands (Lachat Feller 1993, Genovesi and Boitani 1997), and villages and towns (Waechter 1975, Clément and St-Girons 1982, Lucherini and Crema 1993, Tóth 1998). The stone marten is well adapted to humans and continues to expand its range in suburban and urban areas (Bouchardy and Libois 1986, Lachat 1991).

Respondents from France, Switzerland, Denmark, Germany, Italy, Hungary, Romania, Czech Republic, Greece, Lithuana, Croatia, Poland, and Yugoslavia indicated that densities of stone martens were greatest in agricultural, industrial and urban areas. In Portugal and Spain, however, the stone marten is not closely associated with human settlements as in central Europe. When resting, the stone marten prefers mature oaks or riparian vegetation; when foraging, it selects cultivated fields and riparian vegetation (Santos-Reis et al. 2003). In Spain, the stone marten prefers rocky areas and riparian and plain forests to urban and rural habitats (Virgós et al. 2000, E. Virgós, Instituto de Investigación en Recursos Cinegéticos, Spain, personal communication). Likewise, in Albania, stone martens are more frequent in riverine habitat with good riparian vegetation (C. Prigioni, Department of Animal Biology, University of Pavia, Italy, personal communication). Interestingly, a feral population of stone martens was established 20 years ago near Milwaukee, in southeast Wisconsin, USA where the animals inhabit small open and forested deciduous uplands (Long 1995).

3.2.3 Population Status and Trends

In most countries where it occurs, the stone marten is a legally harvested species with stable or increasing populations (Table 2.2). The species is often viewed as a pest and is hunted in response to damages to houses and cars, poultry depredation, smells associated with feces, urine and prey remains, and noise (Waechter 1975, Lachat 1991, Lucherinni and Crema 1993, T. Asferg, National Environmental Research Institute, Department of Landscape Ecology, Rønde, Denmark, personal communication). In Romania, the annual har-

Table 2.2. Responses to questionnaires on the status of stone marten populations since 1995.

Country	Marten Population Status: No Status	Marten Population Status: Furbearer/ Pest	Marten Population Status: Special Status	Marten Harvest: Length of trapping/hunting season (days)	Marten Harvest: Mean Annual Harvest	Data Collected	Population Trend[a]	Reasons for Population Change
Albania		x		181	Unknown		U	
Austria		x		365	1,000		S	
Bulgaria		x			Unknown[b]		S	
Croatia		x		360	Unknown		S-I	
Czech Rep.		x		120	5,000–6,000	Number and location of captures	I	Possible high food supply (pigeons) in urban areas
Denmark		x		133	4,250	Number and location of captures, and questionnaires to hunters	S-I	Unknown
France		x		365	30,000	Questionnaires to trappers	S	
Germany		x		≤140	46,000–52,000	Number of captures	S	
Greece	x							
Hungary		x[c]		168	200–400	Harvest records and questionnaires to hunting clubs	S-I	
Italy	x							
Latvia		x		166	Insignificant		U	
Lithuania		x		275	200–300	Total captures of *M. martes* and *M. foina*	S	

Table 2.2. Continued.

	Marten Population Status			Marten Harvest				
Country	No Status	Furbearer/ Pest	Special Status	Length of trapping/hunting season (days)	Mean Annual Harvest	Data Collected	Population Trend[a]	Reasons for Population Change
Luxembourg		x		137	270	Number and location of captures	S	
Poland		x		181	Unknown	Total captures of *M. martes* and *M. foina*	S	
Portugal	x						U	
Romania		x		196	500	Quotas	D	Poaching
Spain	x						U	
Switzerland		x		168	2,600		S	
The Netherlands			Protected				S-I	
Turkey		x		77	500–1,000		S	
Yugoslavia (Serbia and Montenegro)		x[c]		120	>2,500		S-I	

[a] S = stable; I = increasing; D = decreasing; U = unknown.
[b] Pelt numbers are recorded at the regional level and it is difficult to obtain a figure for the national harvest.
[c] Since 1993 due to increasing urban populations.
[d] Protected in the northern province of Vojvodina.

vest has decreased from 1,000 in 1980 to 500 in 2000. There is a general decline of the population possibly caused by poaching activities (M. Dumitru, "Grigore Antipa" National Museum of Natural History, Bucharest, Romania, personal communication). In Turkey, overharvesting is also a concern (Ö. E. Can, Turkish Society for the Conservation of Nature, Ankara, Turkey, personal communication). In The Netherlands, the species has been protected until recently. However, its status will soon be revised, and it will be possible to catch and kill individuals that cause serious damage (S. Broekhuizen, Wageningen, The Netherlands, personal communication). In Greece and Italy, the species has no special status. In Italy, fewer than 100 animals are thought to be killed each year for damage control (P. Genovesi, National Wildlife Institute, Italy, personal communication).

The stone marten is well adapted to survive in agricultural and urban areas, and most respondents did not identify any population threat. In Spain, however, the stone marten may be threatened by non-selective predator control programs, particularly poisons, and habitat fragmentation (E. Virgós, Instituto de Investigación en Recursos Cinegéticos, Spain, personal communication). In Portugal, stone marten populations are reduced by habitat loss (deforestation, summer fires, afforestation with *Eucalyptus*), poisoning, and trapping.

3.2.4 Research and Management Needs

There is currently little research on the stone marten. Recent studies were conducted or are still underway on habitat preference and food habits in Spain (e.g., Virgós et al. 2000), Germany, Luxembourg, Italy (e.g., Genovesi and Boitani 1997), Hungary, Croatia, and Poland. More research on the distribution of the stone marten in Asia, particularly in mountainous regions, is needed. There is a need to develop cost-effective detection, survey, and monitoring methods. Interspecific relationships of martens inhabiting agricultural and urban areas should be studied. In these areas, stone martens are either protected, legally hunted or controlled as pests. The dynamics of populations subject to different management programs should be investigated in order to better assess the effects of human activities on the viability of populations.

3.3 The Sable (*Martes zibellina)*

3.3.1 Distribution

The sable occurs in 5 countries: Russia, Mongolia, China, North Korea, and Japan (Buskirk et al. 1994) (Fig. 2.4). In Russia, the current distribution is largely the result of mass reintroductions from 1940 to 1965 involving > 19,000 animals (Bakeyev and Sinitsyn 1994). In the nineteenth and early twentieth centuries, sables were intensively harvested over vast areas, and reintroductions and subsequent protection allowed the distribution to recover. The range of the sable extends northward to the limit of trees, reflecting the tolerance of the species for extremely low temperatures. Sables extend southward to 55–60° latitude in western Siberia, and 42° in the mountains of eastern Asia. The sable occurs in the southernmost part of its distribution in mountains that tend to be peninsular or insular. To the west, the sable extends to the Ural Mountains, where it is sympatric with the European pine marten (Geptner et al. 1967, Bakeyev and Sinitsyn 1994, Grakov 1994). The sable is also found on Sakhalin Island (Corbet 1978, Geptner et al. 1967), off the eastern coast of Siberia.

In Mongolia, the sable occurs in the Altai Mountains of the far Northwest, and in forests around Lake Hovsgol. The latter sable habitat is contiguous with the Trans-Baikal boreal forest region, which produces the best-known and most valuable sable pelts. This region has the most sharply continental climate experienced by any *Martes*, with warm summers, but long, severe winters. In China, the sable currently occurs in a small area of the Xinjiang Uygur Autonomous Region, where the southern Altai Mountains enter China from the north (Fig. 2.4). In northeastern China, the sable is now limited to the Daxinganling Mountains of Heilongjiang province and Inner (Nei) Mongolia. In the Xiaoxinganling Mountains of eastern Heilongjiang, the persistence of the sable is suspected, but not confirmed (Ma and Xu 1994, Helin et al. 1999). Sables also occupy the Changbaishan Mountains along the border with, and southward into North Korea (Ma and Xu 1994). Areas of China occupied by the sable have declined drastically over the last 100 years, with the southern margin of the distribution of sables retreating northward by as much as 900 km in some places (Ma and Xu 1994). This contraction of the distribution is attributable to human activities, particularly trapping and hunting, timber harvest, and conversion of land to agriculture.

The sable occurs in Hokkaido, the northernmost major island of Japan, in the main Japanese archipelago, and on the Korean peninsula (Anderson 1970, Corbet 1978, Hosoda et al. 1997) (Fig. 2.4).

3.3.2 Habitat Relations

Sables inhabit taiga forests and their southern montane extensions. Over most of their distribution, they occupy coniferous taiga forest, but in the Daxinganling Mountains and eastward, forests are increasingly deciduous (Ma and Xu 1994). Sables prefer attributes associated with late successional stages: large diameter trees and large diameters and volumes of coarse woody debris (Buskirk et al. 1994). In northern China, these attributes tend to be found on north-facing slopes and in riparian associations. Although mixed coniferous – deciduous forests are suitable habitats, sables avoid pure deciduous stands (V.

Figure 2.4. General distribution of *Martes zibellina* in Asia.

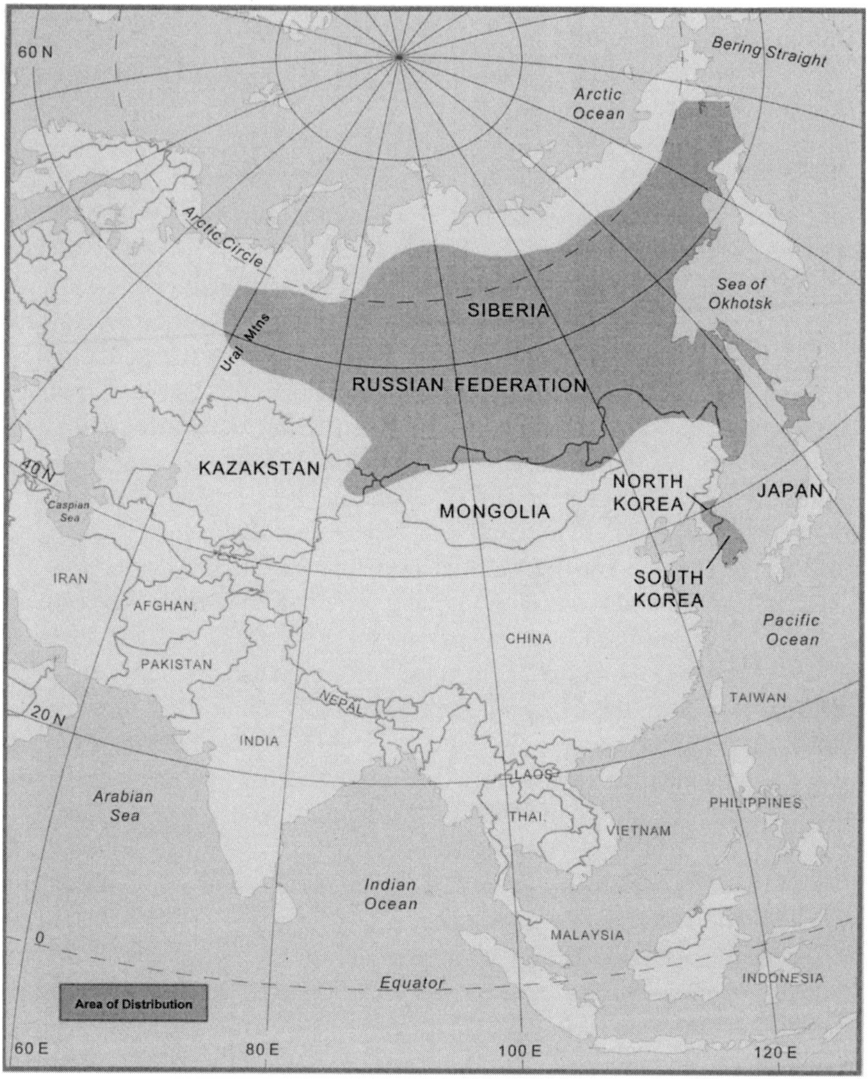

Monakhov, Institute of Plant and Animal Ecology, Ekaterinburg, Russian Federation, unpublished data). Little is known about habitats of sables in Japan.

Bakeyev and Sinitsyn (1994) believed that forest cutting and fire had not yet greatly influenced sable populations in Russia, and that timber harvest had been effective in creating habitat mosaics that support many small mammals and plants that are important foods of sables (Bakeyev and Sinitsyn 1994,

Brzezinski 1994). However, effects of extensive forestry such as large-scale clearcutting have not been studied and could detrimentally affect sables. In Japan, habitat conservation programs are virtually non-existent and the sable is affected by forest destruction and fragmentation.

3.3.3 Population Status and Trends

In Russia, the sable is a furbearer, and is subject to extensive trapping, hunting, and captive propagation, which all contribute to a lucrative fur industry. In the western part of the country, from the Urals to the Yenisey River, 19,600–36,000 pelts are harvested from the wild annually (Minkov 1998; V. Monakov, Institute of Plant and Animal Ecology, Ekaterinburg, Russian Federation, unpublished data). Yet, sable populations have remained stable in the last decade (Minkov 1998, Monakhov 1995, 2000). The most important threats to sable populations in Russia are land conversion, overharvesting, and diseases and parasites (Monakhov 1983, 1999; Valentsev 1996). In China, the sable has been listed as endangered since 1989 (Buskirk et al. 1993). Also, since 1989, all uses of sables, including for research, are under government supervision. One national and 7 provincial nature reserves totaling 812,161 ha have been established for the protection of sables and their habitats (Ma and Xu 1994). However, populations are still threatened by uncontrolled hunting, conversion of forests to other land uses, and logging (Buskirk et al. 1994).

3.3.4 Research and Management Needs

There is an apparent need for more research on sable populations and their habitat relationships. In Russia, several ongoing studies are being conducted at the Institute of Plant and Animal Ecology (Ekaterinburg), All-Russia Institute of Hunting and Fur Farming (Kirov), and Krasnoyarsk State University. In Hokkaido, only one study was carried out on sable food habits (Nitta 1982). More research is needed on sable habitat use, particularly in coniferous forests. A population monitoring program and a better control of harvest activities is required to ensure the future of sables in Japan.

3.4 The Yellow-Throated Marten (*Martes flavigula*)

3.4.1 Distribution

The yellow-throated marten occurs in sub-tropical and tropical forests from the Himalayas to eastern Russian Federation (V. Monakhov, Institute of Plant and Animal Ecology, Russian Federation, unpublished data), south to the Malay Peninsula and Sunda Shelf (Borneo, Sumatra, and Java) to Taiwan (Medway 1978, Buskirk 1994) (Fig. 2.5).

In India, the yellow-throated marten has been reported in the northeast states of Arunachal Pradesh (Choudhury 1997a), Manipur (Ramakantha 1994), and Assam (Choudhury 1997b), and in Indo-Myanmar (Burma) border areas (Ramakantha 1994). In southwest India, the Nilgiri marten (*Martes gwatkinsi*) is regarded by some as a subspecies of *Martes flavigula* (Corbet and Hill 1992). This subspecies is a rare mustelid endemic to the forested tracts of the western Ghat (Madhusudan 1995).

The yellow-throated marten occurs in central and northeast China (Helin et al. 1999), and on the Korean peninsula (Tatara 1994). In Malaya, the species is not common but it is well distributed throughout the mainland in all types of tall forest (Medway 1978). In Taiwan, it occurs in the Central Mountain Range and in southern areas (Lin 2000).

3.4.2 Habitat Relations

In spite of a general lack of data on yellow-throated marten habitat associations, observations indicate that it is associated with forested areas, both tropical and subtropical (Medway 1978, Ramakantha 1994, Choudhury 1997a, Helin et al. 1999). In the Himalayas, the yellow-throated marten inhabits the temperate forest belt between 1220 and 2745 m; it is not found above tree line. It is also found in sub-tropical and tropical forests extending downslope to the edge of the plains (Prater 1971).

3.4.3 Population Status and Trends

There is little information on yellow-throated marten population status and trends. The species is considered to be rare (Helin et al. 1999, Lin 2000). The yellow-throated marten is not typically killed for its fur, but some pelts are sold in Taiwan shops (Wang 1986).

The Nilgiri marten is listed as threatened by the IUCN (Groombridge 1993, Christopher and Jayson 1996). It is well known to the Kani tribals of south-western India. Being hunter-gatherers, the Kanis consume many types of wild animals. However, they avoid eating the Nilgiri marten because they believe its meat to be poisonous. The unpleasant body odor of the marten may be the reason for this belief (Christopher and Jayson 1996).

3.4.4 Research and Management Needs

There is a significant lack of information about yellow-throated martens and Nilgiri martens, and their habitat relationships. More research is required on the reproductive biology, food habits, movements, and behavior in order to develop sound management programs.

Figure 2.5. General distribution of *Martes flavigula* in Asia.

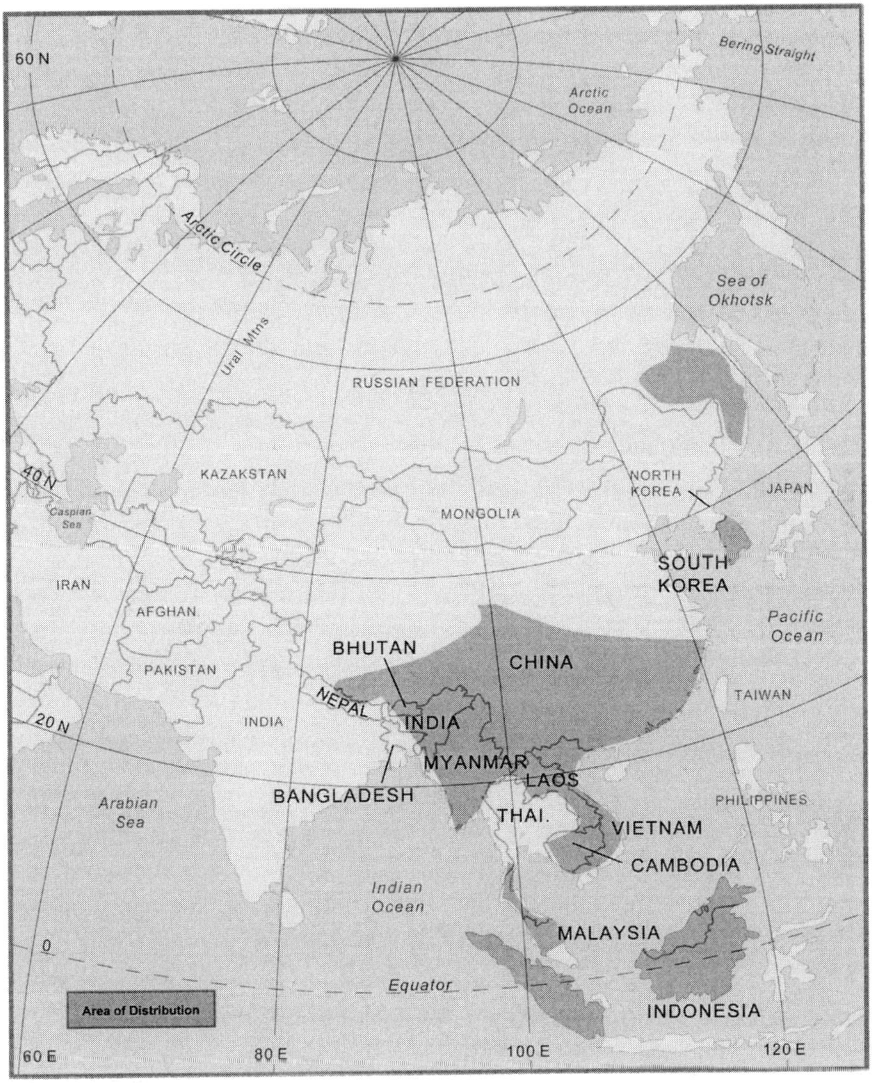

3.5 The Japanese Marten (*Martes melampus*)

3.5.1 Distribution

The Japanese marten occurs in the main Japanese archipelago and the Korean peninsula. Three separate subspecies are recognized on the basis of differences in their coat coloration (Anderson 1970, Corbet 1978): *M. m. melampus* in Honshu, on the islands of Shikoku, Kyushu, Awaji, and Sado (introduced), and in southwestern Hokkaido (introduced); *M. m. tsuensis* on Tsushima Island; and *M. m. coreensis* on the Korean Peninsula (although the identity of this subspecies is subject to controversy).

3.5.2 Habitat Relations

The Japanese marten occurs only in forested areas (Tatara 1994): mainly old deciduous forests in the north, and coniferous forests in the south. Martens avoid plantations and open fields.

3.5.3 Population Status and Trends

In Japan, this marten is trapped for its fur from 1 December to 31 January except on Hokkaido Island, where it is sympatric with the fully protected sable, and on Tsushima Islands where it is designated as a rare species by IUCN and as a species at risk by the Environment Agency of Japan. The annual harvest rate is 5,000–10,000 pelts.

Logging constitutes a serious threat to martens as large tracts of broad-leaved forests are replaced by conifer plantations, which are poor in food resources (Tatara 1994). Other threats include habitat fragmentation from roads, road kills, and mortality caused by feral dogs (Tatara 1994, M. Saeki, Osaka, Japan, unpublished observation). Greater interspecific competition by introduced carnivores (e.g., mongoose, *Herpestes* spp.; civet, *Paguma larvata*; raccoon, *Procyon lotor*; Yamada 1998) may also affect the survival of Japanese martens.

3.5.4 Research and Management Needs

Research on the reproduction and interspecific relationships of the Japanese marten is needed for science-based management. There is a need to establish sound population monitoring programs, including the management of trapping activities, and to designate protected areas.

3.6 The American Marten (*Martes americana)*

3.6.1 Distribution

The American marten occurs in forested habitats of North America north of 35° latitude. It is present in all of the Canadian territories and provinces, except Prince Edward Island. In the United States, it is found in regions west of 105° longitude (except for a re-introduced population in South Dakota), and east of 95° longitude (Fig. 2.6). The following review of marten distribution in North America covers 4 regions: eastern, central, western, and northern.

3.6.1.1 Eastern region

The marten populations of southern Québec (south of St. Lawrence River), New Brunswick, Maine, and New Hampshire are contiguous (Fig. 2.6). In Québec, the marten is absent from the St. Lawrence River valley and Anticosti Island (Newsom 1937). In New Brunswick, martens occur in the central and northwest regions of the province, and in Fundy National Park where they were introduced in the 1980s (C. Libby, Fish and Wildlife Branch, Department of Natural Resources, Fredericton, New Brunswick, Canada, personal communication). Individuals have been seen in the southern and eastern regions of the province, but their numbers are unknown. In Maine, the marten has expanded its range since 1980, due partly to the translocation of 63 martens from northern and western Maine to southeast portions of the state (W. Jakubas, Department of Inland Fisheries and Wildlife, Maine, USA, personal communication). Dispersal movements between northern Maine and Canada may occur in the northern portion of the state where forestlands are contiguous. The marten is also indigenous to New Hampshire, and has also benefited from reintroductions in the early 1970s. Martens occur in northern New Hampshire where most of the species' current habitat is in the White Mountain National Forest (E. Orff, Fish and Game Department, New Hampshire, USA, personal communication).

Disjunct populations of the American marten are found in Newfoundland, Nova Scotia, and New York (Fig. 2.6). In the 1980s, in Newfoundland, the species was found only in the western part of the province (Forsey et al. 1995). However, after a series of reintroductions from 1984–1988, a small population is now present on the east side of the island, in Terra Nova National Park. In Nova Scotia, there is a small remnant population on Cape Breton Island (Anonymous 1998) and a reintroduced population in the south. The marten population in the Adirondack Mountains of New York State is second to Maine's in size, and is disjunct from all other martens in the northeast United States (M. Brown, Department of Environmental Conservation, New York, USA, personal communication).

3.6.1.2 Central region

The marten populations of Labrador (Province of Newfoundland), Québec (north of St. Lawrence River), Ontario, Manitoba, and Minnesota are contiguous (Fig. 2.6). In Labrador, martens are found over most of the territory, except in the northernmost areas dominated by tundra (R. Otto, Inland Fish and Wildlife Division, Labrador, Canada, personal communication). The marten is present in all forest regions of Québec north of the St. Lawrence River (Prescott and Richard 1996, Fortin et al. 1997). Although the species distribution did not include the Ungava Peninsula in 1980, recent research has detected it there

Figure 2.6. General distribution of *Martes americana* in North America (after Gibilisco 1994, Aune and Schladweiler 1997, Groves et al. 1997, Johnson and Cassidy 1997, Zielinski et al. 2001; W. Melquist, Idaho Department of Fish and Game, Idaho, USA, personal communication).

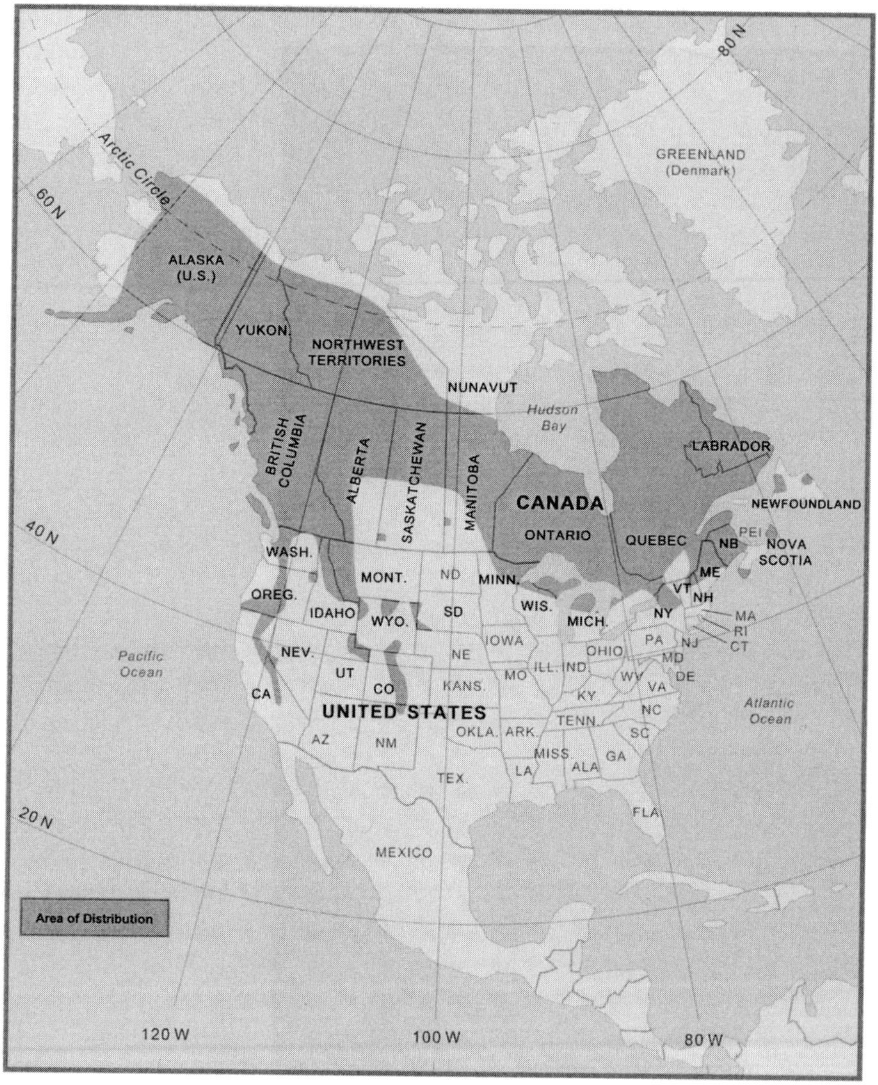

(Fortin et al. 1997). In Ontario, the marten is found throughout most of the province; the Algonquin region is on the southern fringe of the species' current range (Strickland 1989) (Fig. 2.6). The distribution of marten populations in Ontario and Manitoba is contiguous along their common border. In Manitoba, however, martens are mainly located north of Lake Winnipeg (~ 53° latitude)

(Gibilisco 1994). In southwestern Manitoba, a disjunct population has been reintroduced (1991–1993) in Riding Mountain National Park, in the forest-agricultural transition zone; reproduction has been observed since then (Schmidt and Baird 1995). In Minnesota, martens occur only in the northern districts. In northeastern Minnesota, its distribution is contiguous with forested areas in Ontario (Berg and Kuehn 1994). In the northwest, however, the species' distribution extends to the edges of the prairies (W.E. Berg, Minnesota Department of Natural Resources, unpublished report).

In Michigan, following a successful reintroduction in the northern Lower Peninsula in 1985–1986, martens appear to be expanding their distribution widely, but Earle and Reis (1996) believe that they are still below the region's carrying capacity. Martens have been reintroduced to the Upper Peninsula 3 times between 1955–1981. Natural dispersal has also been supplemented by translocations in 1990 and 1992 (Earle and Reis 1996). The presence of marten has recently been confirmed throughout most of the upper peninsula (Earle 1999). In Wisconsin, after several reintroduction programs from 1975–1987 (Kohn and Ashbrenner 1996), martens are found in Nicolet and Chequamegon National Forests and adjacent areas (Anonymous 2000). The Wisconsin population is contiguous with that of Michigan, and is close to the southernmost range of the Minnesota population (Fig. 2.6). A total of 125 martens were released in the Black Hills of South Dakota from 1980–1983. The population is considered to be well established on the basis of documented reproduction, observations, and recoveries (Fredrickson 1995).

3.6.1.3 Western region

In western North America, the marten is found in 3 Canadian provinces (Saskatchewan, Alberta, and British Columbia) and 10 states (Washington, Oregon, California, Nevada, Montana, Idaho, Wyoming, Colorado, Utah, and New Mexico) (Fig. 2.6).

From the eastern border of Saskatchewan to the western border of British Columbia, the distribution is contiguous. In Saskatchewan, the marten occurs in boreal ecoregions. It is rare in the lower portion of the southern boreal ecoregion, but common in the northern and subarctic regions (A. Arsenault, Saskatchewan Environment and Resource Management, Saskatoon, Saskatchewan, Canada, personal communication). It is also present in the southeast, in Cypress Hills Provincial Park, a coniferous and mixed forested area within the mixed-grass prairie, where females with kits and adult males were released in 1986 (Hobson et al. 1989). In Alberta, it is present in boreal, subalpine and montane forest regions (Skinner and Todd 1988, G. Proulx, Alpha Wildlife Research & Management Ltd., Alberta, Canada, unpublished data). In British Columbia, the marten is present throughout the province and on the coastal islands (M. Badry, Ministry of Environment, Lands and Parks, Victoria, British Columbia, personal communication) (Fig. 2.6).

In the western United States, marten populations typically occur in upper-elevation montane habitats, and are geographically disjunct (Gibilisco 1994, Graham and Graham 1994). In Washington, the marten is found in 4 distinct regions corresponding to the Selkirk Mountains in the northeast, the Blue Mountains in the southeast, the Cascade Range in the center, and the Olympic Mountains in the northwest (Johnson and Cassidy 1997) (Fig. 2.6). The range of the marten in coastal areas of Washington contracted substantially during the 20th century, and appears now to be restricted to a small population on the east slope of the Olympic Mountains (Zielinski et al. 2001). In Oregon, the marten is found in the Blue Mountains in the northeast, the Cascade Range in the center, and the southern portions of the Coast Range in the west (Marshall 1992). Coastal populations are restricted in distribution and abundance. In California, the marten is found in several mountain ranges including the Klamath, the Cascades, and the Sierra Nevada, but has drastically decreased in northwestern California within the range of *M. a. humboldtensis* (Zielinski et al. 2001, Slauson 2003). Although the distribution and abundance of inland marten populations in the Pacific States have remained relatively stable, coastal populations have been substantially reduced in distribution and appear to occur at extremely low densities (Zielinski et al. 2001); these populations are particularly vulnerable to extirpation. In Nevada, martens have been documented in the Tahoe Basin and along portions of the Carson Range (S. Espinosa, Department of Wildlife, Nevada, personal communication). In Montana, martens occur on the west side of the state. In the northwest region, they occur in habitats ranging from low forested valley bottoms to the alpine zone (Aune and Schladweiler 1997). In southwestern Montana, martens are restricted to high elevation mountain ranges (Fig. 2.6). Their distribution is interrupted by large open grassland valleys, resulting in naturally fragmented habitats that isolate populations (Gibilisco 1994). In Idaho, on the basis of habitat types (Groves et al. 1997) and capture locations, the distribution of marten populations is likely limited to the northern half of the state, which includes the Bitterroot Range and several groups of mountains. There is also a small population in southern Idaho where 59 martens were re-introduced in 1993 and 1994 (W. Melquist, Idaho Department of Fish and Game, Idaho, USA, personal communication). In Wyoming, the marten is found in the Absaroka Mountains, Bighorn Mountains, Wind River Mountains, Uinta Mountains, and Medicine Bow Mountains. In Utah, martens are most abundant in mature forest stands located in the Uinta Mountains (C. McLaughlin, Division of Wildlife Resources, Utah, personal communication). Scattered sightings also indicate their presence in other high, forested ranges such as the Wasatch Mountains (Parker 2001). In Colorado, the marten is found in the western half of the state, which is characterized by the presence of several mountain ranges such as the Rocky Mountains, the Sangre de Cristo Mountains, and the San Juan Mountains (Fitzgerald et al. 1994, Byrne 1998). The American marten is very rare and restricted in distribution in New Mexico, but its presence has been verified in the San Juan and Sangre de Cristo Mountains (Anonymous 1996).

3.6.1.4 Northern region

The American marten generally occurs throughout forested areas of northern Canada and Alaska (Fig. 2.6). The range encompasses most of the Northwest Territories, from the Mackenzie Delta to the southern border of the territories (R. Mulders and R. Popko, Resources, Wildlife, and Economic Development, Government of the Northwest Territories, Northwest Territories, Canada, personal communication). In Nunavut, the species is limited to narrow zones in the northwest and along the southern border of the Territories. The marten is absent along eastern portions of the arctic coastal region and on the islands of the high arctic. The marten is found throughout most of the Yukon, being absent only in the treeless tundra of the north and in the high mountains of the southwestern region of the province (H. Slama, Yukon Department of Renewable Resources, Whithorse, Yukon, Canada, personal communication). In the Whitehorse area of southwestern Yukon, marten declined during the 1940s and 1950s because of habitat loss and overtrapping. However, martens are now more common (H. Slama, Yukon Department of Renewable Resources, Whithorse, Yukon, Canada, personal communication) following a successful reintroduction in the late 1980s (Slough 1994).

The American marten is common in the central and southern portions of mainland Alaska, south and east of the northern tree line (Fig. 2.6). Insular populations are recorded for Afognak Island in southcentral Alaska, and for several large islands (some with introduced populations from the 1930s and 1940s) in southeast Alaska. The presence of the marten on remote smaller islands in southeast Alaska is uncertain because these islands are typically not trapped (R. Flynn, Department of Fish and Game, Douglas, Alaska, USA, H. Golden, Department of Fish and Game, Anchorage, Alaska, USA, and M. McNay, Department of Fish and Game, Fairbanks, Alaska, USA, personal communication).

3.6.2 Habitat Relations

The American marten is a forest specialist. Martens are associated with areas of overhead cover, especially near the ground, large volumes of large-diameter (> 50 cm dbh) live trees, snags, and coarse woody debris for denning and resting, and small-scale horizontal heterogeneity especially the interspersion of herbaceous vegetation and patches of large old trees (Buskirk and Ruggiero 1994, Raphael and Jones 1997). Particularly in the West, the marten is commonly associated with late-successional mesic coniferous or conifer-dominated mixed forests (Strickland et al. 1982, Buskirk and Powell 1994). Work in Acadian forests of eastern North America has indicated that mid-successional (9–12 m in height) forests and mature forests of deciduous, mixed conifer-deciduous, and conifer compositions are preferred similarly by martens (Chapin et al. 1997, Payer 1999). These mesic forests contain high volumes of the necessary vertical and horizontal cover required by martens (Chapin et al. 1997, Payer and Harrison 2003); forest-maturity thresholds determining marten use of forest stands in the Acadian region have been estimated to be trees and snags >9 m in height with basal areas of >18 m^2/ha (Payer and Harrison 2003). Habitat fragmentation (often measured by the percent of the landscape that is unforested) even at low levels, i.e., 20–30% of a home range area, may have negative effects on martens (Thompson and Harestad 1994, Hargis and Bissonette 1997, Chapin et al. 1998, Potvin et al. 2000).

All respondents reported the importance of late-seral coniferous forests for American marten. In most jurisdictions, logging has been identified as a major threat for the species. Concerns are mainly about the loss of canopy cover and coarse woody debris (e.g., Flynn and Schumacher 1999).

Although some timber harvesting occurs in the Northwest Territories, Yukon and Alaska, the predominant disturbance is fire. While burns with early successional shrub-sapling vegetation may be inhabited by juvenile martens, they are not used by adult females, and they may act as population sinks for nonbreeders (Paragi et al. 1996). In many jurisdictions, insect epidemics, e.g., bark beetles (*Dendroctonus* spp.) and spruce budworm (*Choristoneura fumiferana*), have resulted in intensive timber harvest operations, often with little or no forest retention, that impact significantly on marten habitat. On the other hand, Yeager (1950) reported that, while outbreaks of the Engelmann spruce bark-beetle (*Dendroctonus engelmanii*) created forests of standing dead trees, such outbreaks were not detrimental to martens where preferred small mammals were still present and cover was provided by residual fir (*Abies* spp.) stands. Chapin et al. (1997) also reported that forest stands with significant mortality from spruce budworm were preferred by marten, despite a canopy closure of mature trees that was typically <30%. These naturally disturbed stands were characterized by increased numbers of snags, windfalls, and root mounds. Habitat loss through urban and sub-urban encroachment is an issue in some jurisdictions.

3.6.3 Population Status and Trends

Trapping seasons for American martens occur in 17 of 25 jurisdictions (Table 2.3). Annual harvests range from 30 in Oregon to ≥30,000 in Québec and Ontario (Robitaille 2000; Table 2.3). In most of these areas, marten populations appear stable. Because habitat loss is a general concern, however, careful monitoring will be required to ensure the future of sustainable populations. For example, in the Northwest Territories and the Yukon, increased numbers of seismic trails for oil and gas exploration results in greater access to areas that have received little trapping pressure in the past. The combined effects of habitat degradation and trapping pressure might be detrimental to the future of marten populations (Banci and Proulx 1999). In New Brunswick, Michigan, Washington, and Oregon, harvest seasons are relatively short, and appear to take into account the small populations. In Colorado, the season is closed due to a lack of ecological and population data. Finally, in some jurisdictions (e.g., Michigan, Maine), quotas have been established to better control fur catches.

The status of most protected American marten populations is unknown (Table 2.3). However, respondents have identified serious threats such as demographic and environmental stochasticity in California, South Dakota, and islands in southeast Alaska. In California, the range of the Humboldt marten has been reduced to one small area that probably contains fewer than 20 individuals (Zielinski et al. 2001, Slauson 2003). The decline of the Humboldt marten is probably the result of habitat loss due to excessive logging of the redwood region during the 20th century. In South Dakota, coyotes may be a threat to marten during winter, when access into deep snow areas is facilitated by compacted snowmobile trails (Buskirk et al. 2000). The resumption of trapping has been, and continues to be, an objective for restoring marten to the Black Hills in South Dakota. If trapping occurs, the population could become vulnerable to over-exploitation unless the harvest is strictly regulated and monitored. In Newfoundland, the marten is endangered by habitat loss, but also by incidental capture from snowshoe hare (*Lepus americanus*) snaring and fur trapping (Thompson 1991, Proulx et al. 1994a, B. Hearn, Canadian Forest Service, Corner Brook, Newfoundland, personal communication).

3.6.4 Research and Management Needs

Marten populations are the subject of many investigations throughout North America. Whereas surveys (e.g., remote cameras, track plate surveys, aerial and ground track counts, questionnaires, trappers' logbooks) are being conducted in many jurisdictions (e.g., New Brunswick, Québec, Washington, Oregon, and California), there is a need to identify and quantify regional habitat requirements of martens in order to customize forest management plans. Conservation assessments at the regional level (e.g., Proulx 2001) are needed to

Table 2.3. Responses to questionnaires on the status of American marten populations since 1995.

	Marten Population Status			Marten Harvest				
Country	No Status	Furbearer/ Pest	Special Status	Length of trapping/hunting season (days)	Mean Annual Harvest	Data Collected	Population Trend[a]	Reasons for Population Change
Alaska		x		77 to 120 depending on regions	> 3,000	Number and location of captures in most regions; age-sex composition in select areas	S	
Alberta		x		92	1,100	Number and location of captures		
British Columbia		x		120	18,500	Number and location of captures	S	
California			Protected			Camera and track surveys, roadkills	U	
Colorado	x (non-game)						U	
Idaho		x		92	575	Harvest locations	U	
Labrador		x		150	1,000–5,500	Number and location of captures	S	
Maine		x		61	3,795 (quota of 25 martens/trapper/ season)	Number and location of captures, and voluntary trapper log book	S	
Michigan		x		10	1 per trapper		S-I	
Nevada			Protected			U		
Newfoundland			Protected (endangered)				S	

Table 2.3. Continued.

	Marten Population Status			Marten Harvest				
Country	No Status	Furbearer/ Pest	Special Status	Length of trapping/hunting season (days)	Mean Annual Harvest	Data Collected	Population Trend[a]	Reasons for Population Change
New Brunswick		x		14	1,800–2,200	Number and location of captures, and age-sex composition; camera-track surveys	S	
New Hampshire			Protected (threatened)				I	
New York		x		44	91 (6 martens per trapper/ season)	Number and location of captures, trapping effort, sex-age composition, and voluntary log book		
Northwest Territories		x		135	6,000–11,000	Number and location of captures; age-sex composition in select areas	S	
Nova Scotia			Protected (endangered)				U	
Oregon		x		61–92 depending on regions.	30	Number and location of captures, trapping effort, and sex-age composition, camera and track surveys	S in most regions D in coastal mountains	Non-coastal regions: urban encroachment Coastal:timber harvest

Table 2.3. Continued.

Country	Marten Population Status			Marten Harvest				
	No Status	Furbearer/ Pest	Special Status	Length of trapping/hunting season (days)	Mean Annual Harvest	Data Collected	Population Trend[a]	Reasons for Population Change
Québec		x		128	30,000	Number and location of captures, trapping effort, and sex-age composition	S	
Saskatchewan		x		120	2,400	Number and location of captures	S	
South Dakota			Protected			Agency personnel interviews	U	
Utah		x		126	<100	Number and location of captures, and sex-age composition	S	
Washington		x		48–71 depending on regions.	70	Number and location of captures, and trapping effort, camera and track surveys	S in most regions D in coastal mountains	Non-coastal regions: urban encroachment Coastal:timber harvest
Wisconsin			Protected				U	
Wyoming		x		182	Unknown		U	
Yukon		x		118	5,000	Number and location of captures	S	

[a]S = stable; I = increasing; D = decreasing; U = unknown.

develop effective management programs. The effects of severe forest fires, bark beetle infestations, emerging silvicultural practices (e.g., partial harvests vs. clearcutting), agricultural developments in forested regions, and urban sprawl on habitat use by martens need to be investigated. In British Columbia, habitat selection (e.g., Therrien and Eastman 1999, Proulx and Kariz 2001) and connectivity (Proulx and Verbisky 2001) studies are underway. More work on habitat fragmentation and protection (e.g., importance of refugia) is needed. In many jurisdictions where martens are trapped for their fur, carcasses are being collected in order to better assess the status of population and harvest programs (e.g., Fortin et al. 2003). Annual meetings of state and provincial biologists responsible for managing furbearers at the regional level are needed to identify issues of common concern, track research progress, and when appropriate assemble information to be provided to the public on special management issues.

3.7 The Fisher (*Martes pennanti*)

3.7.1 Distribution

The fisher occurs in all of the Canadian provinces and territories except Newfoundland and Prince Edward Island, and in disjunct areas within the United States, north of 35° N latitude (Fig. 2.7). This review of its distribution encompasses 3 regions of North America: eastern, central, and western.

3.7.1.1 Eastern region

The fisher populations of southeastern Québec (south of St. Lawrence River), New Brunswick, Maine, Massachusetts, New Hampshire, Vermont, New York, Connecticut, Rhode Island, and northeastern Pennsylvania are contiguous (Fig. 2.7). Fishers occur throughout southeastern Québec, except in the Montreal and Laval urban areas, and on Anticosti Island. Fishers occur throughout New Brunswick, with the exception of Grand Manan, Deer, and Campbello Islands.

In Maine, fishers occur statewide, with the highest densities from central Maine southward (Krohn et al. 1995) (Fig. 2.7). Densities of fishers and martens appear inversely related; Krohn et al. (1995) hypothesized that fishers were limited in northern Maine by deep snow, and that martens were excluded from southern Maine by high fisher densities. In New York, the fisher's distribution has expanded, due in part to past reintroductions of animals in southeastern New York and a recent release in northcentral Pennsylvania, with animals moving into southern New York. Between 1957–1967, 124 fishers were translocated to northern Vermont. Today, the species occurs throughout Vermont, even in the Champlain Valley, a region of extensive agriculture. In New

Hampshire, fishers occupy the entire state and reach highest population densities in the southwestern and southern regions. The species occurs in central and western Massachusetts and is expanding into highly populated areas to the east. It inhabits both Connecticut and Rhode Island, except the coastal islands (Fig. 2.7). The eastern Connecticut population has resulted from southward expansion of populations from central Massachusetts, whereas expansion in the western half of the state probably originated from a release of 32 animals in 1989 and 1990. During 1994–1998, 190 fishers were released at 5 primary reintroduction sites in northern Pennsylvania: Fish Dam Wild Area (25 fishers), Quehanna Wild Area (23 fishers), and Pine Creek Valley (37 fishers) in northcentral Pennsylvania; Sullivan and Wyoming counties (40 fishers) in northeastern Pennsylvania; and the Allegheny National Forest (61 fishers) in northwestern Pennsylvania (T. Serfass, Department of Biology, Frostburg State University, Maryland, USA, personal communication).

The fisher populations in West Virginia, southern Pennsylvania, and Maryland are also contiguous (Fig. 2.7). The occurrence of fishers in southern Pennsylvania is undoubtedly the result of a reintroduction of 23 fishers in West Virginia in 1969 (Pack and Cromer 1980, Williams et al. 1999, 2000). Fishers expanded into Maryland (Garrett County) and into southern Pennsylvania (Somerset, Fayette, Westmorland, Bedford, and Cambina Counties) toward central Pennsylvania (T. Serfass, Department of Biology, Frostburg State University, Maryland, USA, personal communication).

Disjunct populations of fishers exist in Nova Scotia (Fig. 2.7). Reintroductions from 1947–1948 and 1963–1966 of 80 fishers from Maine into eastern and western Nova Scotia resulted in 2 geographically separate and expanding populations (Potter 2002).

3.7.1.2 Central region

The fisher populations of Québec (north of St. Lawrence River), Ontario, Manitoba, Minnesota, Wisconsin, and Michigan are contiguous (Fig. 2.6). Pilgrim (1980) reported a first record of a fisher in Labrador; however, to our knowledge, there is no evidence of an established population in this part of Newfoundland. In Québec, it was believed that fisher populations were well established from Labrador to the Canada-USA border (Banfield 1974). However, recent information on fisher ecology and capture locations suggests that their current distributional range is smaller and south of 50° latitude (Fortin et al. 2003). In Ontario, the distribution of the fisher overlaps that of the American marten (Gibilisco 1994), but has expanded eastward in the suburban and agricultural areas of the Ottawa Region (Egan 2003). In Manitoba, the distribution of fisher coincides with that of the boreal forest, north of 50° latitude (Leonard 1986, Gibilisco 1994).

Figure 2.7. General distribution of *Martes pennanti* in North America (after Gibilisco 1994, Aubry and Lewis 2003).

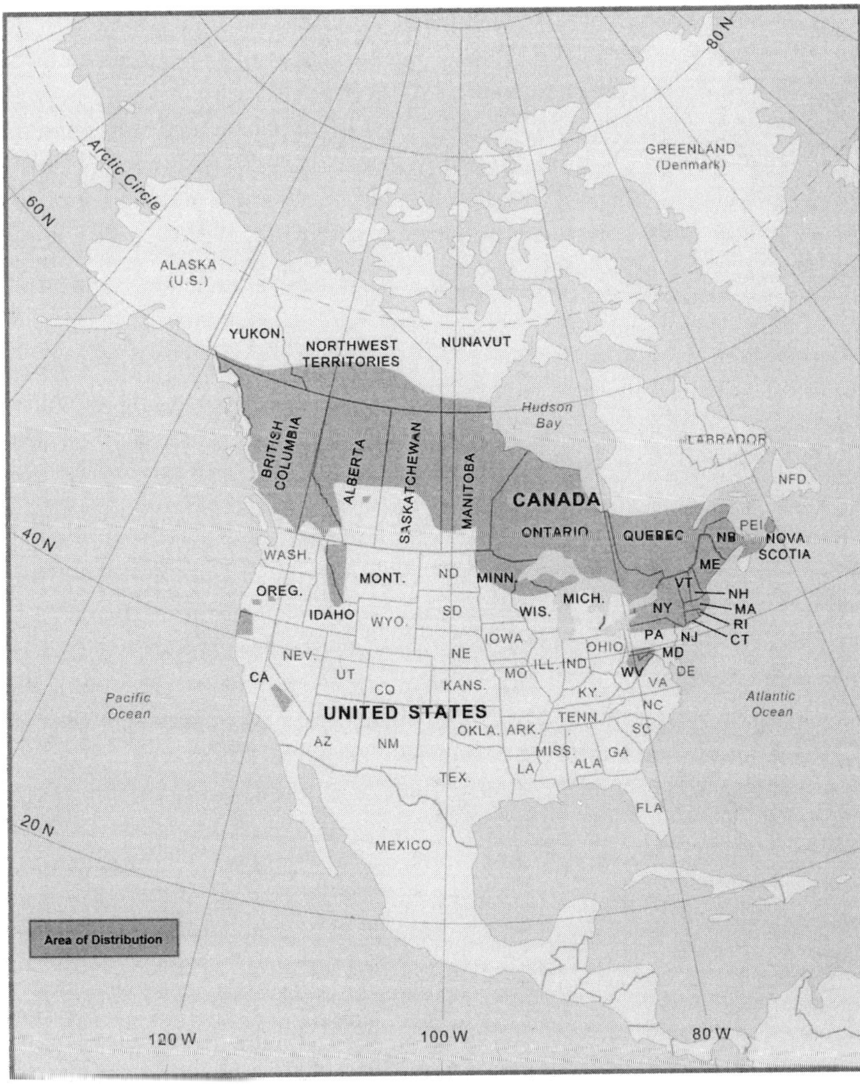

In Minnesota, fisher populations are found in the northeastern corner of the state where they are contiguous with Ontario populations (Berg and Kuehn 1994) (Fig. 2.7). In Michigan, 61 fishers were reintroduced to 3 counties in the western Upper Peninsula from 1961–1963, and these animals, combined with a few immigrants from Wisconsin, populated most of the forested portions of the west and central Peninsula by 1987. Natural dispersal was supplemented by the translocation of 190 fishers to the eastern Upper Peninsula from 1988 to 1992. The fisher is now distributed throughout forested portions of the Upper Peninsula, but has not been reintroduced to the Lower Peninsula (Earle and Reis 1996). In Wisconsin, 120 fishers were successfully reintroduced during 1956–1967. By 1981, fishers occupied all of Wisconsin's Northern Forest Region. There are now approximately 14,000–17,000 fishers in the state and they occupy all suitable habitat (Kohn and Ashbrenner 1996). Gibilisco (1994) reported a recent increase in fisher sightings in North and South Dakota. However, we were unable to confirm the occurrence of resident populations in those 2 states.

3.7.1.3 Western region

In Canada, contiguous populations of fishers occur in Saskatchewan, Alberta, British Columbia, and in a narrow belt in southern Yukon and Northwest Territories tapering off along the Saskatchewan-Nunavut border (Fig. 2.7). In the United States, the fisher occurred historically in most coniferous forest habitats in Montana, Idaho, Wyoming, Washington, Oregon, and California. However, during the 20th century, the range of the fisher in the Pacific states has changed dramatically. The fisher has apparently been extirpated in Washington (Lewis and Stinson 1998) and, in Oregon and California, its range has been reduced to a few disjunct and relatively small areas (Zielinski et al. 1995, Aubry and Lewis 2003). Fisher populations in Montana and Idaho occur over ≤25% of those states; fishers are totally absent from Wyoming.

In Saskatchewan, the fisher is found in the boreal forest, mainly between 52° and 58° N. Although present farther north, it is considered rare in the subarctic boreal region (A. Arsenault, Saskatchewan Environment and Resource Management, Saskatoon, Saskatchewan, Canada, personal communication). In Alberta, Skinner and Todd (1988) reported the presence of fishers in the Rocky Mountains along the British Columbia border, and in the boreal forest, mostly above 54° N. However, they indicated that over most of their range, fisher populations were in decline. In 2000, the distribution of fishers still encompassed boreal and montane forests (G. Proulx, Alpha Wildlife Research & Management Ltd., Sherwood Park, Alberta, Canada, unpublished data). In 1990, Proulx et al. (1994b) released fishers in the parklands of Alberta, near the City of Edmonton. Until recently, the population was thriving (Badry et al. 1997), and reproduction was confirmed in 1993 (G. Proulx, Alpha Wildlife Research & Management Ltd., Sherwood Park, Alberta, Canada, unpublished data). However, the animals were persecuted by local landowners (poisoning and rundown by snowmobiles), and accidentally captured in traps set for beaver (*Castor canadensis*) and other furbearers. The status of this re-introduced population is now uncertain. In British Columbia, the fisher was found throughout the province in the 1980s except for coastal islands (Banci 1989) (Fig. 2.7). Today, fishers are believed to occur at low densities throughout much of the province (Weir 2003); the species is likely extirpated from the Lower Mainland, portions of the Thompson and Okanagan Valleys, and the southeast corner of the province (M. Badry, Ministry of Environment, Lands and Parks, Victoria, British Columbia, Canada, personal communication). Fishers are rare in coastal ecosystems and may be found in boreal forest habitats (Proulx et al. 2003).

The northernmost extent of the range of fishers is the Great Slave Lake region of the Northwest Territories, where the species is found at 63° N latitude (Fig. 2.7). The fisher occurs throughout the southern portion of the Northwest Territories and Nunavut, and the southeast corner of the Yukon Territory. Since 1980, the species' range may have expanded northward in the Northwest Territories and westward in the Yukon (R. Mulders, Resources, Wildlife, and Economic Development, Government of the Northwest Territories, Northwest Territories, Canada, and H. Slama, Yukon Department of Renewable Resources, Whithorse, Yukon, Canada, personal communication). The presence of fishers near Juneau in southeast Alaska (Fig. 2.7) was confirmed by the recovery of a skull in 1993, and the incidental harvest of 4 individuals between 1997–2003. It is likely that these fishers emigrated from British Columbia via the Taku River valley. As of 2003, however, there is no evidence that a viable fisher population occurs in Alaska (R. Flynn, Department of Fish and Game, Douglas, Alaska, USA, personal communication).

In Montana, fishers are rare and found mainly in the northwest portion of the state, in the Swan mountain range (Roy 1990) (Fig. 2.7). In Wyoming, Gibilisco (1994) questioned the presence of fishers in the vicinity of Yellowstone National Park, in the extreme northwestern corner of that state. Uhler (1998) listed fisher as a rare mammal in Yellowstone Park, if present. We were unable to confirm the presence of fishers in Wyoming. The fisher is not common in Idaho despite a reintroduction program in the 1960s (Williams 1963). Its distribution is limited to the northern portion of the state (C. E. Harris, Idaho Department of Fish and Game, Boise, Idaho, USA, personal communication). Fishers are probably extirpated in Washington. Since 1969, documented evidence of their occurrence in the state is limited to 2 records in anomalous habitats along Puget Sound near captive facilities from which fishers are known to have escaped, and 1 marked animal in northeastern Washington that had been translocated into Montana (Lewis and Stinson 1998). There is no evidence that fishers were ever translocated into Washington or California, but planning is currently underway to assess the feasibility of reintroducing fishers to Washington (Lewis 2002). Fishers were translocated from south-central British Columbia and northern Minnesota to several localities in the southern Cascade

Range in southwestern Oregon and the Wallowa Mountains in northeastern Oregon from 1961–1981 (Kebbe 1961, Aubry and Lewis 2003). Although reintroduction efforts in the Wallowa Mountains appear to have failed, translocations to the southern Cascade Range were successful. Currently, fishers occur in Oregon in only 2 small areas in the southwestern portion of the state: the southern Cascade Range and the northern Siskiyou Mountains. The Cascade population is reintroduced and descended primarily from British Columbia fishers, whereas fishers in the northern Siskiyou Mountains are believed to represent the northeastern extension of a relatively large indigenous population in northwestern California (Aubry and Lewis 2003, Aubry et al. 2004, Drew et al. 2003). In California, there are 2 known disjunct populations (Zielinski et al. 1997). One inhabits the Coast Range and Klamath Mountains of the northwest; the other is found in the southern Sierra Nevada (Fig. 2.7).

3.7.2 Habitat Relations

Fishers occur primarily in late-seral coniferous and mixed-coniferous-deciduous forests (Coulter 1966, Powell 1977, Arthur et al. 1989a, Weir and Harestad 2003), but also use younger stands, especially as foraging habitat (Jones 1991, Buskirk and Powell 1994, Powell and Zielinski 1994, Weir and Harestad 2003). In all regions where they occur, fishers inhabit forests with multi-storied and contiguous overhead cover, and complex structure near the ground that typically includes abundant coarse woody debris and a well-developed understory. While trapping can be a limiting factor to fishers (Krohn et al. 1994, Banci and Proulx 1999), especially during periods of high pelt prices, responses to the questionnaires uniformly indicated that loss of forestland habitat from human development is the main long-term threat to fisher populations. For species like the fisher with large spatial requirements (Arthur et al. 1989b, Garant and Crête 1997), the long-term maintenance of extensive forestlands will be a major conservation challenge.

3.7.3 Population Status and Trends

Fisher populations are harvested in 65% of the surveyed jurisdictions, and most of them are stable or increasing (Table 2.4). In Canada, harvest seasons last at least 90 days. In the United States, most seasons are markedly shorter. Approximately 50% of the harvests consist of less than 400 animals per jurisdiction. In 5 jurisdictions where the fisher is protected, 1 population is stable and 1 is increasing. The status of the other populations is unknown. In British Columbia, the fisher has been identified as "imperiled" by the Conservation Data Centre (2003).

Like most carnivores, fisher populations are threatened by habitat loss through fire, logging, oil and gas exploration, and urban encroachment. Many

Table 2.4. Responses to questionnaires on the status of fisher populations since 1995.

Country	Fisher Population Status			Fisher Harvest		Data Collected	Population Trend[a]	Reasons for Population Change
	No Status	Furbearer/ Pest	Special Status	Length of trapping/hunting season (days)	Mean Annual Harvest			
Alberta		x		92	1,100	Number and location of captures, and age-sex composition	S-I	
British Columbia		x[b]		107	275	Number and location of captures, age-sex composition, and trapper effort	D-S	Loss of habitat and trapping
California			Protected[c]			Roadkills, and camera and track surveys	U	
Connecticut			Protected				I	Reintroduction
Idaho			Protected			Sightings and incidental harvest	U	
Maine		x		61	2,091	Number and location of captures, and voluntary trapper log book	S-I	
Massachusetts		x		22	290	Number and location of captures, age-sex composition, and voluntary survey	I	
New Hampshire		x	31	814 (quota of 10 fishers/ trapper/ season)		Number and location of captures, and mandatory trapper log book	S	

Table 2.4. Continued.

Country	Fisher Population Status: No Status	Fisher Population Status: Furbearer/ Pest	Fisher Population Status: Special Status	Fisher Harvest: Length of trapping/hunting season (days)	Fisher Harvest: Mean Annual Harvest	Fisher Harvest: Data Collected	Population Trend[a]	Reasons for Population Change
New York		x		44	1,486	Number and location of captures	I	Reintroductions
Northwest Territories			x[d]	135	22	Number and location of captures	S	
Oregon			Protected			Roadkills, and camera and track surveys	S	
Pennsylvania			Protected				I	Reintroductions
Québec		x		128	4,215	Number and location of captures, age-sex composition, and trapper log book	S	
Rhode Island		x		30	No data – 1st season in November 2000 (quota of 2 fishers/ trapper/season)	Number and location of captures to be collected	S-I	
Saskatchewan		x		120	1,950	Number and location of captures	S-I	
Vermont		x		16	343	Number and location of captures, and mandatory mail survey	I	

Table 2.4. Continued.

Country	Fisher Population Status			Fisher Harvest				
	No Status	Furbearer/ Pest	Special Status	Length of trapping/hunting season (days)	Mean Annual Harvest	Data Collected	Population Trend[a]	Reasons for Population Change
Washington		State endangered[c]				Camera and track surveys	S	May be extirpated
West Virginia		x		90	36 (quota of 1 fisher/ trapper/season)	Number and location of captures	I	Reintroductions
Wyoming			Protected				U	
Yukon		x		135	22	Number and location of captures	S	

[a]S = stable; I = increasing; D = decreasing; U = unknown.
[b]The Government of British Columbia has designated the current status of this "furbearer" as "imperiled".
[c]The fisher has been petitioned 3 times in the Pacific states for listing under the U.S. Endangered Species Act, the most recent petition is pending (Beckwitt 1990, Carlton 1994, Greenwald et al. 2000).
[d]The Government of the Northwest Territories has designated the current status of this "furbearer" as "may be at risk".

of them (e.g., California, Oregon) are vulnerable to demographic and environmental stochasticity. A phylogeographic study of fisher populations in the western United States is currently underway (Buskirk et al. 2002). Fishers have a valuable pelt and they are easily enticed in traps set for other furbearers (Banci and Proulx 1999). Trapping activities must be carefully monitored to ensure the future of exploited populations. Fortunately, most jurisdictions monitor harvest captures and locations, and some of them enforce strict quotas (Table 2.4).

3.7.4 Research and Management Needs

Respondents pointed out that the response of fishers to loss and fragmentation of old forest habitat through natural disturbance agents or human activities is a priority research subject. More research is also needed to improve our understanding of broad-scale ecological factors that may affect the abundance and distribution of fishers, and its relationships with sympatric species such as American martens and lynx (*Lynx canadensis*). Special attention should be paid to climatic changes and snowfall patterns.

In jurisdictions where fishers are trapped, the size and distribution of the harvest, and the sex and age composition of captured populations, should be determined to detect major population changes, and to modify harvest programs through adaptive management. Where fisher populations are endangered, monitoring and modeling of habitats needs to be investigated to improve forest development plans. As for martens, annual meetings of state and provincial biologists responsible for managing furbearers at the regional level are a necessity to track research progress and identify specific management concerns.

4. DISCUSSION

While this review provides up-to-date information on distribution limits of the genus *Martes*, scientific information is lacking in some parts of the world and for some species. For example, more data on the distribution of the pine marten are needed in Portugal, Austria, Hungary and the Balkans. The exact distribution of the sable in Mongolia, North Korea, and Japan still needs to be established. We also know little about the distributional range of the yellow-throated marten.

All *Martes* species (even the stone marten) are associated with forest habitats, preferably late seral conditions in either coniferous or mixed coniferous-deciduous forests. On the basis of today's known distribution records, and the apparent association existing between *Martes* and forest habitats, one can conservatively develop habitat management programs. For the yellow-throated and the Japanese martens, however, baseline research on the species' habitat requirements is required before developing such programs.

This review demonstrated well how valuable reintroduction programs are in the reestablishment of *Martes* species. Today, through repeated release programs, the fisher and the sable reoccupy large portions of their original range. However, the genetic implications of reintroductions are unknown and potentially deleterious (Greig 1979, Templeton 1986, Storfer 1999); unique local adaptations can be disrupted by animals introduced from elsewhere. Reintroductions will not be enough to reestablish *Martes* where habitat has been significantly altered. For example, reintroductions of fishers in many areas of western Washington and Oregon where the species has been extirpated, will be problematic until closed-canopy conditions and key structural elements have been restored. It is therefore essential to develop habitat management programs for landscapes that meet the current needs of *Martes* populations, and that will retain enough interconnected habitats in the future to ensure the long-term viability of populations.

As contradictory as it may seem, *Martes* populations that are annually harvested appear to be the most secure. Data on the distribution of harvested populations are more complete than those of protected populations. Also, because government agencies monitor numbers and locations of captures, changes in population densities or habitat quality are readily determined. Unfortunately, populations that are protected from hunting and trapping are not necessarily better understood. In many cases, these populations have received a special status only after being seriously reduced, and it is difficult to monitor the presence of animals. Data on the dynamics of populations harvested for economic reasons (e.g., pelt value) also are usually more complete than those of populations that are controlled because of damage caused by "pest" animals. It appears that an economically viable *Martes*:human interaction may facilitate the proper management of their populations and habitats.

Martes management programs should take into consideration the impact of global warming on the distribution of species. For example, warmer temperatures and less snow could result in an extension of the geographic range of the stone marten, possibly at the expense of the pine marten and the sable (Lachat Feller 1993, Bakeyev 1994). Likewise, milder winters may benefit fishers over American martens (Krohn et al. 2004).

The limits of distribution of *Martes* species depends on several factors associated with the demographic dynamics of populations and their habitat needs. Scientific studies properly addressing our lack of knowledge on populations and habitats, thorough surveys, and effective monitoring programs will all improve our understanding of the distribution of *Martes* worldwide. We hope that the knowledge gaps identified in this review will be addressed by research organizations and government agencies in the near future so that, in a decade or less, a precise distribution of all *Martes* species and a greater understanding of their population dynamics are available.

5. LITERATURE CITED

Amores, F. 1980. Feeding habits of the stone marten *Martes foina* (Erxleben, 1777) in southwestern Spain. Säugetierkundliche Mitteilungen 28:316–322.

Andera, M., and V. Hanzal. 1996. Atlas of the mammals of the Czech Republic. A provisional version. II. Carnivores (*Carnivora*). National Museum, Praha, Czech Republic.

Anderson, E. 1970. Quaternary evolution of the genus *Martes* (Carnivora, Mustelidae). Acta Zoologica Fennica 130:1–133.

Anonymous. 1996. Threatened and endangered species of New Mexico - 1996 biennial review and recommendations. Unpublished report, New Mexico Department of Game and Fish, Santa Fe, New Mexico, USA.

_____. 1998. Mammalian species of Nova Scotia. Nova Scotia Department of Natural Resources, Kentville, Nova Scotia, Canada.

_____. 2000. Pine marten (*Martes americana*). Wisconsin Department of Natural Resources, Madison, Wisconsin, USA.

Arthur, S. M., W. B. Krohn, and J. R. Gilbert. 1989a. Habitat use and diet of fishers. Journal of Wildlife Management 53:680–688.

_____, _____, and _____. 1989b. Home range characteristics of adult fishers. Journal of Wildlife Management 53:674–679.

Aubry, K. B., and J. C. Lewis. 2003. Extirpation and reintroduction of fishers (*Martes pennanti*) in Oregon: implications for their conservation in the Pacific states. Biological Conservation 114:79–90.

_____, S. M. Wisely, C. M. Raley, and S. W. Buskirk. 2004. Zoogeography, spacing patterns, and dispersal in fishers: insights gained from combining field and genetic data. Pages 201–220 *in* D. J. Harrison, A. K. Fuller, and G. Proulx, editors, Martens and fishers (*Martes*) in human-altered environments: An international perspective. Kluwer Academic Press, Boston, Massachusetts, USA.

Aune, K., and P. Schladweiler. 1997. Age, sex structure, and fecundity of the American marten in Montana. Pages 61–77 *in* G. Proulx, H. N. Bryant, and P. M. Woodard, editors. *Martes*: taxonomy, ecology, techniques, and management. Provincial Museum of Alberta, Edmonton, Alberta, Canada.

Badry, M. J., G. Proulx, and P. M. Woodard. 1997. Home-range and habitat use by fishers translocated to the aspen parkland of Alberta. Pages 233–251 *in* G. Proulx, H. N. Bryant, and P. M. Woodard, editors. *Martes*: taxonomy, ecology, techniques, and management. Provincial Museum of Alberta, Edmonton, Alberta, Canada.

Bakeyev, N. N., and A. A. Sinitsyn. 1994. Status and conservation of sables in the Commonwealth of Independent States. Pages 246–254 *in* S. W. Buskirk, A. S. Harestad, M. G. Raphael, and R. A. Powell, editors. Martens, sables, and fishers: biology and conservation. Cornell University Press, Ithaca, New York, USA.

Bakeyev, Y. N. 1994. Stone martens in the Commonwealth of Independent States. Pages 243–245 *in* S. W. Buskirk, A. S. Harestad, M. G. Raphael, and R. A. Powell, editors. Martens, sables, and fishers: biology and conservation. Cornell University Press, Ithaca, New York, USA.

Balharry, E. A., G. M. McGowan, H. Kruuk, and E. Halliwell. 1996. Distribution of pine marten in Scotland as determined by field survey and questionnaire. Scottish Natural Heritage Report No. 48, Edinburgh, Scotland.
Banci, V. 1989. A fisher management strategy for British Columbia. British Columbia Ministry of Environment, Wildlife Bulletin No. B-63, Victoria, British Columbia, Canada.
_____, and G. Proulx. 1999. Resiliency of furbearers to trapping in Canada. Pages 175–203 *in* G. Proulx, editor. Mammal trapping, Alpha Wildlife Research & Management Ltd., Sherwood Park, Alberta, Canada.
Banfield, A. W. F. 1974. Les mammifères du Canada. Musée National des Sciences Naturelles, Presses de l'Université Laval, Québec, Canada.
Beckwitt, E. 1990. Petition for a rule to list the fisher as endangered. Sierran Biodiversity Project, North San Juan, California, USA.
Berg, W. E., and D. W. Kuehn. 1994. Demography and range of fishers and American martens in a changing Minnesota landscape. Pages 262–271 *in* S. W. Buskirk, A. S. Harestad, M. G. Raphael, and R. A. Powell, editors. Martens, sables, and fishers: biology and conservation. Cornell University Press, Ithaca, New York, USA.
Birks, J. D. S., J. E. Messenger, S. P. Rushton, and P. W. W. Lurz. 2003. Reintroducing pine martens – habitat constraints and enhancement opportunities. *In* C. P. Quine, R. C. Trout, and R. F. Shore, editors. Managing woodlands and their mammals: proceedings of a joint Mammal Society/Forestry Commission symposium, Forestry Commission, Edinburgh, Scotland.
Bouchardy, C., and P. Labrid. 1986. La martre (*Martes martes*). Office National de la Chasse,Bulletin Mensuel No. 104, Fiche No. 33, Paris, France.
_____, and R. Libois. 1986. La fouine (*Martes foina*). Office National de la Chasse, Bulletin Mensuel No. 105, Fiche No. 34, Paris, France.
Brainerd, S. M. 1997. Habitat selection and range use by the Eurasian pine marten (*Martes martes*) in relation to commercial forestry practices in southern boreal Scandinavia. Dissertation, Agricultural University of Norway, Ås, Norway.
_____, J.-O. Helldin, E. Lindström, and J. Rolstad. 1994. Eurasian pine martens and old industrial forest in southern boreal Scandinavia. Pages 343–354 *in* S. W. Buskirk, A. S. Harestad, M. G. Raphael, and R. A. Powell, editors. Martens, sables, and fishers: biology and conservation. Cornell University Press, Ithaca, New York, USA.
_____, J.-O. Helldin, E. Lindström, J. Rolstad, and I. Storch. 1995. Pine marten (*Martes martes*) selection of resting and denning sites in Scandinavian managed forests. Annals Zoologica Fennici 32:151–157.
Broekhuizen, S., and G. J. D. M. Müskens. 1984. Wat is er met de steenmarter *Martes foina* (Erxleben, 1777) in Nederland aan de hand? Lutra 27: 261–273.
Brzeziński, M. 1994. Summer diet of the sable *Martes zibellina* in the Middle Yenisei taiga, Siberia. Acta Theriologica 39:103–107.
Buskirk, S. W. 1994. Introduction to the Genus *Martes*. Pages 1–10 *in* S. W. Buskirk, A. S. Harestad, M. G. Raphael, and R. A. Powell, editors. Martens, sables, and fishers: biology and conservation. Cornell University Press, Ithaca, New York, USA.
_____, K. B. Aubry, S. M. Wisely, G. A. Russell, and W. J. Zielinski. 2002. Population genetic structure of the fisher in the western states. *Martes* Working Group Newsletter 10:10.
_____, Y. Ma, and L. Xu. 1993. Sable ecology in Chinese taiga forests. National Geographic Research and Exploration 9:479–480.
_____, _____, and _____. 1994. Sables (*Martes zibellina*) in managed forests of northern China. Small Carnivore Conservation 10:12–13.
_______, and R. A. Powell. 1994. Habitat ecology of fishers and American martens. Pages

283–296 *in* S. W. Buskirk, A. S. Harestad, M. G. Raphael, and R. A. Powell, editors. Martens, sables, and fishers: biology and conservation. Cornell University Press, Ithaca, New York, USA.

_____, and L. F. Ruggiero. 1994. American marten. Pages 7–37 *in* L. F. Ruggiero, K. B. Aubry, K.B., S. W. Buskirk, L. J. Lyon, and W. J. Zielinski, editors. American marten, fisher, lynx, and wolverine in the western United States. USDA Forest Service, General Technical Report RM-254, Fort Collins, Colorado, USA.

_____, _____, and C. J. Krebs. 2000. Habitat fragmentation and interspecific competition: implications for lynx conservation. Pages 83–100 *in* L. F. Ruggiero, K. B. Aubry, S. W. Buskirk, G. M. Koehler, C. J. Krebs, K. S. McKelvey, and J. R. Squires, editors. Ecology and conservation of lynx in the United States. University Press of Colorado, Boulder, Colorado, USA.

Byrne, G. 1998. A Colorado winter track survey for snowshoe hares and other species. Colorado Division of Wildlife, Glenwood Springs, Colorado, USA.

Carlton, D.C. 1994. Petition for a rule to list the fisher, *Martes pennanti*, as "threatened" in the western United States under the Endangered Species Act, 16 U.S.C. Sec. 1531 *et seq.* (1973) as amended. Biodiversity Legal Foundation, Boulder, Colorado, USA.

Chapin, T. G., D. J. Harrison, and D. D. Katnik. 1998. Influence of landscape pattern on habitat use by American marten in an industrial forest. Conservation Biology 12:1327–1337.

_____, _____, and D. M. Phillips. 1997. Seasonal habitat selection by marten in an untrapped forest preserve. Journal of Wildlife Management 61:707–717.

Chotolchu, N, M. Stubbe, and N. Dawaa. 1980. Der steinmareder *Martes foina* (Erxleben, 1777) in der Mongolei. Acta Theriologica 25:105–114.

Choudhury, A. 1997a. Small carnivores (mustelids, viverrids, herpestids, and one ailurid) in Arunachal Pradesh, India. Small Carnivore Conservation 17:7–9.

_____. 1997b. The distribution and status of small carnivores (mustelids, viverrids, and herpestids) in Assam, India. Small Carnivore Conservation 16:25–26.

Christopher, G., and A. A. Jayson. 1996. Sightings of Nilgiri marten (*Martes gwatkinsi* Horsfield) at Peppara Wildlife Sanctuary and Silent Valley National Park, Kerala, India. Small Carnivore Conservation 15:34.

Clément, R., and M. C. Saint-Girons. 1982. Le régime de la fouine *Martes foina* Erxleben 1777 dans l'agglomération nantaise et en milieu rural. Mammalia 46:550–553.

Clevenger, A. P. 1993. The European pine marten *Martes martes* in the Balearic Islands, Spain. Mammal Review 23(2):65–72.

_____. 1994. Feeding ecology of Eurasian pine martens and stone martens in Europe. Pages 326–340 *in* S. W. Buskirk, A. S. Harestad, M. G. Raphael, and R. A. Powell, editors. Martens, sables, and fishers: biology and conservation. Cornell University Press, Ithaca, New York, USA.

Conservation Data Centre. 2003. *Martes pennanti*. B. C. Species and Ecosystem Explorer Website, Ministries of Sustainable Resource Management and Water, Land, and Air Protection, Victoria, British Columbia, Canada.

Corbet, G. B. 1978. The mammals of the Palaearctic region: a taxonomic review. British Museum Natural History, London, United Kingdom.

_______, and J. E. Hill. 1992. The mammals of the Indomalayan region: a systematic review. British Museum Natural History, London, United Kingdom.

Coulter, M. W. 1966. Ecology and management of fishers in Maine. Dissertation, Syracuse University, Syracuse, New York, USA.

De Marinis, A. M., P. Genovesi, and M. Spagnesi. 2000. Pine marten. Pages 158–159 *in* M. Spagnesi, S. Toso, and A. M. De Marinis, editors. 2000 – Italian mammals. Ministero dell'Ambiente, Servizio Conservazione della Natura and Istituto Nazionale per la Fauna

Selvatica A. Ghigi, Roma, Italy.

_____, and M. Masseti. 1993. Pine marten *Martes martes* on the Island of Elba. Small Carnivore Conservation 8:13.

Drew, R. E., J. G. Hallett, K. B. Aubry, K. W. Cullings, S. M. Koepf, and W. J. Zielinski. 2003.Conservation genetics of the fisher (*Martes pennanti*) based on mitochondrial DNA sequencing. Molecular Ecology 12:51–62.

Earle, R. D. 1999. Furbearer winter track counts survey of 1998. Michigan Department of Natural Resources, Wildlife Division Report No. 3288, Federal Aid in Wildlife Restoration, Michigan Project W-127-R, City, Michigan, USA.

_____, and T. F. Reis. 1996. Fisher and marten management in Michigan. Pages 47–58 *in* Abstracts, 14th Midwest Furbearer Workshop, Ironwood, Michigan, USA.

Egan, K. 2003. Team tags reclusive fishers. The Ottawa Citizen, February 8:D1, Ottawa, Canada.

Fayard (Dir.). 1984. Atlas des mammifères de France. Société Française pour l'Etude et la Protection des Mammifères, Paris, France.

Fedyk, S., Z. Gegczynska, M. Pucek, J. Raczynski, and M. D. Sikorski. 1984. Winter penetration by mammals of different habitats in the Biebrza Valley. Acta Theriologica 29:317–336.

Fitzgerald, J. P., C. A. Meancy, and D. M. Armstrong. 1994. Mammals of Colorado. Denver Museum of Natural History and University Press of Colorado, Denver, Colorado, USA.

Flynn, R. W., and T. Schumacher. 1999. Ecology of martens in Southeast Alaska. Alaska Department of Fish and Game, Federal Aid in Wildlife Restoration Research Progress Report W-27-1, Study 7.16, Juneau, Alaska, USA.

Fornasari, L., L. Bani, I. Bonfanti, E. de Carli, and R. Massa. 2000. A spatial analysis of mustelid distributions in northern Italy. Pages 201–215 *in* H. I. Griffiths, editor. Management and conservation aspects of small carnivore:human interactions. Backhuys Publishers, Leiden, The Netherlands.

Forsey, O., J. Bissonette, J. Brazil, K. Curnew, J. Lemon, L. Mayo, I. Thompson, L. Bateman, and L. O'Driscoll. 1995. National recovery plan for the Newfoundland marten. Report No. 14, Recovery of Nationally Endangered Wildlife Committee, Ottawa, Canada.

Fortin, C., M. Julien, and Y. Leblanc. 1997. Geographic variation of American marten in central and northern Québec. Pages 29–39 *in* G. Proulx, H. N. Bryant, and P. M. Woodard, editors. *Martes*: taxonomy, ecology, techniques, and management. Provincial Museum of Alberta, Edmonton, Alberta, Canada.

_____, Y. Leblanc, and Y. Garant. 2003. Suivi de l'exploitation et des tendances des populations du pékan (*Martes pennanti*) au Québec de 1984 à 1999. Société de la Faune et des Parcs du Québec, Québec, Canada.

Fredrickson, L. F. 1995. History and management plans for the reintroduced Black Hills marten population. Abstracts, 2nd International *Martes* Symposium, Edmonton, Alberta, Canada.

Garant, Y., and M. Crête. 1997. Fisher, *Martes pennanti*, home range characteristics in a high density untrapped population in southern Québec. Canadian Field-Naturalist 111:359–364.

Genovesi, P., and L. Boitani. 1997. Social ecology of the stone marten in central Italy. Pages 110–120 *in* G. Proulx, H. N. Bryant, and P. M. Woodard, editors. *Martes*: taxonomy, ecology, techniques, and management. Provincial Museum of Alberta, Edmonton, Alberta, Canada.

Geptner, V. G., N. P. Naumov, and P. B. Yurgenson. 1967. *Martes* (*Martes zibellina* L.) Pages 507–553 *in* Mammals in the U.S.S.R. Volume 2. Vyskaya Skola, Moscow (in Russian), Russia.

Gibilisco, C. J. 1994. Distributional dynamics of modern *Martes* in North America. Pages 59–71 *in* S. W. Buskirk, A. S. Harestad, M. G. Raphael, and R. A. Powell, editors. Martens, sables, and fishers: biology and conservation. Cornell University Press, Ithaca, New York,

USA.

Godin, J., and E. Vivier. 1995. Les Mustélidés dans les forêts du Nord de la France – leurs fluctuations d'effectifs de 1975 à 1987. Courrier de l'Environnement de l'INRA no. 25, Paris, France.

Graham, R. W., and M. A. Graham. 1994. Late Quaternary distribution of *Martes* in North America. Pages 26–58 *in* S. W. Buskirk, A. S. Harestad, M. G. Raphael, and R. A. Powell, editors. Martens, sables, and fishers: biology and conservation. Cornell University Press, Ithaca, New York, USA.

Grakov, N. N. 1994. Kidus- a hybrid of the sable and the pine marten. Lutreola 3:1–4.

Greenwald, D.N., J. Carlton, and B. Schneider. 2000. Petition to list the fisher (*Martes pennanti*) as an endangered species in its west coast range. Center for Biological Diversity, Tucson, Arizona, USA.

Greig, J. C. 1979. Principles of genetic conservation in relation to wildlife management in southern Africa. South African Journal of Wildlife Research 9:57–78.

Grigorov, G. 1986. Distribution, number and use of the stone marten (*Martes foina* Erxl.), the pine marten (*Martes martes* L.) and the badger (*Meles meles* L.) in Bulgaria. Gorskostopanska nauka 5:59–67.

Groombridge, B. 1993. 1994 IUCN red list of threatened animals. The World Conservation Union, Gland, Switzerland.

Groves, C. R., B. Butterfield, A. Lippincott, B. Csuti, and J. M. Scott. 1997. Atlas of Idaho's wildlife. Integrating Gap analysis and natural heritage information. Idaho Department of Fish and Game, Nongame and Endangered Wildlife Program, Boise, Idaho, USA.

Hargis, C. D., and J. A. Bissonette. 1997. Effects of forest fragmentation on populations of American marten in the intermountain west. Pages 437–451 *in* G. Proulx, H. N. Bryant, and P. M. Woodard, editors. *Martes*: taxonomy, ecology, techniques, and management. Provincial Museum of Alberta, Edmonton, Alberta, Canada.

Hatt, R. T. 1959. The mammals of Iraq. Museum of Zoology, University of Michigan, Miscellaneous Publication No. 106, Ann Arbor, Michigan, USA.

Hayden, T., and R. Harrington. 2000. Exploring Irish mammals. Dúchas The Heritage Service, Dublin, Ireland.

Helin, S., N. Ohtaishi, and L. Houji. 1999. The mammals of China. China Forestry Publishing House, Beijing, People's Republic of China.

Helldin, J. O. 1998. Pine marten (*Martes martes*) population limitation. Dissertation, Swedish University, Agricultural Sciences, Uppsala, Sweden.

______. 2000. Population trends and harvest management of pine marten *Martes martes* in Scandinavia. Wildlife Biology 6:111–120.

Helle, E., P. Helle, H. Lindén, and M. Wikman. 1996. Wildlife populations in Finland during 1990–5, based on wildlife triangle data. Finnish Game Research 49:12–17.

Hobson, D. P., G. Proulx, and B. L. Dew. 1989. Initial post-release behavior of marten, *Martes americana*, introduced in Cypress Hills Provincial Park, Saskatchewan. Canadian Field-Naturalist 103:398–400.

Hosoda, T., H. Suzuki, K. Tsuchiya, H. Lan, L. Shi, and A. P. Kryukov. 1997. Phylogenetic relationships within *Martes* based on nuclear ribosomal DNA and mitochondrial DNA. Pages 3–14 *in* G. Proulx, H. N. Bryant, and P. M. Woodard, editors. *Martes*: taxonomy, ecology, techniques, and management. Provincial Museum of Alberta, Edmonton, Alberta, Canada.

Jędrzejewski, W., A. Zalewski, and B. Jędrzejewska. 1993. Foraging by pine marten *Martes martes* in relation to food resources in Białowieża National Park, Poland. Acta Theriologica 38:405–426.

Jones, J. L. 1991. Habitat use of fisher in northcentral Idaho. Thesis, University of Idaho,

Moscow, Idaho, USA.
Johnson, R. E., and K. M. Cassidy. 1997. Terrestrial mammals of Washington State. Volume 3 *in* K. M. Cassidy, C. E. Grue, M. R. Smith, and K. M. Dvornich, Editors. Washington State Gap Analysis – Final Report. Seattle Audubon Society Publications in Zoology No. 1, Seattle, Washington, USA.
Kauhala, K., and P. Helle. 2000. The interactions of predator and hare populations in Finland – a study based on wildlife monitoring counts. Annals Zoologica Fennici 37:151–160.
Kebbe, C. E. 1961. Return of the fisher. Oregon State Game Communication Bulletin 16:3–7.
King, C. M. 1977. Pine marten. Pages 323–330 *in* G. B. Corbet and H. N. Southern, editors. The handbook of British mammals. Blackwell, Oxford, United Kingdom.
Kohn, B. E., and J. E. Ashbrenner. 1996. Fisher and marten management in Wisconsin. Page 45 *in* Abstracts, 14th Midwest Furbearer Workshop, Ironwood, Michigan, USA.
Krohn, W. B., S. M. Arthur, and T. F. Paragi. 1994. Mortality and vulnerability of a heavily trapped fisher population. Pages 137–145 *in* S. W. Buskirk, A. S. Harestad, M. G. Raphael, and R. A. Powell, editors. Martens, sables, and fishers: biology and conservation. Cornell University Press, Ithaca, New York, USA.
_____, K. D. Elowe, and R. B. Boone. 1995. Relations among fishers, snow, and martens: development and evaluation of two hypotheses. The Forestry Chronicle 71:97–105.
_____, C. L. Hoving, D. J. Harrison, D. M. Phillips, and H. C. Frost. 2004. *Martes* foot-loading and snowfall patterns in eastern North America: implications to broad-scale distributions and interactions of mesocarnivores. Pages 115–132 *in* D. J. Harrison, A. K. Fuller, and G. Proulx, editors. Martens and fishers (*Martes*) in human-altered environments: An international perspective. Kluwer Academic Press, Boston, Massachusetts, USA.
Kryštufek, B. 2000. Mustelids in the Balkans – small carnivores in the European biodiversity hot-spot. Pages 281–294 *in* H. I. Griffiths, editor. Management and conservation aspects of small carnivore:human interactions. Backhuys Publishers, Leiden, The Netherlands.
Kurki, S., P. Helle, H. Lindén, and A. Nikula. 1997. Breeding success of black grouse and capercaillie in relation to mammalian predator densities on two spatial scales. Oikos 79:301–310.
_____, A. Nikula, P. Helle, and H. Lindén. 1998. Abundances of red fox and pine marten in relation to the composition of boreal forest landscapes. Journal of Animal Ecology 67:876–886.
Kyle, C. J., A. Davison, and C. Strobeck. 2003. Genetic structure of European pine martens (*Martes martes*), and evidence for genetic introgression with *M. americana* in England. Conservation Genetics 4:179–188.
Lachat, N. 1991. Stone martens and cars: a beginning war? Mustelid and Viverrid Conservation 5:4–6.
Lachat Feller, N. 1993. Eco-éthologie de la fouine (*Martes foina* Erxleben, 1777) dans le Jura suisse. Dissertation, Université de Neuchâtel, Suisse, Switzerland.
Leonard, R. D. 1986. Aspects of reproduction of the fisher, *Martes pennanti*, in Manitoba. Canadian Field-Naturalist 100:32–44.
Lewis, J. C. 2002. Feasibility study underway for fisher reintroduction in Washington. *Martes* Working Group Newsletter 10:11.
_____, and D. W. Stinson. 1998. Washington State status report for the fisher. Washington Department of Fish and Wildlife, Olympia, Washington, USA.
Libois, R. 1983. Protégeons nos mammifères. Animaux menacés en Wallonie. Editions Duculot, Région Wallone, Gembloux, Belgium.
Lin, L-K. 2000. The status and conservation of Taiwan's mustelids. Pages 331–335 *in* H. I. Griffiths, editor. Management and conservation aspects of small carnivore:human interactions. Backhuys Publishers, Leiden, The Netherlands.

Lindström, E. R., S. M. Brainerd, J.-O. Helldin, and K. Overskaug. 1995. Pine marten-red fox interactions: a case of intraguild predation? Annals Zoologica Fennici 32:123–130.
Long, C. A. 1995. Stone marten (*Martes foina*) in southeast Wisconsin, USA. Small Carnivore Conservation 13:14.
Lucherini, M., and G. Crema. 1993. Diet or urban stone martens in Italy. Mammalia 57:274–277.
Ma, Y., and L. Xu. 1994. Distribution and conservation of sables in China. Pages 255–261 *in* S. W. Buskirk, A. S. Harestad, M. G. Raphael, and R. A. Powell, editors. Martens, sables, and fishers: biology and conservation. Cornell University Press, Ithaca, New York, USA.
Madhusudan, M. D. 1995. Sighting of the Nilgiri marten (*Martes gwatkinsi*) at Eravikulam National Park, Kerala, India. Small Carnivore Conservation 13:6–7.
Mallon, D. 1991. Mustelids in Ladakh, India. Mustelid and Viverrid Conservation 5:9.
Marchesi, P. 1989. Ecologie et comportement de la martre (*Martes martes* L.) dans le Jura suisse. Dissertation, Université de Neuchâtel, Switzerland.
Marshall, D. B. 1992. Threatened and sensitive wildlife of Oregon's forests and woodlands. Audubon Society, Portland, Oregon, USA.
Medway, L. 1978. The wild mammals of Malaya (Peninsular Malaysia) and Singapore. 2nd edition, Oxford University Press, Oxford, United Kingdom.
Messenger, J. E., and J. D. S. Birks. 2000. Monitoring the very rare: pine marten populations in England and Wales. Pages 217–230 *in* H. I. Griffiths, editor. Management and conservation aspects of small carnivore:human interactions. Backhuys Publishers, Leiden, The Netherlands.
_____, _____, and D. J. Jefferies. 1997. What is the status of the pine marten in England and Wales? British Wildlife 8:273–279.
Mickevicius, E., and K. Baranauskas. 1992. Status, abundance, and distribution of mustelids in Lithuania. Small Carnivore Conservation 6:11–14.
Milenkovic, M. 1985. About the distribution of stone marten *Martes foina* Erxleben (Mustelidae, Carnivora) in Vojvodina. Archives Biological Sciences 37 (1–4):3P-4P, Belgrade, Serbia. (in Serbian)
Minkov, S. I., editor. 1998. The sable: Resources conditions and hunting. Kirov City, Russia:VNIIOZ (All-Russia Institute of Hunting and Fur Farming). 207 pages (in Russian).
Mitchell-Jones, A. J., G. Amori, W. Bogdanowicz, B. Kryštufek, P. J. H. Reijnders, F. Spitzenberger, M. Stubbe, J. B. M. Thissen, V. Vohralík, and J. Zima. 1999. The atlas of European mammals. Society Europaea Mammalogica, T. & A. D. Poyser Publishers, London, United Kingdom
Monakhov, V. G. 1983. Filaroises incidence rate in Russian sables of different sex and age. Moscow Bull. MOIP (Bulletin of Moscow Society of Naturalists, Series Biol.), 88 (4):67–69. (in Russian).
_____. 1995. Sable of the Urals, Priobye and the Eniseiskaya Siberia: results and acclimatization. Ekaterinburg: Bank Kul'turnoi Informatsii. (in Russian)
_____. 1999. The ecological structure of sable populations in the nidus of *filaroides* invasion. Russian Journal of Ecology 30:420–427. (in Russian)
_____. 2000. Population analysis of sables in the Ural-Ob Region of the species range. Russian Journal of Ecology 31:422–428.
Moutou, F. 2003. Pests in France. IUCN Small Carnivore Conservation Newsletter 28:18.
Muskens, G. J. D. M., D. J. C. Klees, and S. Broekhuizen. 2000. Dagrustplaatsgebruik van een boommartermannetje *Martes martes* op de zuidoostelijke Veluwe. Lutra 43:81–91.
Newsom, W. 1937. Mammals of Anticosti Island. Journal of Mammalogy 18: 435–442.
Nitta, N. 1982. Winter food habits of the sable in the Hokkaido University forest: the effects on mouse populations and a role of damage prevention in plantations. Unpublished

Dissertation, Hokkaido University, Japan. (in Japanese)
Oli, M. K. 1994. Snow leopards and blue sheep in Nepal: densities and predator:prey ratio. Journal of Mammalogy 75:998–1004.
O'Sullivan, P. J. 1983. The distribution of the pine marten (*Martes martes*) in the Republic of Ireland. Mammal Review 13:39–44.
Ozolinš, J., and V. Pilats 1995. Distribution and status of small and medium-sized carnivores in Latvia. Annals Zoologica Fennici 32:21–29.
Pack, J. C., and J. I. Cromer. 1981. Reintroduction of fisher in West Virginia. Pages 1431–1442 *in* J. A. Chapman and D. Pursley, editors. Worldwide furbearer conference proceedings, Vol. II, Frostburg, Maryland, USA.
Paragi, T. F., W. N. Johnson, D. D. Katnik, and A. J. Magoun. 1996. Marten selection of postfire seres in the Alaskan taiga. Canadian Journal of Zoology 74:2226–2237.
Parker, C. 2001. Marten (*Martes americana*). Utah Division of Wildlife Resources, Wildlife Notebook Series No. 4, Salt Lake City, Utah, USA.
Payer, D. C. 1999. Influences of timber harvesting and trapping on habitat selection and demographic characteristics of marten. Dissertation, University of Maine, Orono, Maine, USA.
_____, and D. J. Harrison. 2003. Influence of forest structure on habitat use by American marten in an industrial forest. Forest Ecology and Management 179:145–156.
Pilgrim, W. 1980. Fisher, *Martes pennanti* (Carnivora: Mustelidae) in Labrador. Canadian Field-Naturalist 94:468.
Pocock, R. I. 1999. The fauna of British India including Ceylon and Burma. Volume II. Taylor and Burma, London, United Kingdom.
Potter, D. N. 2002. Modeling fisher (*Martes pennanti*) habitat associations in Nova Scotia. Thesis, Acadia University, Wolfville, Nova Scotia, Canada.
Potvin, F., L. Bélanger, and K. Lowell. 2000. Marten habitat selection in a clearcut boreal landscape. Conservation Biology 14:844–857.
Powell, R. A. 1977. Hunting behavior, ecological energetics and predator-prey community stability of the fisher (*Martes pennanti*). Dissertation, University of Chicago, Illinois, USA.
_____, and W. J. Zielinski. 1994. Fisher. Pages 38–73 *in* L. F. Ruggiero, K. B. Aubry, S. W. Buskirk, L. J. Lyon, and W. J. Zielinski, editors. American marten, fisher, lynx, and wolverine in the western United States. USDA Forest Service, General Technical Report RM-254, Fort Collins, Colorado, USA.
Prater, S. H. 1971. The book of Indian animals. 3rd edition, Bombay Natural History Society, Delhi, India.
Prescott, J., and P. Richard. 1996. Mammifères du Québec et de l'est du Canada. Editions Michel Quintin, Waterloo, Québec, Canada.
Proulx, G. 2000. The impact of human activities on North American mustelids. Pages 53–75 *in* H. I. Griffiths, editor. Management and conservation aspects of small carnivore:human interactions. Backhuys Publishers, Leiden, The Netherlands.
_. 2001. Characteristics and management of American marten habitat at stand and landscape levels. Unpublished report, Alpha Wildlife Research & Management Ltd., Edmonton, Alberta, Canada.
_____, D. Bernier, J. M. Heron, and K. A. Paige. 2003. A field guide to species at risk in the Coast Forest Region of British Columbia. International Forest Products Ltd. and British Columbia Ministry of Water, Land, and Air Protection, Victoria, British Columbia, Canada.
_____, and R. Khariz. 2001. Occurrence of American marten and other furbearers in TFL # 30, Prince George Forest District. Unpublished report, Alpha Wildlife Research & Management Ltd., Edmonton, Alberta, Canada.

_____, A. J. Kolenosky, M. J. Badry, P. J. Cole, and R. K. Drescher. 1994a. A snowshoe hare snare system to minimize capture of marten. Wildlife Society Bulletin 22:639–643.

_____, _____, _______, R. K. Drescher, K. Seidel, and P. J. Cole. 1994b. Post-release movements of translocated fishers. Pages 197–203 *in* S. W. Buskirk, A. S. Harestad, M. G. Raphael, and R. A. Powell, editors. Martens, sables, and fishers: biology and conservation. Cornell University Press, Ithaca, New York, USA.

_____, and R. Verbisky. 2001. Occurrence of American marten within a connectivity corridor. *Martes* Working Group Newsletter 9:4–5.

Pulliainen, E. 1984. Use of the home range by pine martens (*Martes martes* L.). Acta Zoologica Fennica 171:271–274.

Ramakantha, V. 1994. Natural distribution and ecology of mustelids and viverrids in Manipur, north-eastern India. Small Carnivore Conservation 11:16–17.

Raphael, M. G., and L. L. C. Jones. 1997. Characteristics of resting and denning sites of American martens in central Oregon and western Washington. Pages 146–165 *in* G. Proulx,H. N. Bryant, and P. M. Woodard, editors. *Martes*: taxonomy, ecology, techniques, and management. Provincial Museum of Alberta, Edmonton, Alberta, Canada.

Robitaille, J.-F. 2000. Management and conservation of mustelids in Ontario and eastern Canada. Pages 77–96 *in* H. I. Griffiths, editor. Management and conservation aspects of small carnivore:human interactions. Backhuys Publishers, Leiden, The Netherlands.

Roy, K. 1990. Cabinet Mountains fisher reintroduction study. Interim Progress Report 1990, Montana Cooperative Wildlife Research Unit, University of Montana, Missoula, Montana, USA.

Ryvarden, L. 2001. La faune et la flore en Norvège. Ministry of Foreign Affairs, Oslo, Norway.

Saint-Girons, M. C. 1973. Les mammifères de France et du Bénélux. Doin, Paris, France.

Santos-Reis, M. 1983. Status and distribution of Portuguese mustelids. Acta Zoologica Fennica 172:213–216.

_____, and F. Petrucci-Fonseca. 1999. Carnivora. Pages 135–165 *in* ICN/CBA, editors. A guide for terrestrial mammals of continental Portugal, Azores and Madeira. Instituto de Conservação do Ambiente, Ministerio do Ambiente, Lisboa, Portugal. (in Portuguese)

_____, M. J. Santos, S. Lourenço, T. Marques, I. Pereira, and B. Pinto. 2004. Relationships between stone martens, genets and cork oak woodlands in Portugal. Pages 147–172 *in* D. J. Harrison, A. K. Fuller, and G. Proulx, editors. Martens and fishers (*Martes*) in human-altered environments: An international perspective. Kluwer Academic Press, Boston, Massachusetts, USA.

Schmidt, G., and R. Baird. 1995. The reintroduction of pine marten and fishers to Riding Mountain National Park. Page 52 *in* Abstracts, 2nd International *Martes* Symposium, Edmonton, Alberta, Canada.

Serafini, P., and S. Lovari. 1993. Food habits and trophic niche overlap of the red fox and the stone marten in a Mediterranean rural area. Acta Theriologica 38:233–244.

Serviço Nacional de Parques Reservas e Conservação da Natureza. 1990. Red data book of terrestrial vertebrates. Vol. I – Mammals, birds, reptiles and amphibians. Serviço Nacional de Parques, Reservas e Conservação de Natureza, Lisboa, Portugal. (in Portuguese)

Skinner, D. L., and A. W. Todd. 1988. Distribution and status of selected mammals in Alberta as indicated by trapper questionnaires in 1987. Alberta Fish and Wildlife Division, Wildlife Management Branch, Occasional Paper No. 4, Edmonton, Alberta, Canada.

Slauson, K. M. 2003. Habitat selection by American martens (*Martes americana*) in coastal northwestern California. Thesis, Oregon State University, Corvallis, Oregon, USA.

Slough, B. G. 1994. Translocations of American martens: an evaluation of factors in success. Pages 165–178 *in* S. W. Buskirk, A. S. Harestad, M. G. Raphael, and R.A. Powell, editors. Martens, sables, and fishers: biology and conservation. Cornell University Press, Ithaca,

New York, USA.

Spiridonov, G., and N. Spassov. 1998. Large mammals (Macromammalia) of Bulgaria. Pages 467–483 *in* C. Meine, editor. Bulgaria's biological diversity: conservation and status needs assessment, Vol. I and II, Washington, D.C., USA.

Storch, I., E. Lindström, and J. de Jounge. 1990. Diet and habitat selection of the pine marten in relation to competition with the red fox. Acta Theriologica 35:311–320.

Storfer, A. 1999. Gene flow and endangered species translocations: a topic revisited. Biological Conservation 87:173–180.

Strachan, R., D. J. Jefferies, and P. R. F. Chanin. 1996. Pine marten survey of England and Wales 1987–1988. Peterborough: Joint Nature Conservation Committee.

Strickland, M. A. 1989. Marten management in Ontario. Pages 155–174 *in* R. Lafond, editor. Proceedings of the Northeast Fur Resources Technical Committee Workshop, Beauport, Québec, Canada.

_____, M. Novak, and N. P. Hunziger. 1982. Marten. Pages 599–612 *in* J. A. Chapman and G. A. Feldhamer, editors. Wild mammals of North America: biology, management, and economics. The Johns Hopkins University Press, Baltimore, Maryland, USA.

Stubbe, M. 1993. *Martes martes* (Linné, 1758) – Baum., Edelmarten. Pages 374–426 *in* J. Niethammer and F. Krapp, editors. Handbuch der Säugetiere Europas. Band 5: Raubsäuger - Carnivora (Fissipedia). Teil I: Canidae, Ursidae, Procyoinidae, Mustelidae. Weisbaden, Aula Verlag.

Tatara, M. 1994. Ecology and conservation status of the Tsushima marten. Pages 272–279 *in* S. W. Buskirk, A. S. Harestad, M. G. Raphael, and R. A. Powell, editors. Martens, sables, and fishers: biology and conservation. Cornell University Press, Ithaca, New York, USA.

Templeton, A. R. 1986. Coadaptation and outbreeding depression. Pages 105–116 *in* M. E. Soule, editor. Conservation biology: the science of scarcity and diversity. Sinauer Associates, Sunderland, Massachusetts, USA.

Therrien, S., and D. Eastman. 1999. Analyses of habitat selection by marten in areas of alternative forest practices in the Boreal Mixedwood of Northeastern British Columbia. Interim report – 1998 activities. University of Victoria, British Columbia, Canada.

Thompson, I. D. 1991. Could marten become the spotted owl of eastern Canada? Forestry Chronicle 67:136–140.

_____, and A. S. Harestad. 1994. Effects of logging on American martens, and models for habitat management. Pages 355–367 *in* S. W. Buskirk, A. S. Harestad, M. G. Raphael, and R. A.Powell, editors. Martens, sables, and fishers: biology and conservation. Cornell University Press, Ithaca, New York, USA.

Timm, U. 1991. Sieben eue Saugetierarten in Estland (Eesti). Folia Theriologica Estonica, Tartu:31–34.

Tóth, M. A. 1998. Data to the diet of the urban stone marten (*Martes foina* Erxleben) in Budapest. Opuscula Zoologica Budapest 31:113–118.

Uhler, J. W. 1998. The complete Yellowstone page. U.S. National Park Service Website.

Valentsev, A. S. 1996. Helminth invasion of sable. *Martes* Working Group Newsletter 4:22–24.

Velander, K. A. 1983. Pine marten survey of Scotland, England and Wales 1982–1983. The Vincent Wildlife Trust, London, United Kingdom.

_____. 1991. Pine marten. Pages 368–376 *in* G. Corbet and S. Harris, editors. The handbook of British mammals. Blackwell, Oxford, United Kingdom.

Virgós, E., M. R. Recio, and Y. Cortés. 2000. Stone marten (*Martes foina*) use of different landscape types in the mountains of central Spain. Zeitschrift für Säugetierkunde 63:193–199.

Waechter, A. 1975. Ecologie de la fouine en Alsace. La Terre et La Vie 24:399–457.

Wang, Y. 1986. The study of the consumption of wildlife resources in commercial shops in Taiwan. Council of Agriculture Reports, Taipei, Peoples Republic of China. (in Chinese)

Weir, R. D. 2003. Status of fishers in British Columbia. British Columbia Ministry of Water, Land, and Air Pollution report, Victoria, British Columbia, Canada.

_____, and A. S. Harestad. 2003. Scale-dependent habitat selectivity by fishers in south-central British Columbia. Journal of Wildlife Management 67:73–82.

Williams, R. M. 1963. Trapping and transplanting. Part I – fisher. Federal Aid in Wildlife Restoration, Final Segment Report, Project W 75-D-10, Boise, Idaho, USA.

Williams, R. N., L. K. Page, T. L. Serfass, and O. E. Rhodes. 1999. Genetic polymorphisms in the fisher (*Martes pennanti*). The American Midland Naturalist 141:406–410.

_____, O. F. Rhodes, Jr., and T. L. Serfass. 2000. Assessment of genetic variance among source and reintroduced fisher populations. Journal of Mammalogy 81:895–907.

Yamada, F. 1998. Status of alien mammals and problems caused by them in Japan. Mammalian Science 38: 97–105 (in Japanese).

Yeager, L. E. 1950. Implications of some harvest and habitat factors on pine marten management. Transactions North American Wildlife Conference 15:319–334.

Zalewski, A. 1997. Factors affecting selection of resting site type by pine marten in primeval deciduous forests (Białowieża National Park, Poland). Acta Theriologica 42:271–288.

Zielinski, W. J., J. E. Kucera, and R. H. Barrett. 1995. The current distribution of fisher in California. California Fish and Game 81:104–112.

_____, K. M. Slauson, C. R. Carroll, C. J. Kent, and D. G. Kudrna. 2001. Status of American martens in coastal forests of the Pacific states. Journal of Mammalogy 82:478–490.

_____, R. L. Truex, C. V. Ogan, and K. Busse. 1997. Detection surveys for fishers and American martens in California, 1989–1994: summary and interpretation. Pages 372–392 *in* G. Proulx, H. N. Bryant, and P. M. Woodard, editors. *Martes*: taxonomy, ecology, techniques, and management. Provincial Museum of Alberta, Edmonton, Alberta, Canada.

Chapter 3

GEOGRAPHICAL AND SEASONAL VARIATION IN FOOD HABITS AND PREY SIZE OF EUROPEAN PINE MARTENS

Andrzej Zalewski

Abstract: Although the diet of pine martens (*Martes martes*) has been described in detail from many locations in Europe, the geographical variation in their food habits is unknown. I reviewed the food habits of the pine marten over most of its geographical range, using 43 winter and 23 summer diet studies. Throughout Europe, the most important prey of martens was small mammals, which represented 47% of all prey in winter (range 14–81%), and 42% in summer (range 12–68%). Small mammals were followed in decreasing order of importance by plant (primarily berries) material (16% in winter, 21% in summer), birds (15 and 13%), medium-sized mammals (10 and 4%), and invertebrates (5 and 15%). Plant material and insects were more frequently consumed in southern regions than in northern Europe during winter. Medium-size mammals and large birds were consumed more often at higher latitudes. The proportion of small mammals (mainly rodents) in marten diets increased from the Mediterranean to northern regions, and reached a peak in the temperate deciduous and mixed woodlands; it declined further north in boreal forests. Across all studies, pine martens showed a functional response to fluctuating rodent numbers, but this was much more significant for bank voles (*Clethrionomys glareolus*) than for other rodent species. During winter, there was a trend towards a wider food niche and larger prey in the north compared to the south. Prey size in marten diets was negatively correlated with marten body size, but positively related to the number of days with snow cover. The diet of pine martens varied significantly with latitude and longitude during winter, suggesting that winter is a period of limited food availability.

1. INTRODUCTION

European pine martens (*Martes martes*) are widespread in Europe, inhabiting areas from northern Portugal and Spain to northern Finland and Russia (Grakov 1981). They occupy a wide range of habitats from boreal and temperate forests to Mediterranean forests. The broad habitat niche of martens is reflected in their diverse food habits, which include a variety of mammals, birds, amphibians, insects, fruits, as well as ungulate carcasses and mushrooms (e.g., Grakov 1981, Jędrzejewski et al. 1993, Pulliainen and Ollinmäki 1996). The size of prey utilized by martens varies from 3 g to 4 kg and the ability of pine martens to use such a wide range of habitats and prey causes them to be generalist forest carnivores. Many studies, however, suggest that martens are rodent specialists, which respond functionally to fluctuations in rodent abundance (Jędrzejewski et al. 1993, Pulliainen and Ollinmäki 1996, Helldin 1999). To better understand the interactions among predator and prey populations, we must better understand the variation in food habits of marten across their geopraphic range (Marcström et al. 1988).

The pine marten is a medium-size predator, and its body size varies regionally, but does not follow Bergmann's rule (Reig 1992). An alternative hypothesis for latitudinal size changes in carnivores is based on the assumption of a positive correlation between the size of the predator and available prey (Rosenzweig 1966, Erlinge 1987). According to this hypothesis, martens in southern Europe are larger (Reig 1992) and should feed on larger prey, whereas in northern Europe, martens are smaller and should consume smaller prey. Knowledge of the ratio of prey size to predator body size is critical for understanding adaptations of martens to climatic, latitudinal and altitudinal variation.

Although diets of martens have been described in detail from many localities in Europe, large-scale geographical variation in marten food habits is poorly understood. Reviews of numerous studies from western and central Europe (Clevenger 1994) and the former Soviet Union (Grakov 1981) have failed to reveal geographical trends in the food composition of pine martens. Clevenger's (1994) review was based on only 7 studies from western and central Europe, thus leaving a gap in the data set from north-eastern parts of the species' range. Grakov's (1981) comparisons included data only from the former Soviet Union. The purpose of this chapter is to review the food habits of pine martens over most of their geographical range, to describe geographical patterns in dietary composition to evaluate relationships between prey size in the marten diet and body size of martens, and to describe the extent of variation in food habits of martens when rodent abundance fluctuates.

2. METHODS

Data on the diet of European pine martens were taken from the literature (Table 3.1). Studies were selected based on the following criteria: (1) diet composition was estimated by the analysis of stomachs and/or scats; (2) the study

Table 3.1. Description and results of studies on pine marten (*Martes martes*) diet composition, reviewed in this paper. No.: consecutive number of study locations in both seasons. Latitude: °N, Longitude: positive values °E, negative values °W. Diet composition expressed as percentage of relative frequency of prey items (all prey items identified from scats and/or stomach taken as 100%). Prey groups: 1–small mammals, 2–medium-sized mammals, 3–birds, 4–invertebrates, 5–plant material. The unlisted 6th group ("others") sums to 100% in each locality.

No.	Country, locality	Latitude	Longitude	Mean percentage frequency in marten diet					Source
				1	2	3	4	5	
				Winter					
1	Spain, Mallorca	39	2	19.1	3.4	7.6	3.6	56.2	Clevenger 1995
2	Spain, Menorca	40	3	13.7	0	11.9	0.6	70.6	Ruiz-Olmo and Nadal 1991
3	Spain, Menorca	40	3	22.2	2.8	16.3	14.0	39.6	Clevenger 1993
5	Russia, Caucasus Mts.	41	42	25.2	0	6.5	2.3	65.9	Donaurov et al. 1938
8	Swiss, Jura Mts.	47	7	27.4	2.8	10.0	12.0	44.5	Marchesi 1989
9	Ukraine, western part	50	32	48.5	4.5	6.8	17.4	15.2	Polushina 1957
10	Germany, Oberlausitz district	51	15	35.9	5.6	15.1	17.0	20.7	Ansorge 1989
11	Poland, central part	52	20	38.8	10.0	27.9	2.2	6.6	Rzebik- Kowalska 1972
12	Poland, Białowieża Forest	52	55	60.3	0.6	2.9	10.6	9.0	Jędrzejewski et al. 1993
13	Belarus, Białowieża Forest	52	55	80.7	4.7	5.6	3.9	2.6	Datskevich 1979
14	Russia, Zhiguli Mts.	53	30	55.1	22.6	6.4	12.8	0	Yurgenson 1951
15	Belarus, central part	54	28	78.2	4.7	2.5	0	2.6	Serzhanin 1973
16	Belarus, Berezinski Reserve	54	28	68.4	0.5	3.3	5.7	12.4	Serzhanin 1961
17	Russia, Bashirskii Reserve	54	57	81.2	1.2	3.0	6.6	1.2	Mozgovoi 1971
18	Russia, Ural Mts. (Bashkiria)	54	58	63.7	5.6	12.1	10.5	8.1	Aspisov 1973
19	Lithuania	55	24	51.2	2.9	20.4	1.2	18.3	Maldzhiunaite 1959
20	Russia, Tatarskoi region	55	51	51.7	7.6	20.6	9.3	8.7	Aspisov 1973
21	Belarus, north-eastern part	56	30	31.7	3.0	7.6	13.2	34.3	Sidorovich 1997
22	Russia, Gorkovskoi region	57	44	46.4	14.0	29.1	6.2	0.8	Yurgenson 1951
23	Russia, Mariiskoi region	57	48	72.3	10.9	4.0	4.0	6.9	Aspisov 1973
24	Russia, Pskov region	58	29	55.7	21.3	23.0	0	0	Morozov 1976
25	Russia, Kirov region	58	50	51.1	20.9	12.0	4.0	12.0	Pleshak 1976

Table 3.1. Continued.

No.	Country, locality	Latitude	Longitude	Mean percentage frequency in marten diet					Source
				1	2	3	4	5	
26	Russia, Perm region	58	56	57.0	16.2	9.5	3.3	10.6	Chashchin 1956
27	Russia, Perm region	58	56	53.0	4.1	17.6	4.0	15.8	Grakov 1981
28	Russia, Central Ural Mts.	58	58	55.8	4.4	17.9	19.0	2.9	Bakeev 1966
29	Russia, Novgorod region	59	32	50.5	14.2	17.7	0	7.0	Morozov 1976
30	Sweden, south-central part	60	15	37.1	7.7	10.9	10.6	15.0	Helldin 1999, 2000
31	Sweden, south-central part	60	15	36.5	26.5	9.2	0	0	Storch et al. 1990
32	Russia, Petersburg region	60	32	53.0	12.1	22.9	0	5.7	Morozov 1976
33	Russia, Leningrad region	60	32	64.1	11.4	14.1	0.9	5.7	Novikov et al. 1970
34	Russia, NW Vologodtskoi region	60	37	50.6	7.5	20.6	2.8	11.0	Grakov 1981
35	Russia, Vologodtskoi region	60	40	53.5	13.1	17.7	0.9	12.0	Gribova 1958
36	Norway, southern part	61	10	18.7	10.7	28.0	6.7	26.7	Selas 1992
37	Russia, Pechora river	62	59	32.1	40.7	25.6	1.5	0	Yurgenson 1951
38	Russia, Pechora river	62	59	36.5	14.3	28.8	0	15.9	Yazan 1962
39	Russia, N Dvina river	63	41	34.5	40.4	11.1	0.8	8.9	Grakov 1981
40	Russia, N Dvina river	64	42	38.9	8.6	17.6	5.4	21.4	Parovshchikov 1961
41	Russia, Karelia	65	32	35.2	17.8	23.4	1.4	9.7	Danilov and Ivanov 1967
42	Russia, Karelia	65	32	52.5	12.2	14.2	0	11.8	Morozov 1976
43	Russia, Arkhangelsk region	66	46	44.4	11.4	21.0	1.9	10.9	Grakov 1962
44	Russia, N Ural Mts.	66	60	40.2	5.4	17.7	2.7	33.7	Gashev 1965
45	Finland, Lapland Forest	68	29	54.4	8.2	17.3	1.1	9.3	Pulliainen and Olinmäki 1996
46	Russia, Laplandskii Reserve	68	32	33.9	7.5	30.7	5.6	11.9	Nasimovich 1948
				Summer					
1	Spain, Mallorca	39	2	23.9	0.8	6.9	13.3	48.3	Clevenger 1995
3	Spain, Menorca	40	3	22.0	3.8	20.1	20.5	28.6	Clevenger 1993
4	Spain, Menorca	40	3	11.9	11.0	15.3	34.7	21.2	Moreno et al. 1988

Table 3.1. Continued.

No.	Country, locality	Latitude	Longitude	Mean percentage frequency in marten diet					Source
				1	2	3	4	5	
5	Russia, Caucasus Mts.	41	42	46.1	0.8	13.7	33.5	5.7	Donaurov et al. 1938
6	Spain, Cantabrian Mts.	43	-5	63.7	0.1	1.9	13.7	12.6	Clevenger 1993
7	France, Côte d'Or	47	5	16.0	0.3	4.9	10.2	68.3	Baudvin et al. 1985
12	Poland, Białowieża Forest	52	55	53.1	0.2	13.7	14.8	8.4	Jędrzejewski et al. 1993
13	Belarus, Białowieża Forest	52	55	67.9	5.6	0.8	11.1	9.5	Datskevich 1979
14	Russia, Zhiguli Mts.	53	30	56.0	8.7	1.8	21.7	10.5	Yurgenson 1951
15	Belarus, central part	54	28	67.3	0.6	3.7	9.6	6.0	Serzhanin 1973
17	Russia, Bashirskii Reserve	54	57	67.7	1.2	16.0	7.8	5.9	Mozgovoi 1971
21	Belarus, north-eastern part	56	30	25.1	5.3	23.6	25.6	16.4	Sidorovich 1997
25	Russia, Kirov region	58	50	46.2	10.4	12.6	4.6	23.2	Grakov 1981
30	Sweden, south-central part	60	15	22.0	5.4	11.8	36.7	15.5	Helldin 1999, 2000
36	Norway, southern part	61	10	22.8	11.8	30.6	5.1	28.9	Selas 1992
37	Russia, Pechora river	62	59	35.4	1.5	24.1	20.9	15.1	Yurgenson 1951
38	Russia, Pechora river	62	59	56.0	4.6	11.6	0	27.9	Yazan 1962
39	Russia, N Dvina river	63	41	45.3	0.6	13.8	14.4	17.3	Grakov 1981
40	Russia, N Dvina river	64	42	25.4	4.1	15.5	16.1	35.8	Parovshchikov 1961
41	Russia, Karelia	65	32	43.4	6.3	17.3	4.7	16.5	Danilov and Ivanov 1967
42	Russia, Karelia	65	32	60.1	0.8	8.0	9.5	9.6	Grakov 1981
44	Russia, N Ural Mts	66	60	29.4	2.3	7.4	16.7	42.6	Gashev 1965
46	Russia, Laplandskii Reserve	68	32	56.0	2.9	11.5	8.2	18.4	Nasimovich 1948

covered part of either winter or summer, and the seasons were analyzed separately; (3) the place and time span of the study were described; (4) sample sizes were ≥40 stomachs or scats. These criteria yielded 43 winter and 23 summer diet studies (Table 3.1). For the analysis of diet composition, I standardized occurrences as a percentage of relative frequency, i.e., the total number of occurrences of each food item recovered from scats or stomachs was divided by the total number of items identified across all samples. Food items were classified into 6 food categories: small mammals (<150 g), medium-sized mammals (150–2,500 g), birds, invertebrates, plant material (including fungi), and others (including amphibians and reptiles, ungulate carcasses). The standardized food niche breadth (Krebs 1989) was calculated for these major food groups. A Principal Component Analysis (PCA) with Varimax rotation was performed for relative frequencies of occurrence of each food type to describe the trophic relationships of martens across their geographic range. The PCA factors from relative frequency of occurrence data were regressed against latitude and longitude using simple linear regression. Prior to analyses, all variables were arcsine transformed.

An index of prey size was calculated for 37 winter and 21 summer studies according to Erlinge (1987). Prey size indices were calculated for all locations, where prey had been divided into 9 categories. The following body weight categories were used for assessing prey size: insectivorous mammals: 10 g, small rodents: 25 g, squirrels (*Sciurus vulgaris*): 230 g, hares (*Lepus* spp.) and rabbits (*Oryctolagus cuniculus*): 1,500 g, small birds: 30 g, large birds: 500 g, amphibians and reptiles: 15 g, insects: 3 g, and carrion (ungulate carcasses): 200 g. I assumed that the weight of carrion consumed by martens corresponded to the maximal capacity of their stomach (Grakov 1981). Multiple linear regression analysis was used to evaluate the influence of a series of climatic factors on prey size: monthly temperature (December, January, and February), average winter temperature, number of days with snow cover, average snow depth, and average winter precipitation. Climatic data were taken from Kostin and Pokrovskaya (1961) and Lebedeva et al. (1979). I compared prey size in the marten's diet with body size of martens using average (for males and females) condylobasal length of pine marten skulls (Maldzhiunaite 1957, Anderson 1970, Reig 1989).

For more detailed analysis of the role of rodents in marten diets, percent frequency of occurrence in scats/stomach was used, i.e., the number of scats or stomachs with rodent remains compared with the total number of scats or stomachs sampled. Spearman rank correlation was used to analyze the association between percent frequency of occurrence of rodents in the marten diet and rodent abundance across years. Only studies with ≥4 years of information were included in this analysis. The association between percent occurrence of rodents and occurrence of other food groups was evaluated using Spearman rank correlation.

3. RESULTS

3.1 Geographical Variation in Diet Composition

Small mammals (<150 g) were the most important food for martens throughout their range during both summer ($\bar{x}$ = 42%) and winter ($\bar{x}$ = 47%) (Table 3.2). Small mammals were followed in frequency by birds (13% summer, 15% winter) and plant material (21% summer, 16% winter). Medium-sized (150–2,500 g) mammals were more frequent in diets during winter than during summer, whereas insects and plant material were more frequent in summer (Table 3.2). The PCA generated 3 factors that explained 78% of the total variance in the winter diet, and 76% of the variance in the summer diet (Table 3.3). Factor 1 for the winter season shows a gradient from diets with a high frequency of invertebrates and plant material towards diets dominated by medium-sized mammals and birds. The second factor describes winter diets with a high frequency of plant material towards those with an important contribution of small mammals. Factor 3 describes winter diets with an increasing frequency of others foods (e.g. from footnote a of Table 3.3). In the summer, factor 1 shows a gradient of increasing small mammals and decreasing plant material in the diet (Table 3.3). Factor 2 during summer indicates an increasing frequency of medium-sized mammals and birds. Factor 3 describes summer diets with an in-

Table 3.2. Comparison of diet composition (relative frequency of occurrence) of European pine martens during winter (n = 43 studies) and summer (n = 23) based on data listed in Table 3.1. Seasonal differences were evaluated using a Mann-Whitney U-test (** $P < 0.01$, *** $P < 0.005$, NS = not significant).

Prey group	Winter		Summer		Mann-Whitney *U*-test
	Average	(min-max)	Average	(min-max)	
Small mammals (<150 g)	46.8	(13.7–81.2)	41.8	(11.9–67.9)	NS
Medium-sized mammals (150–2,500 g)	10.3	(0–40.7)	3.9	(0–11.8)	***
Birds	15.3	(2.5–30.7)	12.5	(0.7–30.6)	NS
Invertebrate	5.2	(0.9–19.0)	15.4	(0–36.7)	***
Plant material[a]	15.8	(0.2–70.6)	21.4	(5.7–68.3)	**
Others[b]	6.5	(0–27.8)	5.0	(0–12.8)	NS

[a]Includes fungi.
[b]Includes amphibians, carrion, fish, and reptiles.

Table 3.3. Correlation between prey groups in pine marten diets and factors from a Principal Component Analysis in two seasons. Numbers in bold had loadings >|0.50|.

	Winter			Summer		
Prey groups	Factor 1	Factor 2	Factor 3	Factor 1	Factor 2	Factor 3
Small mammals (<150 g)	-0.15	**-0.93**	-0.09	**0.80**	-0.41	-0.41
Medium-sized mammals (150–2,500 g)	**0.84**	-0.25	0.03	-0.02	**0.81**	-0.14
Birds	**0.75**	0.40	-0.17	-0.13	**0.78**	0.13
Insects	**-0.55**	0.11	-0.43	-0.09	0.09	**0.89**
Plant material	-0.49	**0.81**	-0.01	**-0.95**	-0.06	-0.18
Others[a]	-0.04	0.10	**0.93**	0.42	-0.28	0.57
Eigenvalue	2.03	1.59	1.06	2.17	1.36	1.09
Percent of variation explained	33.8	26.5	17.6	35.1	22.7	18.2

[a]Includes amphibians, carrion, fish, and reptiles.

creasing contribution of insects. The first two principal components for the winter season clearly separated study sites from 3 different forest zones (Fig. 3.1). The first factor separated Mediterranean, temperate deciduous forests from temperate mixed forests and from boreal forests. The second factor distinguished temperate mixed forests from boreal forests (Fig. 3.1).

The first PCA factor for the winter season was positively correlated with latitude ($r = 0.63$, $n = 43$, $P < 0.001$), and the second factor was negatively correlated with longitude ($r = -0.43$, $n = 43$, $P < 0.005$). Plant material and insects were more frequently consumed in southern regions; their proportions in marten diets decreased in northern Europe. In contrast, birds and medium-sized mammals were consumed more often at high latitudes. Martens preyed on small mammals more often in the eastern portion of their geographic range, but they consumed more plant material in the western portion of their range. During summer, there were weaker correlations between PCA factors and latitude or longitude ($r = -0.41$–0.32, $n = 23$, $P > 0.05$).

Based on latitudinal trends in the proportions of the major prey groups in diets of martens during winter, I constructed a graphical model of geographical variation in the food habits of pine martens (Fig. 3.2). Small mammals, medium-sized mammals, birds, and plant material formed 90% of the frequency of prey. Small mammals were most important in marten diets in the temperate zone (on average, 50% of frequency at 50–60°N) and their role became smaller at both lower and higher latitudes. The frequency of medium-sized mammals increased from zero at 35–40°N to 15–17% at 65–68°N and, similarly, the proportion of birds increased from 7% at 40°N to 20% at 65–68°N (Fig. 3.2). In 37

Figure 3.1. Locations of pine marten diet studies, in relation to the first 2 principle components that described 60% of variation in winter diets of martens and 58% of variation in summer diets of martens across 43 winter and 23 summer diet studies conducted in Europe. Numbers correspond with the studies listed in Table 3.1.

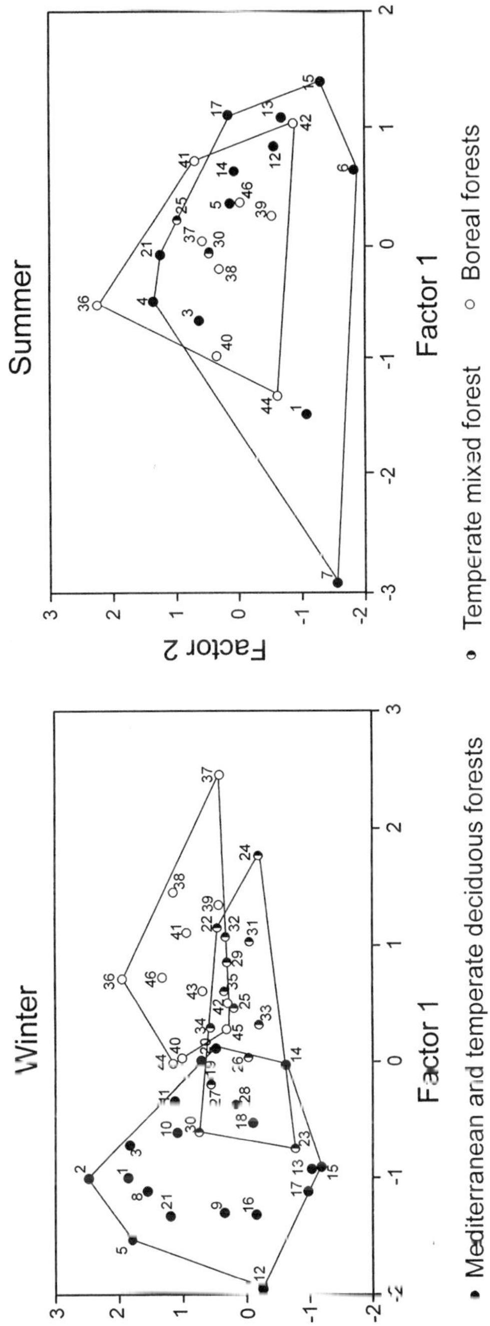

Figure 3.2. Generalized model of latitudinal variation in relative frequency (%) of food categories in winter diets of pine martens (*Martes martes*) in Europe, based on regressions calculated from empirical data (n = 45 localities listed in Table 3.1). The regression equations were as follows: small (<150 g) mammals $Y = -0.14 X^2 + 15.66 X - 379.91$, $r^2 = 0.93$, $P < 0.001$; medium-sized (150–2,500 g) mammals $Y = -22.17 + 0.57 X$, $r^2 = 0.20$, $P = 0.003$; birds $Y = -12.52 + 0.49 X$, $r^2 = 0.21$, $P = 0.002$; invertebrate $Y = 19.36 - 0.25 X$, $r^2 = 0.11$, $P = 0.027$; plant materials (fungi included) $Y = 0.14 X^2 - 16.43 X + 491.03$, $r^2 = 0.83$, $P < 0.001$; others $Y = 0.437 + 0.07 X$, $r^2 = 0.01$, $P = 0.54$.

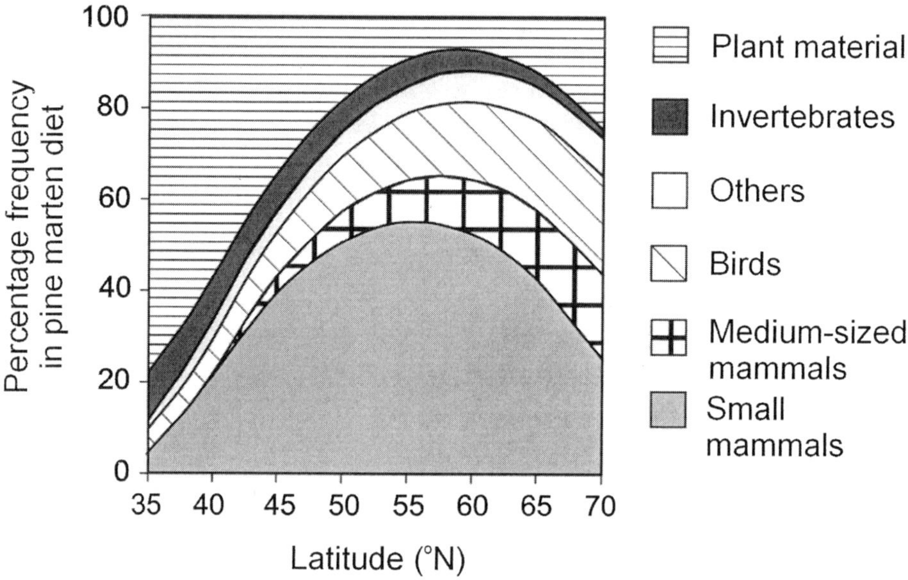

locations, birds were divided into 2 groups: small and large. The latitudinal increase of birds in the marten diet was due to percent occurrence of large birds ($r^2 = 0.44$, $n = 37$, $P < 0.0005$); the share of small birds was not significant ($r^2 = 0.01$, $n = 37$, $P = 0.25$). In northern Europe, large birds consumed by martens were often capercaillie (*Tetrao urogallus*), hazel hen (*Tetrastes bonasia*), black grouse (*Lyrurus tetrix*), and willow grouse (*Lagopus lagopus*).

The frequency of plant material in the marten's diet decreased from southern to temperate regions (on average, 9% at 57–60°N) and increased again in boreal localities (Fig. 3.2). In southern Europe, martens fed on many plant species such as rowanberries (*Sorbus aucuparia*), carob fruit (*Ceratonia siliqua*), myrtle berries (*Myrtus communis)*, juniper (*Juniperus communis)*, cherries (*Prunus* sp.), rose hips (*Rosa* spp.), figs (*Ficus carica*), and citrus (*Citrus* sp.) (Marchesi 1989, Clevenger 1995, Ruiz-Olmo and Lopez-Martin 1996). In Central Europe, *Rubus* spp. and rowanberries were most often reported as vegetable food of pine martens (Ansorge 1989, Jędrzejewski et al. 1993). In north-

Figure 3.3. Latitudinal variation in standardized food niche breadth calculated for 6 major groups of food. Predicted line is calculated based on the generalized data from Figure 3.2. Best-ft line to empirical data according to Lowess methods (Cleveland 1979).

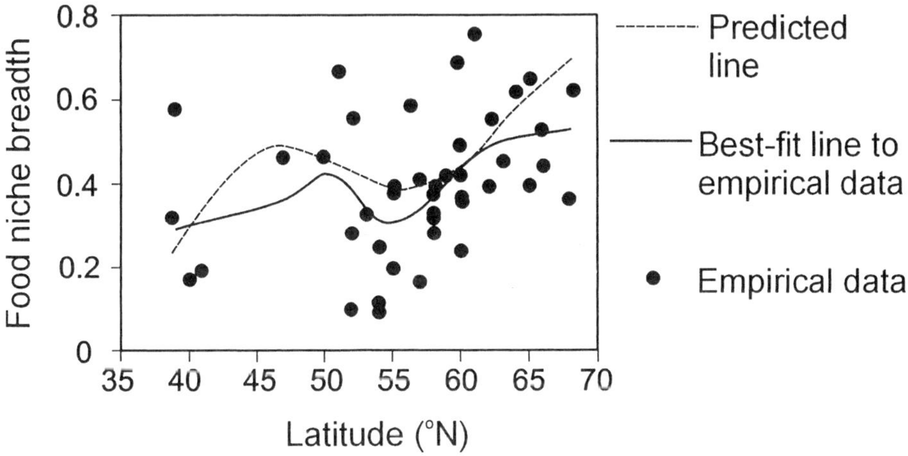

ern Europe, martens consumed mostly blueberries (*Vaccinium myrtillus*), lingonberries (*V. vitis-idaea*) rowanberries, but also mushrooms (Pulliainen and Ollinmäki 1996, Helldin 2000).

The average standardized food niche breadth was 0.34 (SD = 0.15) in winter, and 0.37 (SD = 0.15) in summer. Food niche breadth did not correlate with sample size (winter: $r = -0.15$, $n = 43$, $P > 0.05$; summer: $r = 0.02$, $n = 23$, $P > 0.05$). In winter, food niche breadth was significantly related to latitude ($r^2 = 0.12$, $n = 43$, $P = 0.029$) but not longitude ($r^2 = 0.05$, $P = 0.136$). This indicates a trend towards a wider food niche in northern areas than in southern areas. However, the latitudinal trends in diet were not linear (Fig. 3.3). The food niche was narrow in the south and increased to 50°N. Between 50–60°N food niche breadth decreased, but still further north marten's food niche widened again. Summer values of standardized food niche breadth were not significantly related to latitude or longitude ($r^2 = 0.01$ and $r^2 = 0.07$, respectively, $n = 23$, $P > 0.05$).

3.2 Variation in Prey Size and Marten Size

Within their geographical range, martens consumed prey weighing as much as 4 kg (hares), and consumed very small prey such as shrews or insects. During both winter and summer, the size of marten prey increased with latitude from 2 g (winter) and 4 g (summer) at 40°N to 20 g (winter) and 7 g (summer) at 68°N (Fig. 3.4). Similarly, the frequency of large prey (squirrels, hares, rabbits, and large birds) in marten diets increased with latitude in both seasons (winter: $r^2 = 0.41$, $n = 37$, $P < 0.001$; summer: $r^2 = 0.31$, $n = 21$, $P = 0.004$).

Relative frequency of medium to large sized (>150 g) prey and prey size index were both negatively correlated with marten body size based on condylobasal length of marten skulls (Fig. 3.5). In the south, larger martens consumed smaller prey, but in the north, smaller martens consumed larger prey. Stepwise regression analysis was used to evaluate the influence of 7 climatic factors on prey size; prey size was significantly related to only the number of days with snow cover ($r^2 = 0.35$, $n = 37$, $P < 0.001$).

3.3 Rodents and Alternative Prey in Diets

The composition of rodent species in the diet of martens varied among regions (Fig. 3.6). In the Mediterranean region, mice in genus *Apodemus* comprised the largest proportion of all rodents consumed ($\bar{x}$ = 53%). Frequency of *Apodemus*, however, declined towards the north. Voles in the genus *Clethrionomys* were most prevalent in the temperate and boreal forests ($\bar{x}$ = 46–52%). *Microtus* represented 27–39% of all rodents in diets in the temperate and boreal regions. In the north, martens also consumed lemmings (*Myopus schisticolor* and *Lemmus lemmus*).

Long-term studies demonstrated that pine martens showed a functional response to fluctuations in rodent numbers; the percent occurrence of rodents in the martens diet was positively related to rodent abundance (7 long-term studies; $r_s \geq 0.54$, duration = 4–11 years, $P < 0.05$; calculated from Gribova 1958, Semenov-Tyan-Shanskii 1959, Grakov 1962, Mozgovoi 1971, Helldin and Lindström 1993, Jędrzejewski et al. 1993, A. Zalewski unpubl. data, Pulliainen and Ollinmäki 1996). Three long-term studies conducted within the temperate deciduous to boreal forests analyzed the dietary response of martens in relation to abundance of various species of coexisting rodents. They all demonstrated significant relationships between martens and abundance of bank voles (*Clethrionomys glareolus*), but not with abundances of *Microtus* or *Apodemus* (Jędrzejewski et al. 1993, A. Zalewski unpubl. data, Pulliainen and Ollinmäki 1996, Helldin 1999). Data collected in Białowieża National Park, Poland over an 11-year period clearly elucidate the relationship between occurrence of rodents in the diet and densities of *Clethrionomys*, but not *Apodemus* (Fig. 3.7).

In years of low abundance of rodents, martens utilized different alternative prey types among regions (Table 3.4). The long-term studies showed that in the lowland deciduous forests, martens ate more birds, amphibians and ungu-

Figure 3.4. Relationship between mean weight of prey in diet of pine martens across Europe in the winter ($n = 37$ localities) and summer ($n = 21$) seasons.

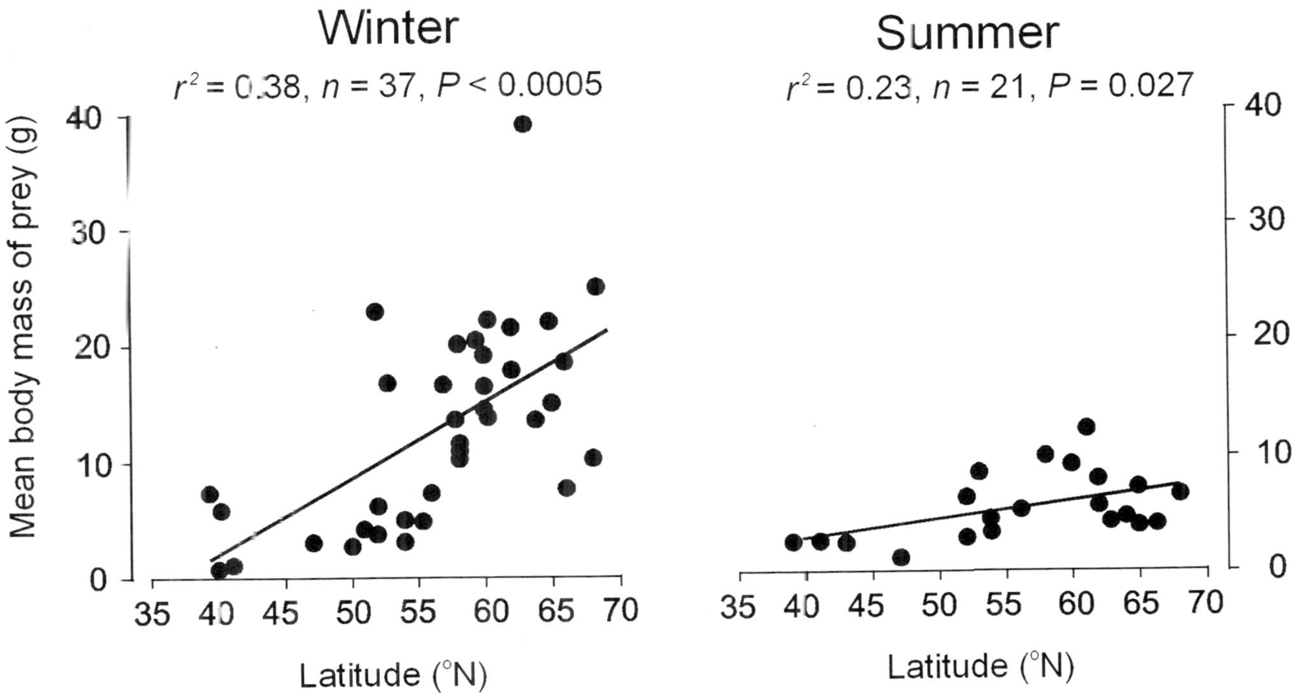

Figure 3.5. Relationship between relative frequency of medium to larger-sized (>150 g) prey (squirrels, hares, rabbits, tetraonid birds) in diets of pine martens and mean body mass of all prey and the condylobasal length of male marten skulls. Data on skull size were from Maldzhiunaite (1957), Anderson (1970), and Reig (1989).

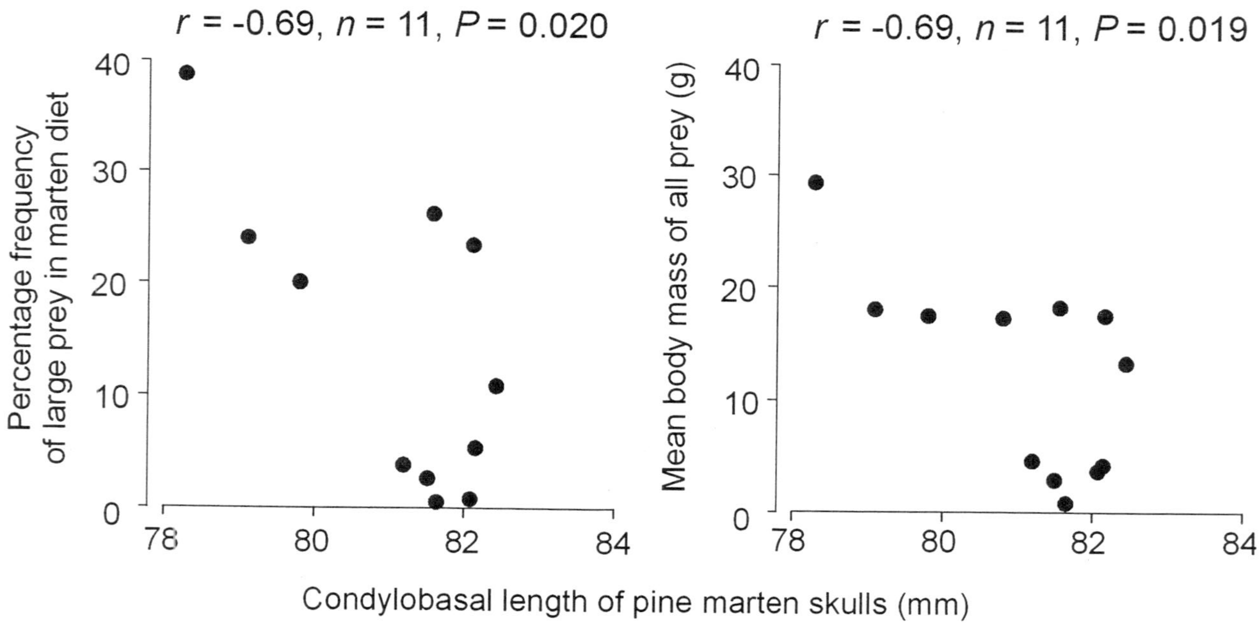

Figure 3.6. Relative frequency of occurrence of 5 groups of rodents in diets of pine martens across 4 biogeographic regions. Sources: Mediterranean forests (Ruiz-Olmo and Nadal 1991, Clevenger 1993, 1995); temperate deciduous forests (Polushina 1957, Maldzhiunaite 1959, Rzebik-Kowalska 1972, Serzhanin 1973, Ansorge 1989, Jędrzejewski et al. 1993); temperate mixed forests (Yurgenson 1951, Gribova 1958, Bakeev 1966, Pleshak 1976, Grakov 1981, Helldin 1999); boreal forests (Nasimovich 1948, Gashev 1965, Parovshchikov 1961, Novikov et al. 1970, Morozov 1976, Pulliainen and Ollinmäki 1996).

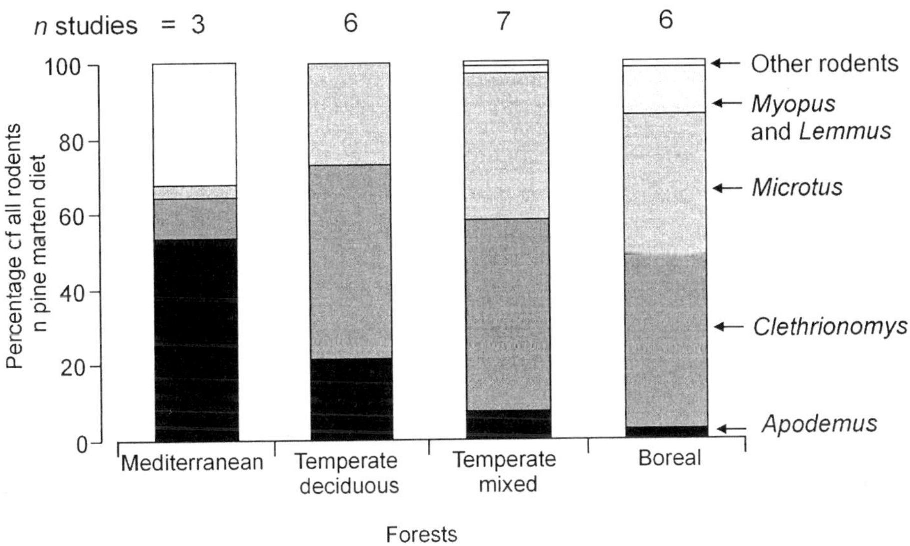

late carcasses, the consumption of which was negatively correlated with consumption of rodents. In the boreal forest, martens consumed more large birds, squirrels, bird eggs, and fruits in years of low rodent abundance. In general, large prey was the alternative prey in Northern Europe.

4. DISCUSSION

I documented a latitudinal variation in diets, food niche breadth, and prey size for pine martens in Europe. The diet of martens varied among years in response to rodent availability and winter conditions (snow cover and temperature) (Jędrzejewski et al. 1993, Pulliainen and Ollinmäki 1996). For example, Helldin's (1999) data were collected during relatively mild winters with a general lack of snow cover; martens ate more berries than in most other studies in this region (Novikov et al. 1970, Morozov 1976, Storch et al. 1990). The percent occurrence of rodents in marten diets varied up to four-fold between years in one study area (Jędrzejewski et al. 1993, Pulliainen and Ollinmäki

Figure 3.7. Eleven-year variations in abundances of bank voles (*Clethrionomys glareolus*) and yellow-neck mice (*Apodemus flavicollis*) during autumn and their percent occurrence in autumn-winter diet of pine martens in Białowieża National Park, Poland. Data on rodent abundance: Pucek et al. (1993), Stenseth et al. (2002); marten diet: Jędrzejewski et al. (1993) and A. Zalewski (unpublished data).

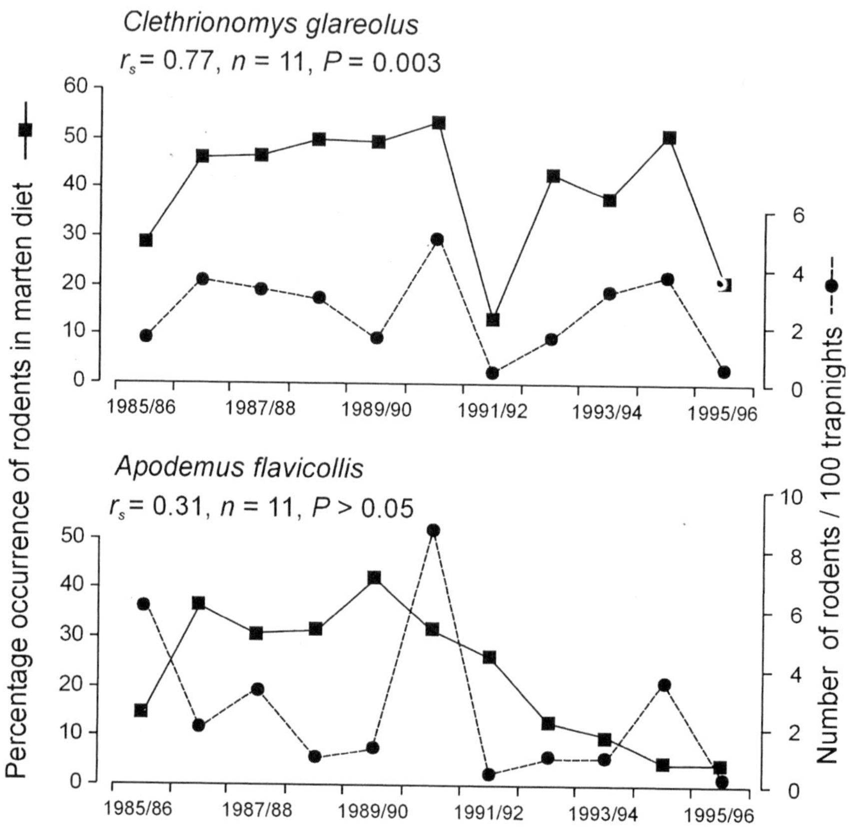

1996). Also, it must be recognized that percent occurrence of food items, although commonly used (Reynolds and Aebischer 1991), overestimates smaller food items (e.g., percent occurrence vs. percent of biomass in insects and fruits) (Jędrzejewski et al. 1993, Helldin 1999). Biomass data would be more informative, but are scarce or have been calculated using different methods among studies.

The latitudinal differences in diet demonstrated the marten's adaptations to varying abundance and availability of food resources. I hypothesize that the most important determinant of dietary composition of martens is the abun-

Table 3.4. Percentage occurrence of alternative prey in winter diet of pine martens and Spearman rank correlations (r_s) between percentage occurrence of rodents and alternative prey in the temperate and boreal regions of Europe.

Country (Source)[a]	*n* years	Alternative prey	Percentage occurrences in marten diet: Average	(min-max)	Correlation with percentage of rodents: r_s	P[b]
Poland (12)	11	Amphibians	21.9	(5.6–47.3)	-0.88	**
		Carcasses	17.8	(5.2–48.2)	-0.73	**
		Birds	9.4	(1.1–29.3)	-0.84	**
Sweden (30)	9	Fruits	23.6	(1.8–52.6)	-0.73	*
Russia (38)	7	Squirrels	19.7	(5.1–41.7)	-0.74	*
		Birds eggs	6.6	(1.4–11.2)	-0.96	***
Russia (43)	8	Large birds	11.2	(2.8–22.5)	-0.76	*
Finland (45)	14	Birds	21.5	(4.0–40.7)	-0.61	*
		Birds eggs	10.9	(0–45)	-0.53	*
		Squirrels	8.1	(0–53.3)	-0.47	*
		Carcasses	6.5	(0–28.6)	-0.52	*

[a]Refers to number of localities in Table 3.1.
[b]* $P < 0.05$, ** $P < 0.01$, *** $P < 0.005$.

dance and availability of rodents (mainly the bank vole). The proportion of rodents in martens diets was largest in the temperate deciduous forests, where densities of forest rodents are high (Jędrzejewski and Jędrzejewska 1996). Availability of rodents probably decreases in northern latitudes with deeper snow cover (Jędrzejewski et al. 1993). Pulliainen and Ollinmäki (1996), however, did not find a significant reduction of consumption of *Clethrionomys* voles by martens during periods of deep snow cover. In southern latitudes, forest rodent communities were dominated by *Apodemus* mice, which are not a preferred prey of martens (Jędrzejewski et al. 1993).

The latitudinal variation of plant material and insects in the diet of martens might be also related to the regional availability of these food resources. Fruits become more available in the southern region of Europe during winter; they are more frequent in the diet of martens during that period. This was also reported for other predators (stone marten, *Martes foina*, Pandolfi et al. 1996; badger, *Meles meles*, Goszczyński et al. 2000). The lower fruit consumption in northern latitudes may be due to lower abundance, but also because snow cover reduces access to fruit. Pulliainen and Ollinmäki (1996) noted a decreased consumption of berries with increasing snow cover. However, in northernmost regions, martens also consumed mushrooms in winter (Pulliainen and Ollinmäki 1996) and the proportion of plant material in their diets increased. As with fruits, insects are more available to martens in southern Europe because insects are not active in cold winters. Also, the number of insect species consumed was much higher in southern than in northern Europe. In southern Europe, Goszczyński et al. (2000) reported a similar observation for badgers.

The greater frequency of birds in the winter diet of martens from northern latitudes was unexpected. In winter, the availability of birds in northern Europe is much lower due to migration of many species to the south. Thus, increased consumption of large birds such as hazel hen, capercaillie, black and willow grouse which are year round residents, was the likely source of this diet change. Similar to birds, squirrels are probably more available during severe winter conditions. Although very agile and difficult to capture, squirrels are less active in winter and are often captured in their dens (Pulliainen and Ollinmäki 1996). In Poland, the proportion of squirrels in the marten's diet increased only in the harshest winters (Jędrzejewski et al. 1993, A. Zalewski unpubl. data).

Pine martens clearly preferred *Clethrionomys* to *Microtus* voles. Thompson and Colgan (1990) reported a similar finding for the American marten (*Martes americana*) in Ontario. A potential reason for preference for *Clethrionomys* may be similar habitat selection by predator and prey. *Clethrionomys* and martens both favor forests, while *Microtus* voles inhabit grasslands, fields, and other open areas (Pucek 1983, Goszczyński 1985, Brainerd et al. 1994, Jędrzejewska and Jędrzejewski 1998).

The marten's diet was flexible across time and space. Predators should have a broader diet in unproductive environments, where prey items are relatively rare and searching time is longer (Begon et al. 1990). Indeed, food niche breadth of martens increased with latitude. In contrast, Martin (1994) recorded the lowest diet diversity for American marten in the subarctic. This may be explained by the fact that a larger prey item provides food for a longer period, hence reducing kills per unit time, and ultimately resulting in a less diverse diet (Martin 1994). In this study, however, broader niches were documented for populations of martens in northern regions, which tended to consume larger prey.

Body size of European pine martens increases from north to south (Reig 1992). For Mustelids, several hypotheses have been proposed to explain this variation: adaptation to winter condition (especially snow cover) (Petrov 1962), and character displacement between competing Muselids (McNab 1971). An alternative hypothesis for latitudinal size trends in carnivores suggests a correlation between the size of predator and prey available (Rosenzweig 1966, Erlinge 1987). However, an inverse relationship was apparent based on the information reported here; size of European pine martens was inversely related to prey size. Perhaps, martens could increase foraging efficiency by hunting larger prey in the north, thus reducing the duration of activity and energy loss at lower temperatures. Compared to larger martens, smaller individuals have lower food requirements, so they could reduce activity by hunting larger prey. Such behavior may enable smaller martens to stay longer in insulated resting sites and to minimize energy expenditure. Thus, the Mustelids' adaptation to cold climates probably involves a reduction in the duration of exposure to low temperature and a behavioral adaptation to prey selection, rather than an increase in body size (morphological adaptation).

In conclusion, this review documented that the European pine marten is a rodent specialist (particularly on *Clethrionomys*) but is opportunistic as well, feeding on various alternative prey in different biogeographic regions. Its diet varies significantly with latitude and longitude, and the variation in winter diet is more pronounced than during the summer. This suggests that winter is the most food limited season for pine martens.

5. ACKNOWLEDGMENTS

I am grateful to J. Birks, J. Goszczyński, B. Jędrzejewska, G. Proulx, and Z. Pucek for helpful comments on a previous draft of this manuscript.

6. LITERATURE CITED

Anderson, E. 1970. Quaternary evolution of the genus *Martes* (Carnivora, Mustelidae). Acta Zoologica Fennica 130:1–132.

Ansorge, H. 1989. Nahrungsökologische Aspekte bei Baummarder, Iltis und Hermelin (*Martes martes*, *Mustela putorius*, *Mustela erminea*) [Aspects of diet ecology of pine marten, polecat and stoat (*Martes martes*, *Mustela putorius*, *Mustela erminea*)]. Populationsökologie marderartiger Säugetiere, Wiss. Beitr. Univ. Halle:494–504.

Aspisov, D. I. 1973. Lesnaya kunitsa: Volzhsko-Kamskii krai [Pine marten: Volga-Kama rivers country]. Pages 161–172 *in* A. A. Nasimovich, editor. Sobol, kunitsy, kharza: razmeshchenie zapasov, ekologiya, ispolzovanie i okhrana [Sable, martens, and yellow-throated marten: distribution of resources, ecology, harvest, and conservation]. Nauka, Moskva. (in Russian)

Bakeev, Y. N. 1966. K pitaniyu lesnoi kunitsy na Srednem Urale [On food of pine marten in the Middle Urals]. Uchenye zapiski Uralskogo Gosudarstvennogo Universiteta, Seriya Biologicheskaya 43(3):58–65. (in Russian)

Baudvin, H., J. L. Dessolin, and C. Riols. 1985. L'utilisation par la martre (*Martes martes*) des nichoirs chouettes dans quelques forets bourguignonnes. Ciconia 9:61–104.

Begon, M., J. L. Harper, and C. R. Townsend. 1990. Ecology. Blackwell Scientific Publications, Cambridge.

Brainerd, S. M., J. O. Helldin, E. Lindström, and J. Rolstad. 1994. Eurasian pine martens and old industrial forest in southern boreal Scandinavia. Pages 343–354 *in* S. W. Buskirk, A. S. Harestad, M. G. Raphael, and R. A. Powell, editors. Martens, sables, and fishers. Biology and conservation. Cornell University Press, Ithaca, New York.

Chashchin, S. P. 1956. Lesnaya kunitsa Kamskogo Preduralya i ee promyslovoe znachene [Pine marten of Kama Predurale and its economic importance]. Dissertation, University of Perm, Perm, Russia. (in Russian)

Cleveland, W. S. 1979. Robust locally weighted regression and smoothing scatterplots. Journal of the American Statistical Association 74:829–836.

Clevenger, A. P. 1993. Pine marten (*Martes martes* Linné, 1758) comparative feeding ecology in an island and mainland population of Spain. Zeitschrift fur Säugetierkunde 58:212–224.

_____. 1994. Feeding ecology of Eurasian pine martens and stone martens in Europe. Pages 326–340 *in* S. W. Buskirk, A. S. Harestad, M. G. Raphael, and R. A. Powell, editors. Martens, sables, and fishers. Biology and conservation. Cornell University Press, Ithaca, New York.

_____. 1995. Seasonality and relationships of food resource use of *Martes martes*, *Genetta genetta* and *Felis catus* in the Balearic Islands. Revue D'écologie - Terre et Vie 50:109–131.

Datskevich, V. A. 1979. Pitanie lesnoi kunitsy v Belovezhskoi Pushche [Food of pine marten in Belovezha Forest]. Zapovedniki Belorussi 3:67–70. (in Russian)

Danilov, P. I., and E. V. Ivanov. 1967. Lesnaya kunitsa v Karelii [Pine marten in Karelia]. Uchenye Zapiski Petrozavodskogo Gosudarstvennogo Universiteta 15:179–197. (in Russian)

Donaurov, S. S., V. P. Teplov, and P. A. Shikina. 1938. The nutrition of the forest marten in the conditions of the Caucasian Reservation territory. Trudy Kavkazskogo Gosudarstvennogo Zapovednika 1:281–316. (in Russian with English summary)

Erlinge, S. 1987. Why do European stoats *Mustela erminea* not follow Bergmann's rule? Holarctic Ecology 10:33–39.

Gashev, N. S. 1965. Nutrition of marten of the *Martes* Genus in North Urals. Byulleten Moskovskogo Obshchestva Ispytatelei Prirody 70 (3):16–21. (in Russian with English summary)

Goszczyński, J. 1985. The effect of structural differentiation of ecological landscape on the predator-prey interactions. Publications of Warsaw Agricultural University SGGW-AR. Treatises and Monographs:1–80.

_____, B. Jędrzejewska, and W. Jędrzejewski. 2000. Diet composition of badgers (*Meles meles*) in a pristine forest and rural habitats of Poland compared to other European populations. Journal of Zoology (London) 250:495–505.

Grakov, N. N. 1962. Rol belki v pitanii kunitsy na Evropeiskom Severe [The role of squirrel in pine marten diets in Northern Europe]. Trudy Vsesoyuznogo Nauchno-Issledovatelskogo Instituta zhivotnogo syrya i pushniny 19:154–163. (in Russian)

_____. 1981. Lesnaya kunitsa [The pine marten]. Nauka, Moskva. (in Russian)

Gribova, Z. A. 1958. Pitanie lesnoi kunitsy v Vologodskoi oblasti [Food of pine marten in Vologda region]. Trudy Vsesoyuznogo Nauchno-Issledovatelskogo Instytuta zhivotnogo syrya i pushniny 17:70–79. (in Russian)

Helldin, J. O. 1999. Diet, body condition, and reproduction of Eurasian pine martens *Martes martes* during cycles in microtine density. Ecography 22:324–336.

_____. 2000. Seasonal diet of pine marten *Martes martes* in southern boreal Sweden. Acta Theriologica 45:409–420.

_____, and E. R. Lindström. 1993. Dietary and numerical responses of pine marten (*Martes martes*) to vole cycles in boreal Fennoscandia. Pages 220–224 *in* I. D. Thompson, editor. Proceeding of the International Union of Game Biologists XXI Congress, Halifax, Nova Scotia, Canada.

Jędrzejewska, B., and W. Jędrzejewski. 1998. Predation in vertebrate communities. The Białowieża Primeval Forest as a case study. Springer-Verlag, Ecological Studies 135.

Berlin, Heidelberg, New York.
_____, and _____. 1996. Rodent cycles in relation to biomass and productivity of ground vegetation and predation in the Palearctic. Acta Theriologica 41:1–34.
_____, A. Zalewski, and B. Jędrzejewska. 1993. Foraging by pine marten *Martes martes* in relation to food resources in Białowieża National Park, Poland. Acta Theriologica 38:405–426.
Kostin, S. I., and T. V. Pokrovskaya. 1961. Klimatologiya [Climatology]. Gidrometeorologicheskoe Izdatelstvo, Leningrad. (In Russian)
Krebs, C J. 1989. Ecological methodology. Harper Collins Publisher, New York.
Lebedeva, A. N., I. S. Borushko, and A. U. Egorovoi. 1979. Climatic Reference book of West Europe. Gidrometeorologicheskoe Izdatelstvo, Leningrad.
Maldzhiunaite, S. 1957. Age determination and age structure of pine marten in Lithuania. Trudy Biologicheskogo Instituta 3:169–177. (in Russian with English summary)
_____. 1959. Biologiya lesnoi kunitsy v Litve [Biology of pine marten in Lithuania]. Trudy Akademii Nauk Litovskoi SSR, Seriya B (17):189–201. (in Russian)
Marchesi, P. 1989. Ecologie et comportement de la martre (*Martes martes* L.) dans le Jura Suisse. Dissertation, Université Neuchâtel, Institut de Zoologie, Switzerland.
Marcström, V., R. E. Kenward, and E. Engren. 1988. The impact of predation on boreal tetraonids during vole cycles: an experimental study. Journal of Animal Ecology 57:859–872.
Martin, S. K. 1994. Feeding ecology of American martens and fishers. Pages 297–315 *in* S. W Buskirk, A. S. Harestad, M. G. Raphael, and R. A. Powell, editors. Martens, sables, and fishers: Biology and conservation. Cornell University Press, Ithaca, New York.
McNab, B. K. 1971. On the ecological significance of Bergmann's rule. Ecology 52:845–854.
Moreno, S., A. Rodriguez, and M. Delibes. 1988. Summer foods of the pine marten (*Martes martes*) in Majorca and Minorca, Balearic Islands. Mammalia 52:289–291.
Morozov, V. F. 1976. Feeding habits of *Martes martes* (Carnivora, Mustelidae) in different regions of the North-west of the USSR. Zoologicheskii Zhurnal 55:1886–1892. (in Russian with English summary)
Mozgovoi, D. P. 1971. O pitanii lesnoi kunitsy [On feeding habits of pine marten]. Sbornik Trudov Bashkirskogo Gosudarstvennogo Zapovednika 3:132–145. (in Russian)
Nasimovich, A. A. 1948. Ekologiya lesnoi kunitsy [Ecology of the pine marten]. Trudy Laplandskogo Zapovednika 3:81–106. (in Russian)
Novikov, G. A., A. E. Airapetyants, Y. B. Pukinskii, P. P. Strelkov, and E. K. Timofeeva. 1970. Zveri Leningradskoi oblasti [Animals of Leningrad district]. Nauka, Leningrad. (in Russian)
Pandolfi, M., A. M. Demarinis, and I. Petrov. 1996. Fruit as a winter feeding resource in the diet of stone marten (*Martes foina*) in east-central Italy. Zeitschrift fur Säugetierkunde 61:215–220.
Parovshchikov, V. Ya. 1961. On feeding habits of *Martes martes borealis* B. Kuztnetz. near Archangelsk. Zoologicheskii Zhurnal 40:1112–1115. (in Russian with English summary)
Petrov, O. V. 1962. The validity of Bergman's rule as applied to intraspecific variation in the ermine. Pages 30–38 *in* C. M. King, editor. Biology of mustelids. Some soviet research. British Library Lending Division.
Pleshak, T. V. 1976. K pitaniyu kunitsy v lesnykh biotopakh, izmenennykh rubkami [On the diet of pine marten in forest biotops affected by logging]. Trudy Kirovskogo Selskokhozyaistvennogo Instituta:38–42. (in Russian)
Polushina, N. A. 1957. Economic significance of some small Mustelidae in the western regions of the Ukrainian SSR. Naukovi zapiski Naukovo-prirodoznavchogo Muzeya Akademii Nauk URSR 6:139–146. (in Ukrainian with English summary)
Pucek, M. 1983. Ecology of bank vole. Habitat preference. Acta Theriologica 28 (Suppl. 1):31–40.

Pucek, Z., W. Jędrzejewski, B. Jędrzejewska, and M. Pucek. 1993. Rodent population dynamics in a primeval deciduous forest (Białowieża National Park) in relation to weather, seed crop, and predation. Acta Theriologica 38:199–232.

Pulliainen, E., and P. Ollinmäki. 1996. A long-term study of the winter food niche of the pine marten *Martes martes* in northern boreal Finland. Acta Theriologica 41:337–352.

Reig, S. 1989. Morphological variability of *Martes martes* and *Martes foina* in Europe. Dissertation, Mammal Research Institute, Białowieża, Poland.

_____. 1992. Geographic variation in pine marten (*Martes martes*) and beech marten (*M. foina*) in Europe. Journal of Mammalogy 73:744–769.

Reynolds, J. C., and N. J. Aebischer. 1991. Comparison and quantification of carnivore diet by faecal analysis: a critique, with recommendations based on a study of the fox *Vulpes vulpes*. Mammal Review 21:97–122.

Rosenzweig, M. L. 1966. Community structure in sympatric Carnivora. Journal of Mammalogy 47:602–612.

Ruiz-Olmo, J., and J. M. Lopez-Martin. 1996. Seasonal food of pine marten (*Martes martes* L., 1758) in a fir forest of Pyrenean Mountains (Northeastern Spain). Pages 189–198. Proceedings of the 1st European Congress of Mammalogy, Lisboa, Portugal.

_____, and J. Nadal. 1991. Régime alimentaire de la martre (*Martes martes* L., 1758) en hiver et taille des portées à Ménorca, Iles Baléares. Mammalia 55:639–642.

Rzebik-Kowalska, B. 1972. Studies on the diet of the carnivores in Poland. Acta Zoologica Cracoviensia 17:415–506.

Selas, V. 1992. Food of pine marten in south Norway. Fauna (Oslo) 45:18–26. (in Norwegian with English summary)

Semenov-Tyan-Shanskii, O. 1959. Ekologiya teterevinykh ptits [Ecology of tetraonid birds]. Trudy Laplandskogo Gosudarstvennogo Zapovednika 5:1–318. (in Russian)

Serzhanin, I. N. 1961. Mlekopitayushchie Belorussii [Mammals of Belarus]. Izdatelstvo Akademii Nauk Belorusskoi SSR, Minsk. (in Russian)

_____. 1973. Lesnaya kunitsa: Belorussiya [Pine marten: Belarus]. Pages 155–158 *in* A. A. Nasimovich, editor. Sobol, kunitsy, kharza: razmieshchenie zapasov, ekologiya, ispolzovanie i okhrana. [Sable, martens, and yellow-throated marten: distribution of resources, ecology, harvest, and conservation]. Nauk, Moscow. (in Russian)

Sidorovich, V. E. 1997. Mustelids in Belarus. Zolotoy uley, Minsk.

Stenseth, N. C., H. Viljugrein, W. Jędrzejewski, A. Mysterud, and Z. Pucek. 2002. Population dynamic of *Clethrionomys glareolus* and *Apodemus flavicollis*: seasonal components of density dependence and density independence. Acta Theriologica 47 (Suppl. 1):39–67

Storch, I., E. Lindström, and J. de Jounge. 1990. Diet and habitat selection of the pine marten in relation to competition with the red fox. Acta Theriologica 35:311–320.

Thompson I. D., and P. W. Colgan. 1990. Prey choice by marten during a decline in prey abundance. Oecologia 83:443–451.

Yazan, Y. P. 1962. Is the marten responsible for a diminishing in squirrel population? Zoologicheskii Zhurnal 41:633–635. (in Russian with English summary)

Yurgenson, P. B. 1951. Ekologo-geograficheskie aspekty v pitanii lesnoi kunitsy i geograficheskaya izmenchivost ekologo -morfologicheskikh adaptatsii ee zhevatelnogo apparata [Ecological-geographical aspects of feeding by pine marten and the geographic variability of ecological-morphological adaptations of its chewing apparatus]. Zoologicheskii Zhurnal 30:172–185. (in Russian)

Chapter 4

TERRITORIALITY AND HOME-RANGE FIDELITY OF AMERICAN MARTENS IN RELATION TO TIMBER HARVESTING AND TRAPPING

David Payer, Daniel Harrison, and David Phillips

Abstract: Timber harvesting and trapping may decrease population density or disrupt sex ratios of American martens (*Martes americana*), potentially affecting fitness by altering spatial relations such as site fidelity and territoriality. We compared home-range fidelity and overlap within and between sexes for 143 (77 M, 66 F) resident, nonjuvenile martens during 1991–1998 among 3 contiguous study sites: (1) an untrapped, unlogged forest reserve (FR) with high marten density (0.63 marten/km^2); (2) an untrapped, extensively clearcut industrial forest (UIF) with moderate marten density (0.31 marten/km^2); and (3) a trapped, extensively clearcut industrial forest (TIF) with low marten density (0.19 marten/km^2). Mean fidelity was 67% for consecutive seasons and 55% for consecutive years, and did not differ among sites ($P \geq 0.33$) or between males and females ($P \geq 0.21$). Extent of same-sex home-range overlap was greater in FR than in either logged site for males ($P <$ 0.01), but did not differ between UIF and TIF for males ($P = 0.16$) or females ($P =$ 0.10). For females, incidence of overlap with male ranges did not differ among sites ($P = 0.07$), although there was a trend of lower incidence in the logged sites, particularly TIF. Incidence of opposite-sex overlap for males was lower in TIF than UIF ($P < 0.01$). In the logged sites, martens established home ranges within residual forest patches that overlapped with ranges of potential mates, were apparently defended against consexuals, and were maintained through consecutive seasons and years similarly to the unlogged reserve. These strategies maintained population social structure and ensured breeding opportunities among females in the trapped and untrapped, logged areas. Higher fur-trapping pressure, greater habitat fragmentation, or isolation of a trapped population from a source population might reduce opposite-sex overlap among females and create social instability.

1. INTRODUCTION

American martens (*Martes americana*), like most mustelids, are intrasexually territorial (Weckwerth and Hawley 1962, Powell 1979). Advantages of maintaining ranges exclusive of conspecifics of the same sex include reduced competition for prey and, in the case of males, exclusive access to mates (Powell 1994). Intrasexual territoriality is not absolute, and several studies have demonstrated home-range overlap between same-sex martens (e.g., Hawley and Newby 1957, Wynne and Sherburne 1984, Katnik et al. 1994). Based on the costs and benefits of territorial defense in relation to resource acquisition, Powell (1994) predicted that martens would display intrasexual territoriality at intermediate levels of prey availability. According to Powell's model, intrasexual territoriality would not occur if resources were not limiting (as might occur in a low-density marten population), or if resources were so limited that home ranges were abandoned and animals became transient. A complete lack of intrasexual territoriality among resident, nonjuvenile martens has not been observed, however, suggesting that intolerance of same-sex conspecifics may be phylogenetically determined (Balharry 1993). This does not preclude the possibility that the degree of territoriality varies across the spectrum of habitat conditions and marten densities, as might occur in landscapes characterized by logging and trapping.

Although martens are intrasexually territorial, they exhibit intersexual tolerance, i.e., males generally maintain ranges that overlap with ≥ 1 female (Balharry 1993, Katnik et al. 1994). Such overlap may be required for breeding to occur because, unlike fishers (*Martes pennanti*) (Arthur et al. 1989) and ermines (*Mustela erminea*) (Erlinge and Sandell 1986), male martens have not been reported to make forays or alter their ranges during the breeding season to access mates (Katnik et al. 1994). Logging and trapping may disrupt opportunities for breeding if differential vulnerability of males and females results in a skewed sex ratio, or if marten density declines to the point where opposite-sex overlap becomes unpredictable. In the absence of over-exploitation, trapped populations typically have a female-biased sex ratio (Strickland and Douglas 1987, Fortin and Cantin 2004) because of greater male vulnerability (Buskirk and Lindstedt 1989). Extensive logging may result in a male-biased sex ratio because energetic demands (Sandell 1989) and habitat requirements associated with raising young (Wynne and Sherburne 1984, Ruggiero et al. 1998, Bull and Heater 2000) may constrain habitat choices of females relative to males. In either case, incidence of opposite-sex home-range overlap may be a useful index for comparing reproductive potential of marten populations among areas with different logging and trapping regimes.

Fidelity to a home range over time may benefit martens by providing familiarity with dispersed resources, such as prey, resting sites, and mates (O'Doherty et al. 1997). Home-range abandonment and subsequent wandering exposes martens to increased mortality risks (Thompson and Colgan 1987, Hodgman et al. 1997) that must be offset by potential benefits of future resource acquisition. Home ranges might shift or be abandoned in response to reduced availability of prey (Thompson and Colgan 1987) or mates (Powell 1994), or habitat alteration associated with natural and anthropogenic disturbances. Seasonal differences in habitat requirements might also precipitate range shifts. In particular, habitat requirements may be more specific in winter versus summer (Soutiere 1979, Steventon and Major 1982, Buskirk et al. 1989), which could lead to seasonal range shifts if resources are unevenly distributed or use of some habitat features is more concentrated in winter. Site fidelity has been examined in unlogged, untrapped landscapes (Phillips et al. 1998), but has not been well-described for extensively logged landscapes, primarily because such areas are often intensively trapped (e.g., Hodgman et al. 1994, Thompson 1994).

We evaluated home-range fidelity over consecutive seasons and years, and determined the extent of same-sex and opposite-sex home-range overlap, among martens on 3 contiguous sites with different management regimes: (1) an untrapped forest reserve with no recent timber-harvesting activity and high marten density; (2) an untrapped, extensively clearcut industrial forest with moderate marten density; and (3) a trapped, extensively clearcut industrial forest with low marten density. Our objectives were to evaluate the individual effects of logging and trapping on home-range fidelity and spatial relations among male and female martens. We discuss our results in relation to habitat selection and demographic characteristics of martens within each forest-management regime.

2. STUDY AREAS

Our forest-reserve study site (FR) was located within Baxter State Park (BSP), north-central Maine (46°4' N, 69°3' W). The 50-km^2 study area was managed as wilderness, and was closed to trapping (>50 yr) and timber harvesting (>35 yr). Prior to protection, some large-diameter red spruce (*Picea rubens*) and eastern white pine (*Pinus strobus*) had been selectively harvested. The reserve was dominated by mature (70–100 yr) forests (73% of the area), consisting of 51% deciduous (>75% deciduous species, e.g., *Acer* spp., *Betula* spp., *Fagus grandifolia*), 17% coniferous (>75% coniferous species, e.g., *Picea* spp., *Abies balsamea*), and 32% mixed (25–75% coniferous species) stands.

Coniferous and mixed stands recovering from a 1974–1984 epidemic of eastern spruce budworm (*Choristoneura fumiferana*) were interspersed throughout the mature-forest matrix. These stands were characterized by reduced overstory canopy closure, abundant coarse woody debris and shrubs, and regenerating coniferous and deciduous species. Topography in the reserve was hilly to mountainous, with elevations 290–735 m. Mean minimum January temperature was –17°C, and mean maximum July temperature was 25°C. Snow cover was usually continuous from late November through mid-April, and snowfall averaged 49 cm/month in December–March (Krohn et al. 1995). A single-lane road accessed the site, yielding a road density of approximately 0.6 km/km^2. During 1991–1997, mean marten density was 0.63 marten/km^2 (0.39 males and 0.24 females/km^2) and marten home ranges occupied an average of 81% of the land area annually. Mammalian predation was the dominant cause of marten mortality (Hodgman et al. 1997).

Our industrial-forest study sites were located in townships T5 R11 WELS (72 km^2) and T4 R11 WELS (58 km^2), Piscataquis County, Maine. These sites were west of BSP and were contiguous with the FR site. Topography was flat to hilly, with elevations 340–500 m. The land was owned and managed primarily for pulpwood by Bowater, Inc. Approximately 55% of the area was harvested during 1974–1998, primarily by clearcutting. Prior to 1974, large-diameter red spruce and eastern white pine had been selectively harvested. The landscape was a mosaic of 39% regenerating ($\leq$6.0 m mean tree height), 12% immature (6.1–9.0 m), and 46% mature (>9.0 m) forest stands. Mature stands with $\geq$50% canopy closure occupied 36% of the area, and were comprised of deciduous (36%), coniferous (43%), and mixed (21%) forest types. Tree-species composition of mature forest types was similar to the forest reserve. Clearcuts were regenerated naturally; tree planting did not occur. Compared to insect-defoliated stands in the reserve, regenerating stands in the industrial forest had reduced volumes of coarse woody debris and less vertical structure (Payer and Harrison 2000). During the 1974–1984 spruce budworm epidemic, susceptible stands were harvested or treated with insecticides; stands recovering from severe insect defoliation did not occur in the industrial forest.

The industrial-forest sites traditionally supported intensive marten trapping. Trapping was responsible for 90% of documented marten mortalities during 1989–1991, and the marten population would have declined ($\lambda = 0.57$) without immigration (Hodgman et al. 1994). High levels of marten harvest were associated with high road density (1.1 km/km^2 of roads passable with a 2-wheel drive vehicle during May–October) and proximity to the forest reserve, which likely served as a reservoir for dispersing martens (Hodgman et al. 1994). During October 1994–December 1998, marten trapping was prohibited in T5

R11 and abutting townships to the north, east, and west. Trapping continued during the legal trapping season (late October–31 December) in T4 R11. We defined 2 industrial-forest management regimes: (1) trapped industrial forest (TIF): T4 R11 WELS May 1994–April 1998, and T5 R11 WELS May–October 1994; and (2) untrapped industrial forest (UIF): T5 R11 WELS, November 1994–October 1998. Mean marten density was 0.19 marten/km^2 (0.10 males and 0.09 females/km^2) in TIF and 0.31 marten/km^2 (0.16 males and 0.15 females/km^2) in UIF. Marten home ranges occupied an average of 36.6% and 64.9% of the landscape annually in TIF and UIF, respectively. Dominant sources of mortality were trapping in TIF and predation in UIF.

3. METHODS

3.1 Trapping, Radiotelemetry, and Home Ranges

We livetrapped and radiocollared martens during 15 May–4 July in FR (1991–1997), TIF (1994–1997), and UIF (1995–1998). We obtained a complete census of resident martens in UIF and TIF by setting pairs of traps on either side of all roads at ≤0.65-km intervals. Traps were placed ≤0.5 km from roads and were checked 1–2 times/day, rebaited at ≤3-day intervals, and maintained for 10 trap nights at each location. We followed the same protocol in FR, but road access was more limited in this site. Trapping effort was sufficient to census resident males, but because females had smaller home ranges (mean home-range radius was 0.7 km for females versus 1.0 km for males), some resident females may have escaped capture. Total trapping effort across study sites was approximately 19,000 trap nights at 390 trap locations.

Captured martens were restrained in a handling cone (Schemnitz 1994:119), immobilized with 10.0–18.0 mg ketamine hydrochloride/kg body weight (Hunter and Clark 1986), and radiocollared (model 070 or configuration 1A, Telonics, Inc., Mesa, Arizona, USA or model SMRC-6, LOTEK Engineering, Inc., Newmarket, Ontario, Canada). We extracted a first premolar from all livetrapped martens for estimation of age via cementum annuli (Strickland et al. 1982). Age estimates were corroborated by examination of second or fourth premolars for martens recovered after death. Animal-handling procedures were approved by the University of Maine's Animal Care and Use Committee.

Marten locations were obtained throughout the year at 18-hr to 10-day intervals. The minimum interval was selected to prevent autocorrelation between consecutive radiolocations (Katnik et al. 1994). Radiolocations were obtained from fixed-wing aircraft with 2 side-facing H-antennas (Gilmer et al. 1981), or by triangulation from ≥3 fixed receiving locations on the road system. Mean error for locations obtained from aircraft was 93.3 m, based on the difference between estimated and true locations for 40 test transmitters. This yielded an estimated location error for telemetry from aircraft of 2.7 ha, assuming a circle with a 93.3-m radius. Angular error for ground-based telemetry was estimated as the difference between 180 actual and estimated bearings for 60 hidden transmitters. Mean angular error (6.0°) was used in the program TRIANG (White and Garrott 1984) to estimate marten locations and error-polygon size for ground-based telemetry.

We used the program CALHOME (Kie et al. 1994) to calculate 95%-minimum convex polygon (MCP) home ranges for resident, nonjuvenile (≥ 1 yr) martens, based on radiolocations with error polygons <25 ha. We assigned resident status to those martens with (1) ≥ 10 radiolocations over ≥ 90 days, and (2) mean minimum distance between consecutive locations (MINDIST; Harrison and Gilbert 1985) $\leq$ mean + 3 SD MINDIST for all martens, stratified by sex, season, and year. The residency-time criterion followed Weckwerth and Hawley (1962), and the MINDIST criterion ensured that we did not assign residency to dispersing or transient martens.

Home ranges were calculated separately for summer (1 May–31 October) and winter (1 November–30 April). Each home range was year-specific; we did not pool locations over multiple years. Based on examination of area-observation curves (Laundre and Keller 1984), a stable estimate of home-range area required ≥ 25 locations during summer and ≥ 23 locations during winter. Mean number of radiolocations used to estimate home ranges was 44 during summer and 37 during winter. We did not estimate home ranges for martens in FR during 1997 because logistical constraints prevented us from obtaining sufficient radiolocations. We did determine residency status for radiocollared animals, however.

3.2 Intrasexual and Intersexual Territoriality

We used percent overlap among summer home ranges of resident, same-sex martens as an index of intrasexual territoriality (Katnik et al. 1994). For each marten with a defined home range, we estimated the percent of its range shared with ≥ 1 same-sex resident(s). We used permutation tests for matched pairs (PTMP) (BLOSSOM statistical software, Slauson et al. 1994) with α = 0.05 to compare the extent of same-sex overlap for males monitored as both ≤ 2-yr olds and ≥ 3-yr olds. If an individual was monitored >1 yr within an age class, we used mean intrasexual overlap within age class in our paired analysis. We chose these age groups because, although males may achieve sexual maturity at 1 yr (Mead 1994), the size of the baculum increases to 3 yr and may be insufficient in younger males to induce ovulation (Strickland and Douglas 1987). Among females, we used the same approach to compare extent of same-sex overlap for individuals monitored as 1-yr olds and ≥2-yr olds. We selected these age groups because females are known to breed as yearlings, but breeding success may be lower for 2-yr olds (bred as yearlings) versus ≥3-yr olds (Strickland and Douglas 1987, Thompson and Colgan 1987).

We used multiresponse permutation procedures (MRPP) (Slauson et al. 1994) to compare extent of intrasexual home-range overlap during summer among forest-management regimes for males. If $P \leq 0.05$, we used MRPPs for *post hoc* pairwise comparisons with $\alpha = 0.017$ ($0.05/k$) (Miller 1981:67–69). We used a similar approach for female martens, but only compared between industrial-forest sites because the census of females in FR may have been incomplete, resulting in a potential negative bias in same-sex overlap. We also used MRPPs to test if extent of intrasexual overlap differed between males and females. We did not estimate same-sex overlap during winter because transmitter batteries tended to fail in late winter, resulting in incomplete information on winter home ranges for some residents.

We used incidence of opposite-sex overlap during summer as an index of reproductive opportunity. Mating occurred during June–August on our study sites. We did not estimate opposite-sex overlap for males in FR because of the possible incomplete census of females on that site. We used Fisher's exact tests with $\alpha = 0.05$ to evaluate whether incidence of opposite-sex overlap differed between age classes for females (1 yr versus ≥2 yr) and males (≤2 yr versus ≥3yr). We compared incidence of opposite-sex overlap between forest-management regimes for both males and females with Chi-square contingency-table analyses. We also used chi-square tests to compare incidence of opposite-sex overlap between males and females.

3.3 Home-Range Fidelity

We estimated home-range fidelity for martens monitored in consecutive seasons as percent of radiolocations (≥10 locations with error polygon <10 ha) during a season that occurred within the boundary of an individual's 95%-MCP home range from a previous season. This index is highly correlated ($r = 0.88$, $n = 23$, $P < 0.001$) with percent overlap of 95%-MCP home-range areas for marten (Phillips et al. 1998). Use of this index increased our sample size for fidelity comparisons that included a winter season because for several martens we obtained ≥10 locations but fewer than the 23 locations needed for reliable estimation of winter home-range area.

We calculated 2 estimates of seasonal fidelity: (1) percent of winter locations within the previous summer's home range, and (2) percent of summer locations within the previous winter's home range. Similarly, we calculated 2 estimates of annual fidelity: (1) percent of summer locations within the home range of the previous summer, and (2) percent of winter locations within the home range of the previous winter. Several martens yielded ≥ 2 measures of fidelity; we treated each value independently in subsequent analyses (Phillips et al. 1998).

Because we obtained sufficient locations to estimate winter home ranges for relatively few martens, we used winter-previous summer fidelity to represent seasonal home-range fidelity and summer-previous summer fidelity to represent annual home-range fidelity in statistical analyses. We used MRPPs with $\alpha = 0.05$ to compare sex-specific seasonal and annual fidelity among forest-management regimes. If $P > 0.05$, we pooled fidelity indices across sites and used MRPPs to compare seasonal and annual fidelity between males and females.

4. RESULTS

4.1 Intrasexual and Intersexual Territoriality

We calculated extent of same-sex overlap for 133 male and 58 female home ranges (Table 4.1). Among 16 males that were monitored for multiple years, we did not observe a difference between the $\leq$2-yr and $\geq$3-yr age classes ($P = 0.29$). Similarly, there was no difference between yearlings and $\geq$2-yr-olds ($P = 0.45$) for 12 females. Percent of same-sex overlap differed among forest-management regimes for males ($P < 0.01$); overlap was greater in FR than in UIF ($P < 0.01$) or TIF ($P < 0.01$), but did not differ between the industrial-forest sites ($P = 0.16$). Same-sex overlap was also similar for females in UIF and TIF ($P = 0.10$), and did not differ between males and females in the combined industrial-forest sites ($P = 0.13$).

Incidence of opposite-sex home-range overlap did not differ ($P = 0.34$) between yearling ($n = 42$) and $\geq$2 yr-old females ($n = 45$). Similarly, no difference was observed ($P = 0.09$) between males in age groups $\leq$2 yr ($n = 45$) and $\geq$3 yr ($n = 15$). We therefore combined age groups in subsequent analyses. We observed a lower incidence of opposite-sex overlap in TIF than in UIF for male martens ($\chi^2 = 9.31$, 1 df, $P < 0.01$) (Table 4.2). Among females, incidence did not appear to differ among sites ($\chi^2 = 5.33$, 2 df, $P = 0.07$), although there was a trend of lower incidence in the industrial-forest sites, particularly TIF (Table 4.2). A higher proportion of males than females maintained 95%-MCP home

Table 4.1. Mean percent (*n*, *SE*) of home-range area shared with resident, nonjuvenile (≥1 yr) martens of the same sex for martens in an untrapped forest reserve (1991–1996), an untrapped industrial forest (1995–1998), and a trapped industrial forest (1994–1997) during May–October in northcentral Maine, USA.

Forest-management regime	Males	Females
Forest reserve	33.5 (70, 3.3)	—[a]
Untrapped industrial forest	8.7 (38, 2.5)	5.9 (39, 1.3)
Trapped industrial forest	15.7 (25, 4.1)	9.7 (19, 5.5)

[a]Same-sex home-range overlap was not estimated for females in the forest reserve because census of females may have been incomplete.

Table 4.2. Percent of resident, nonjuvenile (≥1 yr) martens (*n*) sharing a portion of their home range with ≥1 opposite-sex marten(s) during May–October in an untrapped forest reserve (1991–1996), an untrapped industrial forest (1995–1998), and a trapped industrial forest (1994–1997) in northcentral Maine, USA.

Forest-management regime	Males	Females
Forest reserve	—[a]	100.0 (29)
Untrapped industrial forest	89.5 (38)	94.9 (39)
Trapped industrial forest	56.0 (25)	84.2 (19)

[a]Opposite-sex home-range overlap was not estimated for males in the forest reserve because census of females may have been incomplete.

ranges that did not include opposite-sex marten in TIF ($\chi^2 = 3.96$, 1 df, $P = 0.05$). We detected no such difference in UIF ($\chi^2 = 0.78$, 1 df, $P = 0.38$), where density of potential mates was higher.

4.2 Home-Range Fidelity

The proportion of summer locations that occurred within the 95%-MCP home range of the previous summer (annual home range fidelity) did not differ among FR, UIF, and TIF for male martens ($P = 0.41$) (Table 4.3). Similarly, annual fidelity did not differ among forest-management regimes for females ($P = 0.75$). We therefore pooled data across sites within sex, and tested for differences between males and females. Annual fidelity did not differ ($P = 0.21$) between males ($\bar{x} = 54\%$, $n = 63$) and females ($\bar{x} = 62\%$, $n = 19$).

The proportion of winter locations that occurred within the 95%-MCP home range of the previous summer (seasonal fidelity) did not differ among forest-management regimes for males ($P = 0.99$) or females ($P = 0.33$) (Table 4.3).

Table 4.3. Mean (n, SE)[a] percent of radiolocations that occurred within the 95%-MCP home range of the previous season (seasonal home-range fidelity) or year (annual home-range fidelity) for martens in a forest reserve (1991–1997), a trapped industrial forest (1994–1997), and an untrapped industrial forest (1995–1998) in northcentral Maine, USA.

	Seasonal home-range fidelity		Annual home-range fidelity	
Sex	Winter–previous summer	Summer–previous winter	Summer–previous summer	Winter–previous winter
Forest reserve				
Male	69.5 (48, 2.9)[a]	56.4 (3, 8.9)	56.4 (39, 3.3)	— (0)
Female	60.4 (20, 7.0)	— (0)	74.7 (3, 12.1)	— (0)
Untrapped industrial forest				
Male	67.3 (27, 3.9)	41.1 (11, 5.7)	47.9 (21, 5.2)	34.2 (1)
Female	70.0 (20, 4.7)	65.0 (5, 11.1)	58.6 (12, 5.0)	59.0 (1)
Trapped industrial forest				
Male	69.3 (10, 6.5)	52.2 (2, 29.0)	57.3 (3, 7.5)	6.7 (1)
Female	64.3 (6, 8.5)	76.0 (2, 5.8)	60.5 (4, 19.3)	— (0)
Forest-management regimes and sexes combined[b]				
Both	67.4 (131, 1.9)	52.3 (23, 4.8)	55.5 (82, 2.4)	33.3 (3, 15.1)

[a]n represents number of individual martens; SE represents standard error in percent fidelity of martens.
[b]Winter–previous summer and summer–previous summer home-range fidelity did not differ ($P \geq 0.21$) between forest-management regimes or sexes.

Pooled across sites, we did not observe a difference ($P = 0.44$) in seasonal home-range fidelity between males ($\bar{x} = 69\%$, $n = 85$) and females ($\bar{x} = 65\%$, $n = 46$).

5. DISCUSSION

Based on simulations of random home-range placement, Katnik et al. (1994) determined that expected same-sex overlap in our industrial-forest sites (pre-trapping closure) was 36% under the null hypothesis of no intrasexual territoriality. Even at higher overall marten densities in this study resulting from the trapping closure in UIF, which would lead to a higher expected value under the null hypothesis, we observed a mean of only 9.4% ($SE = 1.5\%$) same-sex overlap for all martens during summer in the industrial-forest sites. Further, at marten densities in FR that were 4.5x greater than Katnik et al. (1994) reported for the industrial forest, we observed only 33% overlap among males during summer. We did not directly evaluate intrasexual overlap during winter. Martens exhibited high fidelity to home ranges through consecutive seasons, however, suggesting that home ranges largely exclusive of other same-sex marten were maintained throughout the year. Our results suggest that martens demonstrated intrasexual territoriality across a wide range of resource conditions and population densities, although we were unable to evaluate this for females in FR.

Powell (1994) predicted that intrasexual overlap would be greater in low-density marten populations because of greater per-capita resource availability. Our finding of greater overlap among sympatric males martens in the high-density reserve versus the lower-density industrial forest do not necessarily conflict with this prediction, because the low-density populations occurred within extensively clearcut landscapes. When establishing home ranges, martens tended to avoid early successional areas and selected mature stands (Payer 1999). Relative to regenerating clearcuts, mature stands had greater abundance of small mammals (Lachowski 1997) and possibly lower risk of avian (Pulliainen 1981, Hargis and McCullough 1984) and mammalian (Hodgman et al. 1997) predation. Habitat selection therefore concentrated martens in patches of mature forest where they likely competed for food and mates, and tended to maintain territories exclusive of same-sex conspecifics.

Our observations are consistent with Balharry's (1993) hypothesis that group living (i.e., complete lack of intrasexual territoriality) in marten is prevented by phylogenetically determined intolerance of conspecifics. In Scotland, male pine martens in breeding condition (≥3 yr) tolerated nonbreeding subadult (≤2yr) males within their territories, although Balharry (1993) suggested that this represented protection of offspring from infanticide. We did not observe differences in extent of intrasexual overlap between 1–2-yr and ≥3-yr males, although we did not assess male reproductive condition or monitor juveniles (<1 yr). Therefore, it is possible that juvenile males were present within the ranges of nonjuveniles. Juvenile martens in our region generally disperse from their natal ranges during late summer and fall (Phillips 1994), suggesting that breeding males do not provide prolonged protection to their offspring. Among females, extent of intrasexual overlap did not differ between yearlings and ≥2-yr olds, and lactation rate did not differ between 2-yr and ≥3-yr olds (Payer 1999). We conclude that yearling females were reproductively active and established intrasexual territorial relationships similar to adults.

We did not observe wandering behavior among resident martens during the breeding season, corroborating Katnik et al.'s (1994) conclusion that intersexual overlap is required for mating. Low marten density in TIF may have been associated with lower incidence of opposite-sex overlap among males and females, although the latter was not statistically significant. Our results suggest that some males in TIF lacked breeding opportunities. Among females, however, the trend of reduced intersexual overlap in the trapped area did not appear to reduce breeding success relative to the higher-density sites, i.e., incidence of lactation in adults captured during early summer did not differ among forest-management regimes (Phillips 1994, Payer 1999). We expected that trapping would bias the sex ratio towards females in addition to reducing marten density (Strickland and Douglas 1987), and that the combined effect would be reduced breeding opportunities for females. The trapped area was contiguous with the reserve, however, which provided a source of immigrants to the trapped site (Hodgman et al. 1994). This probably prevented a more pronounced reduction in density and a biased sex ratio in TIF, which could have depressed reproductive success. Overall, similar patterns of habitat selection by males and females in the industrial forest tended to concentrate martens in mature forest stands (Payer 1999), thereby ensuring that most females were bred.

Fidelity to home ranges may enhance reproductive success and survival for marten (O'Doherty et al. 1997). In our study, a high proportion of winter locations occurred within the boundaries of summer home ranges for both males and females in all forest-management regimes. Based on simulations reported by O'Doherty et aL (1997), our observed mean seasonal fidelity (67.4%) represents little or no shift of winter home ranges outside of summer home-range boundaries. Habitat requirements of marten may be more specific during winter. In particular, near-ground forest structure provided by coarse woody debris and shrubs is important for thermoregulation (Buskirk et al. 1988) and access to subnivean prey (Sherburne and Bissonette 1994). Regardless of differences in trapping and timber harvesting among study sites, martens apparently selected and maintained ranges that were stable across seasons, thus ensuring access to resources required during winter.

Annual home-range fidelity was also high across sites for both males and females. Some shifting of range boundaries did occur between consecutive summers, perhaps in response to death of adjacent consexuals (Katnik et al. 1994). Annual fidelity did not differ between our study sites, however, despite marked differences in habitat conditions and marten densities, and trapping mortality in TIF. In summary, martens selected high-quality home ranges (Payer 1999) that were apparently defended against other same-sex martens, maintained a high degree of fidelity to those ranges, and generally exhibited a high incidence of intersexual home-range overlap during summer. These strategies probably ensured access to prey, cover, and mates, and minimized predation risks associated with major home-range shifts or transience (Hodgman et al. 1997).

6. CONCLUSIONS AND MANAGEMENT IMPLICATIONS

Extensive clearcutting decreased marten density in the industrial forest relative to the forest reserve. Martens responded to habitat loss via landscape-scale habitat selection, thereby establishing high-quality territories that were defended against consexuals and maintained across seasons and years similarly to territories in the unlogged landscape. Further, both males and females in the untrapped population within the logged landscape maintained breeding opportunities via a high incidence of intersexual home-range overlap within patches of mature forest. Therefore, the level of forest harvesting that occurred in our industrial-forest study site did not disrupt conspecific spatial relationships, although population-level performance (i.e., density of lactating females) declined relative to the unlogged reserve because marten territories were discontinuous across the landscape (Payer 1999).

Marten density declined further in the trapped industrial forest because of additive mortality of males (Payer 1999). Both males and females still established territories that were defended against consexuals and were maintained across seasons and years. Opposite-sex overlap declined in males but remained high in females, thereby ensuring that most females had mating opportunities. The reserve was contiguous with the trapped area, and provided a source of dispersing marten (Hodgman et al. 1994). Immigration likely prevented greater declines in marten density or a female-biased sex ratio in the trapped industrial forest. Higher trapping pressure than we observed or isolation of a trapped population from a source population could further reduce population density or bias sex ratios, leading to lower incidence of intersexual overlap. Possible sequelae under this scenario include reduced breeding success among females, reduced intrasexual territoriality (Powell 1994), and transience (Katnik et al. 1994). Additional timber harvest resulting in greater fragmentation of marten habitat could also disrupt social systems and reduce reproductive success if patches of mature forest were insufficient in size or proximity to allow martens to establish stable territories (Chapin et al. 1998, Hargis et al. 1999).

7. ACKNOWLEDGMENTS

Funding for this work was provided by the Maine Department of Inland Fisheries and Wildlife, Federal Aid in Wildlife Restoration Project W-82-R-11-368, the Maine Cooperative Forestry Research Unit at the University of Maine, the Maine Forest Service, the National Council of the Paper Industry for Air and Stream Improvement, and the Maine Agricultural and Forest Experiment Station. Bowater, Inc. and the Baxter State Park Authority provided logistical support. S. Becker, J. Dodge, A. Drake, M. Estabrook, T. Foster, A. Fuller, T. Hodgman, A. Isaacs, R. Kelshaw, S. Kendrot, H. J. Lachowski, J. Martin, A. Matz, S. McLellan, S. Pastva, K. Rogers, M. Saeki, G. Thomas, D. Wrobleski, D. Wroe, and E. York assisted with data collection. J. McPhee skillfully piloted aircraft, and J. Hepinstall assisted with spatial analyses. We thank G. Proulx and 3 anonymous reviewers for helpful comments on an earlier version of the manuscript. This is Scientific Contribution No. 2713 of the Maine Agricultural and Forest Experiment Station.

8. LITERATURE CITED

Arthur, S. M., W. B. Krohn, and J. R. Gilbert. 1989. Home range characteristics of adult fishers. Journal of Wildlife Management 53:674–679.

Balharry, D. 1993. Social organization in marten: an inflexible system? Symposium of the Zoological Society of London 65:321–345.

Bull, E. L., and T. W. Heater. 2000. Resting and denning sites of American martens in northeastern Oregon. Northwest Science 74:179–185.

Buskirk, S. W., and S. L. Lindstedt. 1989. Sex biases in trapped samples of Mustelidae. Journal of Mammalogy 70:88–97.

_____, S. C. Forrest, M. G. Raphael, and H. J. Harlow. 1989. Winter resting site ecology of marten in the central Rocky Mountains. Journal of Wildlife Management 53:191–196.

_____, H. J. Harlow, and S. C. Forrest. 1988. Temperature regulation in American marten (*Martes americana*) in winter. National Geographic Research 4:208–218.

Chapin, T. G., D. J. Harrison, and D. D. Katnik. 1998. Influence of landscape pattern on habitat use by American marten in an industrial forest. Conservation Biology 12:1327–1337.

Erlinge, S., and M. Sandell. 1986. Seasonal changes in the social organization of male stoats, *Mustela erminea*: an effect of shifts between two decisive resources. Oikos 47:57–62.

Fortin, C., and M. Cantin. 2004. Harvest status, reproduction, and mortality in a population of American martens in Québec, Canada. Pages 221–234 *in* D. J. Harrison, A. K. Fuller, and G. Proulx, editors. Martens and fishers (*Martes*) in human-altered environments: An international perspective. Kluwer Academic Publishers, Boston, Massachusetts, USA.

Gilmer, D. S., L. M. Cowardin, R. L. Duval, L. M. Mechlin, C. W. Shaiffer, and V. B. Kuechle. 1981. Procedures for the use of aircraft in wildlife biotelemetry studies. U.S. Department of the Interior, Fish and Wildlife Service, Washington, D.C., USA.

Hargis, C. D., and D. R. McCullough. 1984. Winter diet and habitat selection of marten in Yosemite National Park. Journal of Wildlife Management 48:140-146.

_____, J. A. Bissonette, and D. L. Turner. 1999. The influence of forest fragmentation and landscape pattern on American martens. Journal of Applied Ecology 36:157–172.

Harrison, D. J., and J. R. Gilbert. 1985. Denning ecology and movements of coyotes in Maine during pup rearing. Journal of Mammalogy 66:712–719.

Hawley, V. D., and F. E. Newby. 1957. Marten home ranges and population fluctuations. Journal of Mammalogy 38:174–184.

Hodgman, T. P., D. J. Harrison, D. D. Katnik, and K. D. Elowe. 1994. Survival in an intensively trapped marten population in Maine. Journal of Wildlife Management 58:593-600.

_____, _____, D. M. Phillips, and K. D. Elowe. 1997. Survival of American marten in an untrapped forest preserve in Maine. Pages 86–99 *in* G. Proulx, H. N. Bryant, and P. M. Woodard, editors. *Martes*: taxonomy, ecology, techniques, and management. Provincial Museum of Alberta, Edmonton, Alberta, Canada.

Hunter, B., and B. Clark. 1986. Wildlife restraint handbook. U.S. Department of the Interior, Fish and Wildlife Service, Rancho Cordova, California, USA.

Katnik, D. D., D. J. Harrison, and T. P. Hodgman. 1994. Spatial relations in a harvested population of marten in Maine. Journal of Wildlife Management 58:600-607.

Kie, J. G., J. A. Baldwin, and C. J. Evans. 1994. CALHOME – Home range analysis program electronic user's manual. U.S. Department of Agriculture, Forest Service, PSW Research Station, Albany, California, USA.

Krohn, W. B., K. D. Elowe, and R. B. Boone. 1995. Relations among fishers, snow, and martens: development and evaluation of two hypotheses. Forestry Chronicle 71:97–105.

Lachowski, H. J. 1997. Relationships among prey abundance, habitat, and American marten in northern Maine. Thesis, University of Maine, Orono, Maine, USA.

Laundre, J. W., and B. L. Keller. 1984. Home-range size of coyotes: a critical review. Journal of Wildlife Management 48:127–139.

Mead, R. A. 1994. Reproduction in *Martes*. Pages 404–422 *in* S. W. Buskirk, A. S. Harestad, M. G. Raphael, and R. A. Powell, editors. Martens, sables and fishers: biology and conservation. Cornell University Press, Ithaca, New York, USA.

Miller, R. G. 1981. Simultaneous statistical inference. Second edition. Springer-Verlag, New York, New York, USA.

O'Doherty, E. C., L. F. Ruggiero, and S. E. Henry. 1997. Home range size and fidelity of American martens in the Rocky Mountains of southern Wyoming. Pages 123–134 *in* G. Proulx, H. N. Bryant, and P. M. Woodard, editors. *Martes*: taxonomy, ecology, techniques, and management. Provincial Museum of Alberta, Edmonton, Alberta, Canada.

Payer, D. C. 1999. Influences of timber harvesting and trapping on habitat selection and demographic characteristics of American marten. Dissertation, University of Maine, Orono, Maine, USA.

_____, and D. J. Harrison. 2000. Structural differences between forests regenerating following spruce-budworm defoliation and clear-cut harvesting: implications for marten. Canadian Journal of Forest Research 30:1965–1972.

Phillips, D. M. 1994. Social and spatial characteristics, and dispersal of marten in a forest preserve and industrial forest. Thesis, University of Maine, Orono, Maine, USA.

Phillips, D. M., D. J. Harrison, and D. C. Payer. 1998. Seasonal changes in home-range area and fidelity of martens. Journal of Mammalogy 79:180–190.

Powell, R. A. 1979. Mustelid spacing patterns: variations on a theme by *Mustela*. Zeitscrift fur Tierpsychologie 50:153–165.

_____. 1994. Structure and spacing of *Martes* populations. Pages 101–121 *in* S. W. Buskirk, A. S. Harestad, M. G. Raphael, and R. A. Powell, editors. Martens, sables and fishers: biology and conservation. Cornell University Press, Ithaca, New York, USA.

Pulliainen, E. 1981. Winter habitat selection, home range, and movements of the pine marten (*Martes martes*) in a Finnish lapland forest. Pages 1068–1087 *in* J. A. Chapman and D. Pursley, editors. Worldwide Furbearer Conference Proceedings, Frostburg, Maryland, USA.

Ruggiero, L. F., D. E. Pearson, and S. E. Henry. 1998. Characteristics of American marten den sites in Wyoming. Journal of Wildlife Management 62:663–673.

Sandell, M. 1989. The mating tactics and spacing patterns of solitary carnivores. Pages 164–182 *in* J. L. Gittleman, editor. Carnivore behavior, ecology, and evolution. Cornell University Press, Ithaca, New York, USA.

Schemnitz, S. D. 1994. Capturing and handling wild animals. Pages 106-124 *in* T. A. Bookhout, editor. Research and management techniques for wildlife and habitats. Fifth edition. The

Wildlife Society, Bethesda, Maryland, USA.

Sherburne, S. S., and J. A. Bissonette. 1994. Marten subnivean access point use: response to subnivean prey levels. Journal of Wildlife Management 58:400-405.

Slauson, W. L., B. S. Cade, and J. D. Richards. 1994. User manual for BLOSSOM statistical software. Midcontinent Ecological Science Center, National Biological Survey, Fort Collins, Colorado, USA.

Soutiere, E. C. 1979. Effects of timber harvesting on marten in Maine. Journal of Wildlife Management 43:850-860.

Steventon, J. D., and J. T. Major. 1982. Marten use of habitat in a commercially clear-cut forest. Journal of Wildlife Management 46:175-182.

Strickland, M. A., and C. W. Douglas. 1987. Marten. Pages 530-546 *in* M. Novak, J. A. Baker, M. E. Obbard, and B. Malloch, editors. Wild furbearer management and conservation in North America. Ontario Ministry of Natural Resources, Toronto, Ontario, Canada.

_____, _____, M. K. Brown, and G. R. Parsons. 1982. Determining the age of fisher from cementum annuli of the teeth. New York Fish and Game Journal 29:90-94.

Thompson, I. D. 1994. Marten populations in uncut and logged boreal forests in Ontario. Journal of Wildlife Management 58:272-280.

Thompson, I. D., and P. W. Colgan. 1987. Numerical responses of martens to a food shortage in northcentral Ontario. Journal of Wildlife Management 51:824–835.

Weckwerth, R. P., and V. D. Hawley. 1962. Marten food habits and population fluctuations in Montana. Journal of Wildlife Management 26:55–74.

White, G. C., and R. A. Garrott. 1984. Portable computer system for field processing biotelemetry triangulation data. Colorado Division of Wildlife, Game Information Leaflet 110, Fort Collins, Colorado, USA.

Wynne, K. M., and J. A. Sherburne. 1984. Summer home range use by adult marten in northwestern Maine. Canadian Journal of Zoology 62:941-943.

Zar, J. H. 1999. Biostatistical analysis. Fourth Edition. Prentice-Hall, Inc., Upper Saddle River, New Jersey, USA.

Chapter 5

MARTES FOOT-LOADING AND SNOWFALL PATTERNS IN EASTERN NORTH AMERICA:

Implications to Broad-Scale Distributions and Interactions of Mesocarnivores

William Krohn, Christopher Hoving, Daniel Harrison, David Phillips, and Herbert Frost

***Abstract*:** American martens (*Martes americana*) and fishers (*M. pennanti*) require large areas and live in complex, interacting communities of medium and large size carnivores. Nevertheless, habitat studies of these species continue to emphasize mid- to fine-scale habitat relationships, and rarely examine interspecific relations. Based on our data on foot-loading (ratio of body mass to total foot area) for fishers and martens, and snowfall patterns across eastern North America, we conclude that broader-scale habitat and interspecific relations of these 2 species may affect their regional distributions. Although foot-loading was influenced by sex ($P < 0.001$) and age ($P = 0.08$), foot-loading in fishers was >2 times greater ($P < 0.001$) than martens across all of the 4 age-sex classes. Relative to other large- and medium-sized carnivores in the forests of eastern North America, the 2 *Martes* species have the shortest legs, and thus are most dependent on low foot-loading for mobility in soft snow. To assess temporal and spatial variation in snowfall as related to potential *Martes* distributions, we used a snowfall threshold reported for Maine to define the mid-point of a zone with overlapping populations of fishers and martens and applied this threshold (240 cm mean annual snow) to regional snowfall data for 1970–90. Regression analyses of weather data in conjunction with data on latitude, longitude, and elevation were used to model mean annual snowfall. Since the late 1700's, there has been a general warming trend across eastern North America. If previously proposed hypotheses that snow limits fishers, and large populations of fishers limit martens are true, then one would predict that martens historically occurred south of where they do today. Further, if snowfall continues to decline in the region, fisher populations may expand and martens may decline. To test these predicted broad-scale distribution patterns, we suggest that past and modern occurrence data for fishers, martens, and other forest carnivores be examined across the historic range of both species to evaluate the hypothesis that interactions among morphology and climate affect distribution and degree of sympatry in North American *Martes*.

1. INTRODUCTION

Biogeographers have noted a widespread pattern in the factors limiting species at their geographic edges: the southern range limits tend to be determined by biotic interactions, whereas northern range limits are often caused by abiotic stress (Brown and Lomolino 1998). This pattern may hold for the 2 North American *Martes,* fishers *(M. pennanti)* and American martens (*M. americana*). Krohn et al. (1995, 1997) hypothesized that the energetic costs associated with frequent and deep snows reduce fitness and reproduction of fishers, and that where snow is less frequent and accumulations lower, large populations of fishers may directly increase mortality of martens and ultimately affect their distribution. Snowfall has been hypothesized to be a factor affecting interspecific competition and the allopatric distributions of not only fishers and martens, but also bobcats (*Lynx rufus*) and Canada lynx (*L. canadensis*) (Parker et al. 1983), and some species of weasels (*Mustela* spp.) (Simms 1979).

Sinking depth in snow is influenced by snow depth and structure, leg length, and foot-loading (Peek 1986). Thus, we predicted that martens have an advantage relative to fishers for moving on soft snow because of lighter foot-loading (i.e., a lower ratio of body mass/total foot area). Further, fishers and martens exhibit high degrees of sexual dimorphism (Holmes 1987, Powell 1993) that could contribute to differences in foot-loading and associated energetic constraints between sexes. If martens are better equipped to travel over soft snow than fishers, and the hypotheses of Krohn et al. (1995, 1997) are true, then the broad-scale distributions of the 2 *Martes* species should be predictable. Fishers should be associated with regions of lower snowfall and martens should be largely confined to regions of deepest snow because of reduced competition from fishers.

A second physical adaptation in mammalian predators that enhances mobility in snow is leg length, which is generally longer in larger animals. Principally as a result of selective pressures imposed by climate and competition, size structuring is prevalent in carnivore communities (Rozenweig 1966, Dayan and Simberloff 1996). Because of the high potential for interference and exploitation competition (Case and Gilpin 1974) among medium-sized mammalian carnivores (i.e., mesocarnivores) (Carbyn 1982, Sargeant et al. 1987, Litvaitis and Harrison 1989, Harrison et al. 1989, Arjo and Pletscher 1999, Fedriani et al. 1999, 2000), we were also interested in the potential effects of snow on the mobility of other predators that often occur sympatrically with fishers and martens. Carnivores of special interest, in addition to the 2 *Martes*, were the gray wolf (*Canis lupus*), coyote (*C. latrans*), red fox (*Vulpes vulpes*), Canada lynx, and bobcat.

We evaluated broad-scale spatial distributions of martens and fishers in eastern North America by: (1) testing for species, sex- and age-specific differences in foot-loading between fishers and martens; (2) examining the relations between foot-loading and leg length for a variety of northern forest mesocarnivores; and (3) determining the spatial variation in total mean annual snowfall across eastern North America to estimate the potential affects of snowfall variation on the regional distributions of *Martes,* 1970–90.

2. METHODS

2.1 *Martes* Foot-loading

Foot-loading was calculated by dividing the total body weight (g) of an individual animal by the total area of the 4 feet (cm^2) (Peek 1986, Murray and Boutin 1991). Fishers used in this study were captive animals held at the University of Maine, Orono. Animals were live-trapped in eastern and central Maine, or were the offspring of these wild-caught fishers (Frost and Krohn 1994). The body masses of wild-caught fishers were determined within a few days of initial capture in November or December, 1992; body masses of juveniles were determined when feet of all fishers were measured during mid-January, 1993. Wild martens were live-trapped as part of a telemetry study in north-central Maine (Phillips 1994). Body masses and foot areas of captured martens were measured in the field during August–September, 1993. We assessed the repeatability of our foot-area measurements for martens by comparing our late summer-fall measurements of each foot on several martens that were subsequently recaptured the following spring (May–early June).

Because juvenile fishers were in captivity for months prior to weighing, the body weights of these animals may not represent animals in the wild (i.e., could be either over- or under-weight due to the effect of captivity or captive feeding). However, body weight (g) of the juvenile male and female fishers we studied (3,936 + 568 g and 2,415 ± 263 g, respectively) (Table 5.1) were similar (3,940 + 590 g and 2,170 ± 390 g, respectively) to mid-winter weights reported by Douglas and Strickland (1987), suggesting that foot-loading of our study animals was comparable to free-ranging fishers.

Body mass of fishers was measured with a sliding beam scale in the laboratory, and mass of martens was taken with a spring scale in the field. Measurements were recorded to the nearest gram for both species. Foot measurements were taken from anesthetized animals by tracing the outline of each foot on paper. Toes were compressed and we attempted to keep even pressure on the

Table 5.1. Mean foot area (cm²) and body mass (g) of fishers and martens by sex and age (juvenile <1.0 years of age; adult ≥1.9 years) class in Maine, USA.

	Males							Females					
Species and age-class (n)	Front feet Lt.	Front feet Rt.	Rear feet Lt.	Rear feet Rt.	Total foot area	Body mass	Species and age-class (n)	Front feet Lt	Front feet Rt	Rear feet Lt	Rear feet Rt	Total foot area	Body mass
Fishers							**Fishers**						
Juvenile (3)							Juvenile (8)						
$\bar{x}$	38.1	38.8	47.1	45.6	169.6	3,936	$\bar{x}$	30.2	29.3	36.7	37.3	133.5	2,415
SD	3.4	3.5	9.3	2.6	18.6	568	SD	3.0	3.4	3.0	3.6	10.8	263
Adult (5)							Adult (7)						
$\bar{x}$	37.7	37.3	47.1	47.4	169.5	5,447	$\bar{x}$	28.8	28.2	35.5	35.1	127.6	2,693
SD	7.7	3.5	2.0	1.9	13.2	999	SD	5.4	3.3	2.1	2.1	10.3	417
Pooled (8)							Pooled (15)						
$\bar{x}$	37.9	37.9	47.1	46.7	169.6	4,880	$\bar{x}$	29.5	28.8	36.1	36.3	130.8	2,545
SD	6.1	3.3	5.2	2.2	14.1	1,129	SD 4.2	3.2	2.6	3.1	10.6	360	
Martens							**Martens**						
Juvenile (3)							Juvenile (6)						
$\bar{x}$	12.1	12.7	19.9	20.0	64.6	648	$\bar{x}$	10.0	9.9	15.4	15.2	50.5	457
SD	2.4	0.7	3.9	1.8	8.6	63	SD	0.6	1.0	1.7	1.7	4.5	36
Adult (6)							Adult (3)						
$\bar{x}$	12.9	12.6	20.2	19.8	65.5	798	$\bar{x}$	10.1	11.0	15.0	13.9	49.9	501
SD	1.6	1.5	1.7	1.1	3.0	111	SD	0.8	0.6	0.6	1.3	3.2	22
Pooled (9)							Pooled (9)						
$\bar{x}$	12.6	12.6	20.1	19.9	65.2	748	$\bar{x}$	10.0	10.3	15.3	14.8	50.3	471
SD	1.8	1.3	2.3	1.2	4.9	120	SD	0.6	1.0	1.4	1.6	3.9	37

foot so that the paw remained stationary but the toes were not splayed. Measurements of martens were recorded in the field on the individual's capture form. When a marten was recaptured, use of this form ensured proper placement of the heel when subsequent foot measurements were taken. The area of each tracing (i.e., each foot) was measured with an electronic planimeter to the nearest 0.1 cm^2; we used the average of 3 measurements for each tracing as our estimate of foot area. For each animal, these average values for all 4 feet were summed to estimate total foot area.

Ages of both martens and fishers were classified as juvenile (<1.0 yr old) or adult (≥1.9 years). Ages were determined by counting cementum annuli in first premolars (Strickland et al. 1982, Arthur et al. 1992) in combination with tooth radiographs (Dix and Strickland 1986).

Within each species the overall effect of sex and age and sex*age interactions on foot area was evaluated using a 2-way analysis of variance (ANOVA) for unbalanced sampling designs (MGLH procedure, SYSTAT Inc., Evanston, Illinois, USA; use of trade names does not imply endorsement) followed by Tukey *post hoc* comparisons to test for pairwise differences among the 4 age-sex categories (i.e., juvenile male, juvenile female, adult male, adult female). We evaluated the effects of species, sex, and age on foot-loading using a 3-way ANOVA and tested the interactions among species*sex, species*age, and species*sex*age. Significant results from the ANOVA were used to subset the data for subsequent pairwise comparisons using Tukey *post-hoc* tests. We used a square root transformation on our estimates of foot area and foot-loading for all statistical analyses.

2.2 Data for Other Mesocarnivores

Foot-loading data for wolves, coyotes, red foxes, and bobcats were taken from the literature, whereas data on the lengths of the hind limbs of North American mesocarnivores (including fishers and martens) came from Harris and Steudel (1997). These authors measured hind-limb length as the total lengths of the femur, tibia, and longest metatarsus.

We were interested primarily in species differences, and because species-specific data were not consistently available by age-sex classes, comparisons were made with data combined across sexes within a species. Our approach assumed that foot-loading was consistent across age and sex classes within a species, except for martens and fishers.

2.3 Snowfall Distribution

Climate data (1970–90) for the United States was extracted from the Cooperative Summary of the Day, published by the U. S. National Oceanic and Atmospheric Administration (NOAA), National Climatic Data Center (NCDC). Canadian Monthly Climate Data was purchased from Environment Canada, Atmospheric, Climate and Water Systems Branch.

Monthly sums of snowfall for the northeastern U. S. and eastern Canada, hereafter called the Northeast, were then summed for each weather station by year from November through the following March. Months missing more than 3 consecutive or 5 total days of data were omitted, as were years missing ≥1 month of data. Stations with less than 7 years of data per decade were also omitted.

Using the latitude and longitude provided with the data, stations were mapped using ARC/INFO 7.2.1 (Environmental Systems Research Institute, Redlands, California, USA). To avoid variations caused by ocean currents and lake effects (Wallace and Hobbs 1977), we excluded weather stations within 70 km of Lake Erie and Lake Ontario, those north of the St. Lawrence Seaway in Quebec, stations on Newfoundland (no historical or recent occurrences of fishers) and Prince Edward Island (currently supports neither *Martes* species), and those stations on small islands more than 20 km from the mainland.

We joined 1:250,000 scale Digital Elevation Models (DEMs), downloaded from U. S. Geological Survey (USGS), Geospatial Data Clearinghouse and DEMs purchased from the Centre for Topographic Information Sherbrooke, Geomatics Canada in ARC/INFO. From these DEMs we calculated elevations (m above mean sea level) for each weather station.

2.4 Snowfall Trends and Potential Effects

To determine recent trends in annual snowfall, we performed multiple linear regression with 10-yr mean snowfall, over 2 time periods 1970–80 and 1980–90, as the dependent variable. Elevation, latitude, and longitude were the independent variables. The coefficients from the regression models were then used to map predicted mean snowfall in the Northeast for each decade at 1 km^2 resolution.

To retrospectively examine the potential effects of measured change in snowfall on the distributions of *Martes*, we evaluated which snowfall depth was associated with a change in the high density distribution of fishers versus martens. We could not use the mean level of 48 cm of snowfall per month identified by Krohn et al. (1995) because this threshold was calculated from snowfall amounts by interpolation among weather stations as a monthly mean. Snowfall in this study, in contrast, was calculated by regression as a yearly mean. However, by overlaying the 48-cm snowfall line in the mean monthly snowfall, 1980–87 from Krohn et al. (1995) onto our mean annual snowfall data, 1980–90, we identified 240 cm as coinciding to the midpoint in Maine where the distributions of martens and fishers overlapped. The 240 cm threshold was mapped in relation to snowfall during 2 time periods, 1970–80 and 1980–90, with areas above the threshold presumed to be primarily potential marten habitat, and areas lower than the threshold presumed to be primarily fisher habitat.

3. RESULTS

3.1 Physical Adaptations

3.1.1 Foot Area

Repeatability of planimeter measurements from foot tracings was excellent. The 3 measurements of individual paws were often identical and were always within 0.1 cm^2. Differences in foot areas of 5 individual marten (1 female, 4 males) measured during spring 1993 and remeasured in fall 1993 were insignificant (t-tests, $P > 0.10$ for all pairwise comparisons), suggesting that measurement errors in the field had insignificant effects on our results.

Data were obtained from 23 fishers (adult male = 5, adult female = 7, juvenile male = 3, and juvenile female = 8) and 18 martens (adult male = 6, adult female = 3, juvenile male = 3, and juvenile female = 6) (Table 5.1). There was a significant effect of sex on foot area of fishers ($P < 0.001$, $F = 49.3$) and martens ($P < 0.001$, $F = 38.9$); however, age was not a significant effect on foot area for fishers ($P = 0.58$, $F = 0.31$) or martens ($P = 0.93$, $F = 0.01$). Further, there was no significant age and sex interaction with foot area for either species ($P = 0.58$, $F = 0.32$ for fisher; $P = 0.75$, $F = 0.11$ for martens). Total foot area was greater for males than females in both fishers ($P < 0.001$, $T = 6.96$) and martens ($P < 0.001$, $T = 7.08$) (Table 5.1), which was consistent with the pronounced sexual dimorphism in both species.

3.12 Species Effects

Foot-loading in fishers and martens differed significantly by species ($P < 0.001$, $F = 365.9$), sex ($P < 0.001$, $F = 41.1$), and age ($P < 0.001$, $F = 21.2$). Species * sex ($P < 0.001$, $F = 12.5$) and species * age ($P = 0.08$, $F = 3.23$) interactions were significant, but species *sex*age interactions ($P = 0.20$, $F = 1.68$) were not. The significance of species * sex and species * age interactions confounded interpretation of main effects; therefore, we developed statistical conclusions based solely on *post-hoc* comparisons.

3.1.3 Age and Sex Effects

Pairwise comparisons indicated that in fishers, foot-loading was greater for adult males compared to juvenile males ($P = 0.004$) and for adult females compared to juvenile females ($P = 0.02$). The difference ratio in fishers was 1.4:1 for adult males versus juvenile males and 1.2:1 for adult females versus juvenile females (Table 5.2). In martens, however, foot-loading did not differ significantly between adult males and juvenile males ($P = 0.99$) nor between adult females and juvenile females ($P = 0.34$), reflecting faster growth to adult size in the smaller bodied martens.

In fishers, adult males had greater foot-loading than either juvenile or adult females ($P < 0.001$); however, differences in foot-loading between juvenile male and adult female fishers were not significant ($P = 0.64$). Difference ratios for foot-loading in adult males versus adult female fishers were 1.5:1 (Table 5.2). In martens, adult males had greater foot-loading than juvenile females ($P = 0.04$), but no other differences between sexes were significant ($P \geq 0.31$). Foot-loading of fishers was approximately twice (range: 2.0–2.6) that of martens within each of the 4 age-sex classes ($P < 0.001$) (Table 5.2).

3.1.4 Mesocarnivore Foot-loading and Leg Length

Based on foot-loading alone, the relative ability of mammalian carnivores of medium size to move over soft snow was as follows (from best to worst): marten, lynx/fisher, red fox, bobcat, coyote, and wolf (Table 5.3). In contrast, leg length was longest in the wolf, followed by lynx, coyote, bobcat, red fox, fisher, and marten (Fig. 5.1). When considering both foot-loading and leg length, the Canada lynx was the most specialized of the species for mobility in deep, soft snow, whereas fishers and martens relied on foot area more than limb length for mobility in soft snow. Relative to the other carnivores in this study, fishers and martens had the shortest leg length, with fishers having only slightly longer legs than martens (Fig. 5.1).

3.2 Snowfall Distribution and *Martes*

Mean annual snowfall was estimated from 1,321 weather stations (Hoving 2001). These stations were well distributed along the north-south axis through the middle of the study area. Mapping of the residual errors were similar between the 2 periods, and were generally located near mountains (e.g., the Adirondacks) and areas of large bodies of water (i.e., Atlantic ocean), suggest-

Table 5.2. Comparison ($\bar{x}$ ± SD) (n) of foot-loading (g/cm^2) of fishers and martens during fall-winter by age-sex class[a].

Species and foot-loading[b]	Females		Males	
	Juvenile	Adult	Juvenile	Adult
Fishers[c]	18.2 ± 2.0 (8)	21.1 ± 2.6 (7)	23.2 ± 1.7 (3)	32.0 ± 4.2 (5)
Martens[d]	9.1 ± 0.4 (6)	10.1 ± 0.9 (3)	10.1 ± 1.3 (3)	12.2 ± 1.5 (6)
Ratios	2.0	2.1	2.3	2.6

[a]Juvenile < 1.0 years of age; adults ≥ 1.9 years.
[b]Body mass (g) divided by total area of feet (cm^2). Analysis of variance showed significant ($P < 0.001$) main effects by species, sex, and age, and significant interactions for species * sex ($P = 0.001$), and species * age ($P = 0.08$). Species * sex * age interaction was not significant ($P = 0.20$).
[c]Age differences in mean foot-loading within sexes were significant ($P \leq 0.02$).
[d]Age differences in mean foot-loading within sexes were not significant ($P \geq 0.31$).

Table 5.3. Average foot-loading (g/cm^2) for adult, large and medium-sized mammalian carnivores that historically occurred in eastern North America.

Species	Sex	Foot loading	Source
Fisher	Female	21.1	This study (Table 5.2)
	Male	32.0	
Marten	Female	10.1	This study (Table 5.2)
	Male	12.2	
Lynx	Female	25.6	Murray and Boutin (1991)
	Male	31.6	
Bobcat	Sexes	(57.2)[a]	Parker et al. (1983)
	Combined	(57.2)[a]	
Red Fox	Sexes	40-50	Nasimovich (1955)
	Combined		(cited in Peek [1986:Table 7.2])
Coyote	Female	106.0	Murray and Boutin (1991)
	Male	136.8	
Wolf	Sexes	160.0	Murray and Boutin (1991)
	Combined		

[a]Lynx = $\frac{25.6 + 31.6}{2} = 28.6$; bobcat = 28.6 x 2 = 57.2

ing local effects (Hoving 2001). Both snowfall models had relatively high coefficients of determination ($r^2 = 0.65$ and 0.67; 1970–80 and 1980–90, respectively), and the coefficients for the descriptor variables elevation and latitude were significant ($P < 0.0001$) for both models. Longitude contributed significantly to the 1980–90 model, but not the 1970–80 model; storm tracks likely

Figure 5.1. Foot-loading (g/cm^2) and hind limb length (cm) for large and medium-sized carnivores that historically occurred in eastern North America. Foot-loading data from Table 5.3 and data on hind leg length from Harris and Steudel (1997). Because this figure illustrates differences among species, species-specific data were averaged across the sexes.

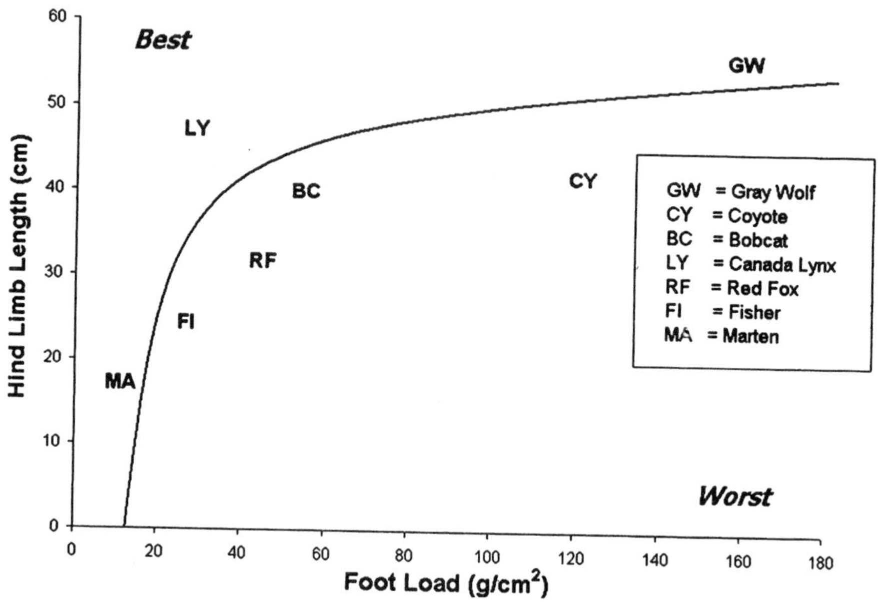

differed between the 2 decades. Snowfall predicted from elevation and latitude (and longitude in 1980–90) showed higher snowfall in the north and at higher elevations. Areas of heaviest snowfall were greatly reduced throughout the study area from 1970–80 to 1980–90 (Fig. 5.2). By applying the 240 cm threshold, an expansion in the northern range limit for fishers, and a corresponding northward contraction in the southern range limit for martens, was predicted between the 2 time periods (Fig. 5.2).

4. DISCUSSION

Total foot area was 2.6 times greater in fishers than martens, whereas the ratio of body masses between the species ranged from 5.3 for juvenile females to 6.8 for adult males. Thus, most of the difference in foot-loading between the 2 species is due to the relatively more massive body of fishers versus martens. Weight (W) of organisms is a direct function of volume, whereas foot size is a

Figure 5.2. Mean annual snowfall (cm/yr) for two ten-year periods illustrating the potential geographic affects of declining snowfall trends on geographic ranges of fishers and martens in eastern North America. Areas of deep snow (>240 cm) presumed to be primarily marten habitat shown in dark gray; areas of light snowfall (<240 cm) that are presumed fisher habitats are shown in light gray.

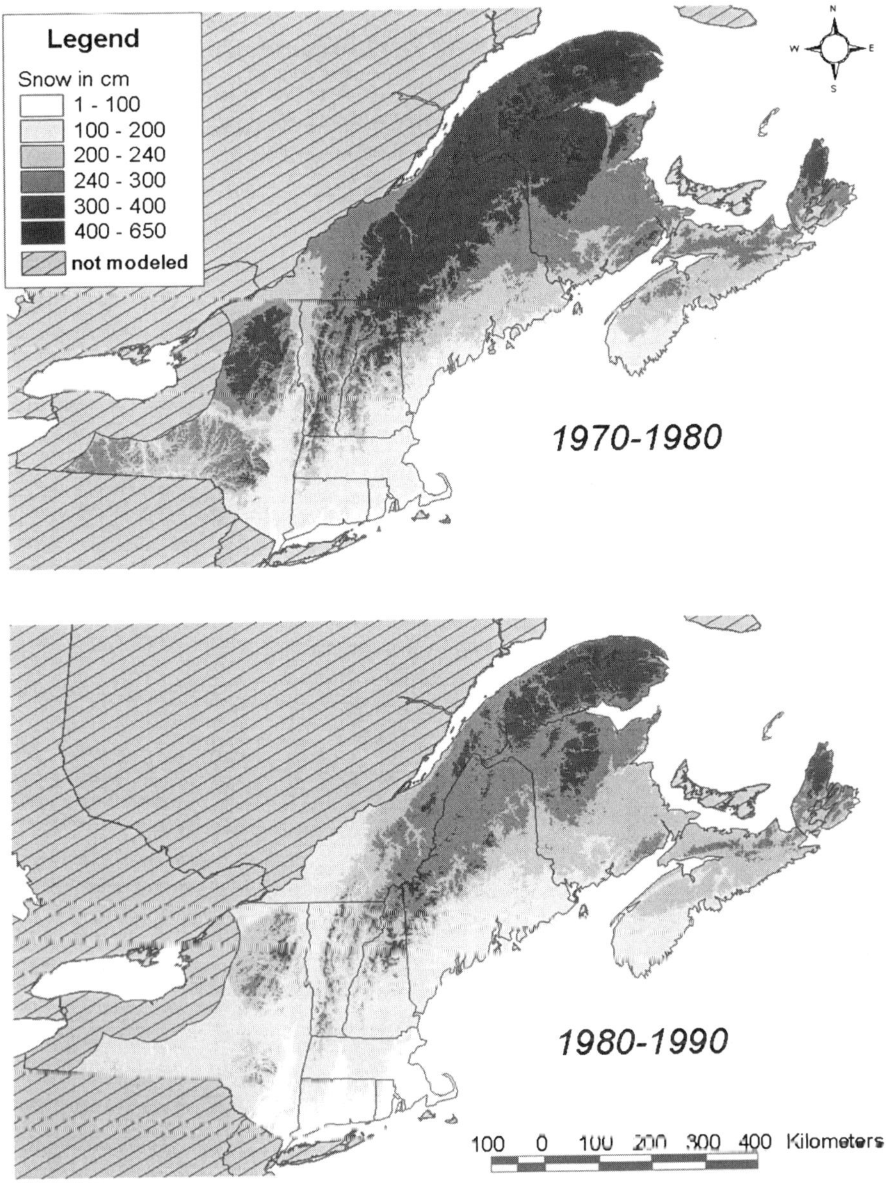

function of area. The allometric relationship of surface area (S) to volume (V) is $S = V^{2/3}$ (Ricklefs 1990), hence, foot-loading would be expected to scale as a function of $W^{3/2}$. The significance of this relationship is that animals with similar morphology should experience greater foot-loading on snow as weight increases. Thus, the trends in greater foot-loading among males versus females in both fishers and martens, as well of the greater foot-loading of fishers compared to martens, are ecological tradeoffs of sexual and interspecific dimorphism.

We found that mean foot area did not change with age in either fishers or martens. However, body masses of both species increase with age (Strickland and Douglas 1987, Douglas and Strickland 1987), and thus we believe that there is an age as well as sex-specific variation in foot-loading, especially in fishers. We cannot say, however, whether this age-specific difference has a survival significance or is merely a consequence of animals becoming larger as they grow into adulthood (Frost and Krohn 2004) and become reproductively active.

Powell (1993; see also Holmes and Powell 1994) discussed a number of hypotheses regarding the evolution of sexual dimorphism in mustelids, and noted that "...other selective pressures, as of yet not studied, may be involved." Our foot-loading data suggest that female fishers and martens are more capable of moving on snow than males. A selective pressure that could be operating, but not previously discussed (see Powell 1993), is the pressure to stay small enough so that movement over snow is facilitated. This constraint could be particularly important for female fishers, which give birth in early spring (Powell 1993, Frost 1994). Body condition during the latter third of active gestation affects birth rates in most larger mammals (Kirkpatrick 1988), and thus the ability to move over snow and hunt successfully during late-winter through early-spring could translate into strong selective pressure to reduce weight loading in the sex with highest reproductive costs (Erlinge 1979).

Our finding that mean foot-loading of fishers was greater than for martens is consistent with the results of Raine (1983), who concluded that fishers were hindered more than martens by soft snow. In a snow-tracking study of both species in Manitoba, Raine (1983) observed that fishers sank in snow to greater depths than martens regardless of gait. He also reported that during the soft snow period of midwinter, fishers traveled more upon their own tracks and those of snowshoe hares (*Lepus americanus*) compared to early winter when snow was shallower, or during late winter when snow was denser and harder. During midwinter, fishers often left furrows in the snow, whereas martens did not (Raine 1983).

Fishers and martens on our study area are part of a complex of mammalian predators including coyotes, red foxes, Canada lynx, and bobcats; historically, wolves were also present. Competition for food among sympatric predators can be expected to be greatest among similar sized animals (Gittleman 1985), particularly those with similar morphology (Dayan and Simberloff 1966, VanValkenburgh 1989). Snowfall, by affecting the mobility and distribution of some species of carnivores, could thus also be expected to have indirect effects on a carnivore community.

Responses to snowfall have also been proposed to influence the distribution of several mesocarnivores. Simms (1979) noted that long-tailed weasels (*Mustela frenata*) probably had poorer access to subnivean prey than the ermine (*M. erminea*) and least weasel (*M. nivalis*). The distribution of these weasels coincides with a snowfall gradient (Simms 1979). Parker et al. (1983) noted that Canada lynx distribution on Cape Breton Island contracted to deep snow areas as bobcats invaded the island in about 1955. They hypothesized that the low foot loading of Canada lynx allowed them to hunt in deep snow. In an analysis of 1,150 lynx occurrences, Hoving (2001) found mean annual snowfall to be the strongest predictor variable of lynx occurrences across eastern North America.

McNab (1971) argued that the northward increase in body size of North American *Mustela* and *Martes* is from competitive release. He presented data indicating that the ermine and least weasel increase in body size only north of the range of the long-tailed weasel, a close competitor. However, these conclusions were based on small samples (especially for *M. nivalis*) and have been questioned by many (see Ralls and Harvey 1985 for arguments and citations). With substantially larger samples, McNab (1971) also noted that in western Canada martens increase in size only north of about 62° N. latitude, the fisher's northern range limit. In Labrador, however, where fishers are absent above approximately 50° N latitude (Proulx et al. 2004), martens are considerably larger than in the southern part of the province inhabited by both species. Thus, size structuring in North American *Martes* could be the result of complex evolutionary trade-offs involving behavioral and physical adaptations to competition, as well as to climate.

The predicted distributions suggested by Figure 5.2 must be viewed with caution. First, the 2 species of *Martes* do live sympatrically in parts of this region, and thus there is no sharp line that separates 1 species from the other. Krohn et al. (1995) delineated a zone of approximately 50 km in width where the 2 species occurred regularly in northern Maine, 1980–87 Further, the threshold level of snow depth that may influence interspecific relationships of martens and fishers is likely to vary with forest canopy closure, prey abundance, and other environmental conditions. For example, the extent of the fisher/marten overlap zone is smaller in California versus Maine, possibly due to sharper elevation-induced snowfall gradients in western versus eastern North America (Krohn et al. 1997).

If snowfall is an important factor in the broad-scale distribution of some mesocarnivores, then these trends have 3 important implications for the distribution of these mesocarnivores in the Northeast. First, historical distributions during periods of high snowfall (Lamb 1977, Baron 1992) may not accurately reflect current potential distribution with low snowfall. This could be an important factor to consider when determining the feasibility of managing or recovering a species within its historic range. Second, given the rapidly decreasing snowfall of 1970–90, current distributions of snow-adapted mesocarnivores may be in flux. This could complicate habitat studies of species at the edge of their range. For example, if fishers are expanding their range north, then forested habitats currently considered "unsuitable" may be occupied in the future. Third, many regions in the Northeast, especially the Maritime Provinces south of the St. Lawrence River have barriers to north-south movements, which could reduce the opportunity for northern species to recolonize range to the south as conditions become more favorable (e.g., wolves; Harrison and Chapin 1998, Wydeven et al. 1998). Further, geographic ranges of mesocarnivores such as martens in northern Maine or Canada lynx on the Gaspé Peninsula or Cape Breton Island, could contract to the north and subsequently become geographically isolated and at greater risk of regional extinction.

In summary, we believe that the abundance and distribution patterns of *Martes* observed today are merely snapshots of a continually changing set of factors and events. One of the key factors underlying this dynamic system that determines the distribution of *Martes*, and other mesocarnivores, appears to be snowfall, as well as interspecific interactions among some carnivore species. Because the potential effects of snowfall on the distribution of fishers and martens have been explicitly stated (see Krohn et al. 1995,1997), and we have now demonstrated the feasibility of mapping snowfall distribution over extensive areas, it appears possible to test these predictions across time over a large geographic region. To do this, we suggest a cooperative study to examine the broad-scale distributions of *Martes,* using different regions as experimental replications. For example, modern data on the occurrence of fishers and martens from New England states, the Maritime Provinces, and southern Quebec should be pooled and examined as a unit, and compared to patterns in the Great Lakes Region and western North America. To the extent possible, historic distribution data on these 2 species from this same region should also be assembled and examined, along with historic data on snowfall. Analyses of spatial and temporal patterns of change in selected environmental factors as related to changes in the abundance and distribution of *Martes* could greatly improve our understanding of the relationships between and among these species and their habitats. Without an understanding of these dynamics, our ability to effectively manage habitats and populations today, or to respond to future conditions may be diminished.

5. ACKNOWLEDGMENTS

We thank the Maine Department of Inland Fisheries and Wildlife (MDIFW) for supporting both our fisher and marten studies through Federal Aid in Wildlife Restoration Project W-69-R. Field work with martens was further supported by the Maine Agricultural Experiment Station, the Maine Cooperative Forestry Research Unit, and the Maine Forest Service. Snowfall data were generated as part of a lynx habitat relations study, funded by the MDIFW, U. S. Fish and Wildlife Service, and The National Council for Air and Stream Improvement. This is a contribution of the Maine Cooperative Fish and Wildlife Research Unit (MDIFW, U.S. Geological Survey's Biological Resources Division, University of Maine, and Wildlife Management Institute) and is Publication Number 2647 of the Maine Agricultural and Forest Experiment Station. K. D. Elowe and T. P. Hodgman reviewed an early version of this manuscript, and we appreciate the thoughtful reviews of three referees.

6. LITERATURE CITED

Arjo, W. M., and D. H. Pletscher. 1999. Behavioral responses of coyotes to wolf recolonization in northwestern Montana. Canadian Journal of Zoology 77:1919–1927.

Arthur, S. M., R. A. Cross, T. F. Paragi, and W. B. Krohn. 1992. Precision and utility of cementum annuli for estimating ages of fishers. Wildlife Society Bulletin 20:402–405.

Baron, W. R. 1992. Historical climate records for the northeastern United States, 1640 to 1900. Pages in 74–91 *in* R. S. Bradley, and P. D. Jones, editors. Climate since A. D. 1500. Routledge, New York

Brown, J. H., and M. V. Lomolino. 1998. Biogeography. Sinauer Associates. Sunderland, Massachusetts, USA.

Carbyn, L. N. 1982. Coyote population fluctuations and spatial distribution in relation to wolf territories in Riding Mountain National Park, Manitoba. Canadian Field-Naturalist 96:176–183.

Case, T. J., and M. R. Gilpin 1974. Interference competition and niche theory. Proceedings from National Academy of Science 71:3073–3077.

Dayan, T., and D. Simberloff. 1996 Patterns of size separation in carnivore communities. Pages 243–266 *in* J. L. Gittleman, editor, Carnivore behavior, ecology and evolution, Volume 2. Comstock Publications Association, Ithaca, New York, USA.

Dix, L. M., and M. A. Strickland. 1986. Use of tooth radiographs to classify martens by sex

and age. Wildlife Society Bulletin 14:275–279.
Douglas, C. W., and M. A. Strickland. 1987. Fisher. Pages 510–529 *in* M. Novak, J. A. Baker, M. E. Obbard, and B. Malloch, editors. Wild furbearer management and conservation in North America. Ontario Trappers Association, North Bay, Canada.
Erlinge, S. 1979. Adaptive significance of sexual dimorphism in weasels. Oikos 33:233–245.
Fedriani, J. M., F. Palomares, and M. Delibes. 1999. Niche relations among three sympatric Meiterranean carnivores. Oecologia 121:138–148.
_____, T. K. Fuller, R. M. Sauvajot, and E. C. York. 2000. Competition and intraguild predation among three sympatric carnivores. Oecologia 125:258–270.
Frost, H.C. 1994. Reproductive biology of captive fishers. Dissertation, University of Maine, Orono, Maine, USA.
_____, and W.B. Krohn. 1994. Capture, care, and handling of fishers (*Martes pennanti*). Maine Agricultural and Forestry Experiment Station, Technical Bulletin 157, University of Maine, Orono, Maine, USA.
_____, and _____. 2004. Post-natal growth and development in fishers (*Martes pennanti*). Pages 253–264 *in* D. J. Harrison, A. K. Fuller, and G. Proulx, editors. Martens and fishers (*Martes)* in human altered environments: An international perspective. Kluwer Academic Publishers, Boston, Massachusetts, USA.
Gittleman, J. L. 1985. Carnivore body size: ecological and taxonomic correlates. Oecologia 67:540–554.
Harris, M. A., and K. Steudel. 1997. Ecological correlates of hind-limb length in Carnivora. Journal of Zoology (London) 241:381–408.
Harrison, D. J., J. A. Bissonette, and J. A. Sherburne. 1989. Spatial relationships between coyotes and red foxes in eastern Maine. Journal of Wildlife Management 53:181–185.
_____, and T. G. Chapin. 1998. Extent and connectivity of habitat for wolves in eastern North America. Wildlife Society Bulletin 26:767–775.
Holmes, T., Jr. 1987. Sexual dimorphism in North American weasels with a phylogeny of the Mustelidae. Dissertation, University of Kansas, Lawrence, Kansas, USA.
Holmes, T., Jr. and R. A. Powell. 1994. Morphology, ecology, and the evolution of sexual dimorphism in North American *Martes*. Pages 72–84 *in* S. W. Buskirk, A. S. Harestad, M. G. Raphael, and R. A. Powell, editors. Martens, sables, and fishers: Biology and conservation. Cornell University Press Ithaca, New York, USA.
Hoving, C. L. 2001. Historical occurrence and habitat ecology of Canada lynx (*Lynx canadensis*) in eastern North America. Thesis, University of Maine, Orono, Maine, USA.
Kirkpatrick, R. L. 1988. Comparative influence of nutrition on reproduction and survival of wild birds and mammals—an overview. Caesar Kleberg Wildlife Research Institution, Texas A&M University, Kingsville, Texas, USA.
Krohn, W. B., K. D. Elowe, and R. B. Boone. 1995. Relations among fishers, snow, and martens: development and evaluation of two hypotheses. The Forestry Chronicle 71:97–105.
_____, W. J. Zielinski, and R. B. Boone. 1997. Relations among fishers, snow, and martens in California: results from small-scale spatial comparisons. Pages 211–232 *in* G. Proulx, H. N. Bryant, and P. M. Woodard, editors. *Martes*: Taxonomy, ecology, techniques, and management. Provincial Museum of Alberta, Edmonton, Alberta, Canada.
Lamb, L. L. 1977. Climate history and future, volume 2 of Climate: Present, past, and future. Methuen, London.
Litvaitis, J. A., and D. J. Harrison. 1989. Bobcat-coyote niche relationships during a period of coyote population increase. Canadian Journal of Zoology 67:1180–1188.
McNab, B. K. 1971. On the ecological significance of Bergmann's rule. Ecology 52:845–854.
Murray, D. L., and S. Boutin. 1991. The influence of snow on lynx and coyote movements: does morphology affect behavior? Ocelogia 88:463–469.

Nasimovich, A. A. 1955. The role of the regime of snow cover in the life of ungulates in the USSR. Translation from Russian by Canadian Wildlife Service, Ottawa, Canada. 371 pages typed (original not seen; cited in Peek [1986:162–164]).

Parker, G. R., J. W. Maxwell, L. D. Morton, and G. E. J. Smith. 1983. The ecology of lynx (*Lynx canadensis*) on Cape Breton Island. Canadian Journal of Zoology 61:770–786.

Peek, J. M. 1986. A review of wildlife management. Prentice-Hall, Englewood Cliffs, New Jersey, USA.

Phillips, D. M. 1994. Social and spatial characteristics, and dispersal of marten in a forest preserve and industrial forest. Thesis, University of Maine, Orono, Maine, USA.

Powell, R. A. 1993. The fisher: Life history, ecology, and behavior. Second edition. University of Minnesota Press, Minneapolis, Minnesota, USA.

Proulx, G., K. B. Aubry, J. Birks, S. W. Buskirk, C. Fortin, H. C. Frost, W. B. Krohn, L. Mayo, V. Monakhov, D. Payer, M. Saeki, M. Santos-Reis, R. Weir, and W. J. Zielinski. 2004. World distribution and status of the genus *Martes* in 2000. Pages 21–76 *in* D. J. Harrison, A. K. Fuller, and G. Proulx, editors. Martens and fishers (*Martes)* in human altered environments: An international perspective. Kluwer Academic Publishers, Boston, Massachusetts, USA.

Raine, R. M. 1983. Winter habitat use and responses to snow cover of fishers (*Martes pennanti*) and marten (*Martes americana*) in southeastern Manitoba. Canadian Journal of Zoology 61:25–34.

Ralls, K., and P. H. Harvey. 1985. Geographic variation in size and sexual dimorphism of North American weasels. Biological Journal of the Linnean Society 25:119–167.

Ricklefs, R. E. 1990. Ecology. Third edition. W. H. Freeman and Company. New York, New York, USA.

Rosenzweig, M. L. 1966. Community structure in sympatric Carnivora. Journal of Mammalogy 47:602–612.

Sargeant, A. B., S. H. Allen, and J. O. Hastings. 1987. Spatial relations between sympatric coyotes and red foxes in North Dakota. Journal of Wildlife Management 51:285–293.

Simms, D. A. 1979. North American weasels: resource utilization and distribution. Canadian Journal of Zoology 57:504–520.

Strickland, M. A., C. W. Douglas, M. K. Brown, and G. R. Parsons. 1982. Determining the age of fisher from cementum annuli of the teeth. New York Fish and Game Journal 29:90–94.

_____, and C. W. Douglas. 1987. Marten. Pages 531–546 *in* M. Novak, J. A. Baker, M. E. Obbard, and B. Malloch, editors. Wild furbearer management and conservation in North America. Ontario Trappers Association, North Bay, Canada.

VanValkenburgh, B. 1989. Carnivore dental adaptions and diet: a study of trophic diversity within guilds. Pages 410–436 *in* J. L. Gittleman, editor. Carnivore behavior, ecology and evolution, Volume 2. Cornell University Press, Ithaca, New York, USA.

Wallace, J. M. and P. V. Hobbs. 1977. Atmospheric science. Academic Press, New York, New York, USA.

Wydeven, A. P., T. K. Fuller, W. Weber, and K. MacDonald. 1998. The potential for wolf recovery in the northeastern United States via dispersal from southeastern Canada. Wildlife Society Bulletin 26:776–784.

Part II

Habitat Relationships

Chapter 6

HOME RANGES, COGNITIVE MAPS, HABITAT MODELS AND FITNESS LANDSCAPES FOR *MARTES*

Roger Powell

Abstract: Martens (members of the genus *Martes*) maintain home ranges because the long-term benefits (food, access to mates, rest sites, information, etc.) from doing so exceed the long-term costs (travel, risk of predation, competitors, etc.). Martens appear to have cognitive maps of their home ranges, which one might envision as an integration of contour maps, perhaps one for each benefit and cost. Habitat provides and affects the benefits and costs for martens and, therefore, contributes to each marten's fitness. By integrating how habitat characteristics contribute to a marten's fitness, one can construct a fitness landscape, a map of how much each site on a landscape can contribute to a marten's fitness. I propose that energy might be used as a crude index for modeling cognitive maps and fitness landscapes for fishers. That model would form a basis for testing my *a priori* hypothesis that fishers choose home ranges that minimize the area needed to meet their requirements, thus maximizing their fitness. Combining the concepts of fitness landscapes, habitat modeling, and *a priori* hypothesis testing would enhance our understanding of the interactions among a marten's habitat, home range characteristics, and fitness.

The wild things that live on my farm are reluctant to tell me, in so many words, how much of my township is included within their daily or nightly beats. I am curious about this, for it gives me the ratio between the size of their universe and the size of mine, and it conveniently begs the much more important question, who is the more thoroughly acquainted with the world in which he lives? Aldo Leopold (1949:78)

1. HOME RANGES

Adult martens (meaning all members of the genus *Martes*) confine their day to day activities to areas we call "home ranges". Burt (1943:351) provided the definition and concept of a mammal's home range that we use today. "that area traversed by the individual in its normal activities of food gathering, mating, and caring for young. Occasional sallies outside the area, perhaps exploratory in nature, should not be considered part of the home range." Burt's definition is clear and accurate conceptually, though vague in ways that are consistent with the difficulties of quantifying animals' home ranges, such as defining boundaries (White and Garrott 1990). The vague wording implicitly and correctly allows a home range to include areas used in diverse ways for diverse behaviors. Members of 2 different species may use their home ranges very differently with very different behaviors but, for both, the home ranges are recognizable as home ranges, not something different for each species.

Animals establish and maintain home ranges because doing so provides long term benefits that exceed the costs (Powell 2000). Benefits include (but are not limited to) food, access to mates, rest sites, escape cover and information (knowledge), while costs include travel, risk of predation, competitors for food and mates, and learning new information. Some carnivores, and probably all martens, have cognitive maps of where they live (Peters 1978). Fishers (*M. pennanti*), for example, do not use the space within their home ranges randomly (Powell 1978, 1979, 1981, 1994b). I documented that fishers in Upper Peninsula Michigan directed travel in roughly straight lines through areas of low prey availability (Powell 1978, 1994b). Such straight line travel sometimes took fishers directly from one porcupine (*Erethizon dorsatum*) winter den to another. Fishers also approached hollow trees used as rest sites directly and ran directly to and through an old culvert deep in the forest. These data suggest strongly that fishers have cognitive maps of their home ranges and that those maps incorporate aspects of prey availability, juxtaposition of habitats, and proximity of habitat features. Spencer (1992) showed that knowledge of resources in an area may, in and of itself, be the major benefit of establishing a home range. Indeed, knowledge of resources and cognitive maps are inherent in Burt's (1943) definition of home range.

To know how martens conceive their home ranges would provide tremendous insight into their lives. Over 50 years ago, Leopold wrote that "...animals frequently disclose by their actions what they decline to divulge in words." Today, we still use animals' actions to learn about their lives but we also use models, especially models built to mimic biological functions, to provide additional insight. From optimal foraging models (Charnov 1976a, Pyke et al. 1977, Pyke 1984), for example, we know that animals often rank resources in some manner. Consequently, *we* might envision a marten's cognitive map of its home range as an integration of contour maps: one (or more) for food resources, one for escape cover, one for travel routes, one for known home ranges of members of the other sex, and so forth. Cognitive maps are probably sensitive to where a marten finds itself within its home range or to its nutritional state; resources that the marten perceives to be close at hand or resources far away that balance the diet may be more valuable than others.

A marten's cognitive map changes as new resources develop or are discovered and as old ones disappear. Such changes may occur quickly, because the marten has an instantaneous concept of its cognitive map. A researcher, in contrast, can learn of the changed cognitive map only through studying changes in the marten's behavior over time. With extensive data, one can build sequential images (and analyses) of a marten's use of space and thereby gain insight into how its concept of its habitat and environment changes over time (Doncaster and Macdonald 1991). To gain this insight, however, we must understand what habitat is.

2. HABITAT

The word "habitat" is defined too much and too little (Mitchell and Powell 2003): too much because the word is defined differently by different people (e.g., Morrison et al. 1992, Hall et al. 1997, Corsi et al. 2000, Garshelis 2000) and too little because the word is often used critically without being defined at all (Forman and Godron 1986). Garshelis (2000:112) noted two distinct suites of definitions of "habitat". The first is "the type of place where an animal normally lives or, more specifically, the collection of resources and conditions necessary for its occupancy." The second is "a set of specific environmental features that, for terrestrial animals, is often equated to a plant community, vegetative association, or cover type." Originally, "habitat" was a species-specific property (the first definition; Leopold 1933) but, with the development of habitat mapping, land-based definitions (the second definition) have become common (Corsi et al. 2000). Garshelis argued that the prevalence of the second definition confers legitimacy and is consistent with the normally accepted concept of "habitat use." Vegetative communities, however, vary across animal species' ranges and habitat needs of martens change across space as multiple factors vary, e.g., predominant prey, habitat structures that affect foraging, and snow. Thus, to be useful a habitat model must explain or predict how animals associate with the vegetation and physical structure of a particular area. To do this, a model must relate habitat to an animal's fitness or to correlates of fitness, emphasizing the superiority of the first definition.

The present predominance of land-based definitions of habitat derives, at least in part, because habitat types are easy to map with today's technology. This ease, in turn, has contributed to a fixation on patches, especially within landscape ecology. Plants are not distributed uniformly in space (Gleason 1926) and, therefore, distributions of plants and animals are heterogeneous, or patchy, within populations and ecological communities. Such patchiness, however, is not equivalent to vegetation being arranged in clearly outlined patches (Gleason 1926), or polygons. In patchy distributions, organisms of the same species are clumped in space, rather than being randomly or evenly distributed. The world, in contrast, seldom fits into easily discernable patches or polygons of different vegetative types. Vegetative types often intergrade gradually into each other. In northern landscapes, for example, spruce bogs often support thinner and thinner densities of black spruce (*Picea mariana*) as they grade into open muskeg. In Appalachian landscapes, white oak (*Quercus alba*) forests gradually become red oak (*Q. rubra*) forests that become cove forests going from ridgetop to cove (Powell et al. 1997a). The boundaries we place on these vegetative types are arbitrary and may have no relationship to how animals view habitat.

Even when a landscape can be divided reasonably unambiguously into patches (at least for us), the patch patterns for different landscape characteristics seldom coincide. In Appalachian landscapes, stands of understory shrubs such as mountain laurel (*Kalmia latifolia*), rhododendrons and azaleas (*Rhododendron* spp.) and berries (*Vaccinium* spp, *Gaylusaccia* spp) cross boundaries of forest types. In addition, patches of the same type have different values to animals depending on their size and juxtaposition with patches of other types. Black bears (*Ursus americanus*) leave patches of good habitat faster if they are imbedded in a local landscape of poor quality as opposed to a local landscape of high quality (Powell, unpublished data). Similarly, large patches of lowland conifers close to porcupine dens were undoubtedly more valuable to the fishers I studied in Upper Peninsula Michigan than were small patches isolated in a sea of even-aged, second growth hardwood forest.

The fixation on patches has its origins in optimal foraging (Fretwell and Lucas 1970, Charnov 1976b) and treating landscapes as a spatial array of arbitrary, unambiguous patches has contributed tremendously to our understanding of animals' foraging patterns and spacing patterns. Nonetheless, little empirical evidence suggests that this model of landscapes is realistic enough to continue to be used so universally (Kareiva 1990). The time is overdue to return to a "habitat" as a species-specific property.

3. HABITAT AND FITNESS

If habitat is important to an animal because it provides critical resources and costs, then our habitat models should reflect how habitats provide those resources and costs. Ultimately, a model must show how those resources and costs affect an animal's fitness, a task that is easier to state than to do. Fitness is a relative measure that increases with increasing survival and number of offspring (Fisher 1930). A more precise definition (Stearns 1992) is the expected contribution of an individual to future generations, homologous to the intrinsic rate of increase, *r*, for a population. Like "habitat", "fitness" has become jargon in animal ecology and is often used without explicit definition. Relatively few studies have addressed linkages between habitat and components of fitness (survival, reproduction), undoubtedly due to the challenges of measuring fitness in field studies (Garshelis 2000). A complete field measure of fitness must include lifetime reproduction for an individual organism *and* its descendants, essentially an impossible task. In lieu of this complete measure, two approaches exist for indexing the fitnesses of animals. First, one can measure components or supposed correlates of fitness for individuals (annual survival and reproduction, energy gain and expenditure) and assume that these measures extrapolate accurately to fitness. This approach is only as accurate as the extrapolation. The second approach is to infer fitness indirectly from animals' behavior, assuming that animals prefer environments that enhance their fitness. The theoretical foundation of this approach lies in optimality theory and in research showing that natural selection has molded foraging decisions, patch selection and patch use to maximize fitness or indices of fitness (Stephens and Krebs 1986). Field studies use this approach extensively, largely because behavioral data (usually telemetry locations) are easier to collect than data on vital rates. This approach is only as accurate as the assumption that a researcher knows how all short- and long-term habitat components affect an animal's optimality decisions.

Calls to understand better the relationship between fitness and habitat focus largely on better ways of indexing fitness (see Garshelis 2000) and not on understanding how habitat contributes to fitness (Mitchell and Powell 2003). Given that the only definition of habitat with a functional, biological basis ties resources within an area to the fitnesses of animals, understanding how habitat affects fitness is critical. A proven approach to solving biological problems is to incorporate present biological knowledge into models that predict animal behavior and then test the models. This modeling approach should be productive for evaluating the utility of fitness landscapes as a tool to understand better the habitat relationships of martens.

4. MODELS OF FITNESS LANDSCAPES

Habitat models that incorporate functional, biological relationships between animals and resources produce fitness landscapes. To understand fitness landscapes, one begins with the concept of an adaptive habitat landscape. Wright (1932, 1978) introduced the concept of an adaptive genetic landscape (Fig. 6.1), showing fitness valleys and peaks based on an organism's potential genome. Note that the axes in Figure 6.1 have no labels; they are two representative axes of an organism's genome, somewhat like two axes from a principal components analysis. Alternatively, each axis might represent (in Wright's terms) the alleles at a single locus. Similarly, Figure 6.1 can represent fitness peaks and valleys based on the habitat within the home range of an organism. Viewed in this manner, each axis represents a habitat or ecological variable and the adaptive habitat landscape shows how that variable affects fitness through its effects on prey or resting sites or escape cover. Each home range has a different location on the adaptive habitat landscape and, therefore, has different effects on fitness, just as each genotype has a different location on the adaptive genetic landscape and leads to a different fitness. Here is where the analogy stops, however, because an animal can change its home range but cannot change its genotype.

To construct a fitness landscape, one might take each axis of the adaptive habitat landscape in turn. Knowing the important axes requires solid knowledge of the biology of the organism in question. For example, one axis for the fishers I studied in Upper Peninsula Michigan might be the extent of dense, lowland habitats that support high densities of snowshoe hares (*Lepus americanus*). One presumes (hypothesizes) that greater extent of hare habitat leads to higher fitness for fishers. To incorporate the relationships of this axis with other habitat axes, one must incorporate juxtaposition of hare habitat with porcupine dens and resting sites and other habitat variables. For a site, a point, in a fisher's home range, the potential contribution of that site to the fisher's fitness includes whether the site has hare habitat, if so how extensive the hare habitat is, the proximity of the site to porcupine dens, the proximity to rest sites, the quality of the rest sites, and so forth. Mapping the fitness values for all points across an area produces the fitness landscape for the area. Each axis of the adaptive habitat landscape becomes a variable in an *n*-dimensional habitat model for the fitness landscape.

Fitness landscapes will change over time as real landscapes change. During summer, habitats with high densities of ground squirrels (*Spermophilus* spp.) can be important for American martens (*M. americana*; Zielinski et al. 1983) and make peaks on a marten's summer fitness landscape. In winter, those habitats make valleys on its winter fitness landscape because ground squirrels hibernate and become inaccessible (Zielinski et al. 1983). On a longer time scale, as a snowshoe hare population grows and hares begin using marginal habitats, the contributions of those habitats to the fitness of American martens and fishers may increase.

Figure 6.1. Seen as an Adaptive Genetic Landscape (Wright 1932, 1978): Each axis represents a dimension of an animal's potential genome. The axes might be considered as the alleles at a locus or abstractly as two genome variables, perhaps as two major axes from a principal components analysis of the entire genome. Different gene combinations contribute differently to the animal's fitness. Areas marked with +s are adaptive peaks, or gene combinations (genotypes) that contribute positively to the animal's fitness, while areas marked with -s are adaptive valleys. *Seen as an Adaptive Habitat Landscape:* Each axis represents how a habitat or ecological variable (vegetation and physical structure) affects different prey or resting sites or escape cover, which affect an animal's fitness. Each axis becomes a component of an *n*-dimensional habitat model. At certain combinations of habitat variables, an animal finds itself on a fitness peak. At other combinations of habitat variables, an animal finds itself in a fitness valley (Figure drawn after Wright 1932).

5. USING FITNESS LANDSCAPES

Most research on habitats used by martens has parsed landscapes into habitat polygons (e.g., Brainerd et al. 1994; Herrmann 1994; Jones and Garton 1994; Powell 1978, 1994a 1994b; Buck et al. 1994; Badry et al. 1997; Coffin et al. 1997; Gilbert et al. 1997; Powell et al. 1997b; Sturtevant and Bissonette 1997). To their credit, most researchers consider habitat variables on more than one scale and some base their research on continuous variables (e.g., Thompson and Harestad 1994, Krohn et al. 1997). Such recent research has provided critical insight into the biology of martens, yet it highlights the common practice of viewing landscapes as patches or polygons and not as a continuous landscape. Even though most researchers know that habitat patches and forest types are not truly discrete, and are ultimately arbitrary, the very fact that we call them "patches" leads us to treat them as if they were discrete and real. An additional value of viewing a landscape as a fitness landscape is that resource clumps, juxtaposition and proximity of those clumps, and landscape features are incorporated into a continuous map of the landscape's contributions to an animal's fitness.

Many studies correlate location data for animals with forest or habitat types. Most studies misrepresent their findings as conclusions rather than identifying them appropriately as hypotheses for future testing. Until correlations between use and habitat classes are tested with independent data sets, they constitute only the first few step of the scientific method and remain untested hypotheses. The wildlife literature is replete with habitat models based on *a posteriori*, correlation analyses, suggesting that we are more interested in generating new hypotheses of habitat relations than we are in testing our hypotheses. Extensive theory and empirical literature allow us to construct habitat hypotheses, habitat models, *a priori* for martens; that is, construct them *before* we collect data. We can then test them, and test our knowledge and understanding, with the data we collect.

Allen (1982, 1983) developed models of habitat suitability for American martens and fishers. These indices of habitat suitability were built from general requirements of martens and fishers and how general forest characteristics were believed to index the ability of a forest to meet those requirements. These indices are biologically pertinent insofar as the information used to build them documented accurately the habitat variables that provide information on prey, rest sites, and other biological requirements of martens or fishers. The indices are simple and lack site specific resources such as rest sites or prey or, for fishers, porcupine dens, and they lack information on interspersion and juxtaposition of habitats with different resources. Allen's (1983) index for fishers, however, was tested by Thomasma et al. (1991, 1994) and found to predict use of space by fishers.

I also tested Allen's model of habitat suitability for fishers, using habitat data and fishers' tracks in the snow (my study site, the track data and the habitat data were detailed by Powell 1994b). Both distance travelled and time spent increased with increasing habitat suitability (both distance and time: $P < 0.001$, General Linear Model, SAS). Blocking the analyses showed that individual tracks were indistinguishable in their use of areas with different habitat suitability and, hence, could be treated as independent samples. Even though Allen's model does not specify direct relationships between its habitat variables and relative fitness of fishers, it does appear to be biologically pertinent.

If an animal's critical resources are food and its critical costs are travelling to find food, then the animal's costs and benefits can be indexed by energy. In this simplistic case, the animal's cognitive map of its home range and the fitness landscape on which it lives might be indexed with energy contours. When the benefits, such as rest sites and escape routes, are not neatly tied to energy and costs include risk of predation, then energy is not an appropriate unit to index either cognitive maps or fitness landscapes. I have argued that fishers' major activities can be indexed by energy because fishers spend most of their time running on the ground hunting or resting (Powell 1979, 1981, 1993). In habitats where fishers can catch prey easily, one can model higher energy gain and lower energy expenditure than for other habitats. The value of rest sites can be modeled by their seasonally relevant insulating qualities. One can model seasonal home ranges by changing the potential energy gain and expenditure for habitats whose prey differ seasonally. Modeling short-term changes in cognitive maps due to nutritional conditions or distances to resources is more problematic but is conceivable. Such changes could probably be ignored at first. I believe that energy based models are a good way to make the first attempt at modeling cognitive maps and fitness landscapes for fishers.

The area that a marten considers to be within its home range may be based on an annual cognitive map that amalgamates information across seasons as well as across space. Alternatively, martens may conceive of their home ranges differently in different seasons. Mitchell (1997) recently developed and tested models for the optimal choice and placement of home ranges on a fitness landscape. Black bears in the Southern Appalachian Mountains appear to chose area-minimizing home ranges, which are home ranges that incorporate the best, local sites that are sufficient to meet minimum requirements. Old, dominant females appear to have priority access to the best home ranges (Mitchell 1997). From appropriate, large fitness landscapes for martens, one should be able to predict optimal home ranges for martens. I predict that martens, too, have area-minimizing home ranges and that when an individual marten incorporates a site into its home range, it reduces the value of that site enough that other martens avoid incorporating that site into their home ranges (Powell 1994a). By testing the goodness of fit of predicted home ranges to true home ranges, we can evaluate our understanding of the underlying relationships among habitat characteristics, habitat quality, use of space and fitness across landscapes. Then, we may be able to answer Leopold's (1949:78) question of "who is more thoroughly acquainted with the world in which he lives?"

6. ACKNOWLEDGMENTS

I thank David Payer, Angela Fuller, Daniel Harrison and 2 anonymous reviewers for comments on a very different, previous draft of this paper. Mike Mitchell made significant contributions through extensive discussions we have had developing other papers and nearly deserves coauthorship.

7. LITERATURE CITED

Allen, A. W. 1982. Habitat suitability index models: Marten. U.S. Fish and Wildlife Service, FWS/OBS-82/10.11.

_____. 1983. Habitat suitability index models: Fisher. U.S. Fish and Wildlife Service, FWS/OBS-82/10.45.

Badry, M. J., G. Proulx, and P. M Woodard. 1997. Home-range and habitat use by fishers translocated to the aspen parkland of Alberta. Pages 233–251 *in* G. Proulx, H. N. Bryant, and P. M. Woodard, editors. *Martes*: taxonomy, ecology, techniques, and management. The Provincial Museum of Alberta, Edmonton, Alberta, Canada.

Brainerd, S. M., J.-O. Helldin, E. Lindström, and J. Rolstad. 1994. Eurasian pine martens and old industrial forest in southern boreal Sweden. Pages 343–354 *in* S. W. Buskirk, A. S. Harestad, M. G. Raphael, and R. A. Powell, editors. Martens, Sables and Fishers: Biology and Conservation. Cornell University Press, Ithaca, New York, USA.

Buck, S. G., C. Mullis, A. S. Mossman, I. Show, and C. Coolahan. 1994. Habitat use by fishers in adjoining heavily and lightly harvested forest. Pages 368–376 *in* S. W. Buskirk, A. S. Harestad, M. G. Raphael, and R. A. Powell, editors. Martens, sables and fishers: biology and conservation. Cornell University Press, Ithaca, New York, USA.

Burt, W. H. 1943. Territoriality and home range concepts as applied to mammals. Journal of Mammalogy 24:346–352.

Charnov, E. L. 1976a. Optimal foraging: Attack strategy of a mantid. American Naturalist 110:141–151.

_____. 1976b. Optimal foraging: The marginal value theorem. Theoretical Population Biology 9: 29–136.

Coffin, K. W., Q. J. Kujala, R. J. Douglass, and L. R. Irby. 1997. Interactions among marten prey availability, vulnerability, and habitat structure. Pages 199–210 *in* G. Proulx, H. N. Bryant, and P. M. Woodard, editors. *Martes*: taxonomy, ecology, techniques, and management. The Provincial Museum of Alberta, Edmonton, Alberta, Canada.

Corsi, F., J. De Leeuw, and A. Skidmore. 2000. Modeling species distributions with GIS. Pages 389–434 *in* L. Boitani and T. K. Fuller, editors. Research techniques in animal ecology. Columbia University Press, New York, New York, USA.

Doncaster, C. P., and D. W. Macdonald. 1991. Drifting territoriality in the red fox *Vulpes vulpes*. Journal of Animal Ecology 60:423–439.

Fisher, R. A. 1930. The genetical theory of natural selection. Clarendon Press, Oxford, England.

Forman, R. T. T., and M. Godron. 1986. Landscape Ecology. John Wiley and Sons, New York, USA.

Fretwell, S. D., and H. L. Lucas, Jr. 1970. On territorial behavior and other factors influencing habitat distribution in birds. I. theoretical development. Acta Biotheoretica 19:16-36.

Garshelis, D. L. 2000. Delusions in habitat evaluation: measuring use, selection, and importance. Pages 111–164 *in* L. Boitani and T. K. Fuller, editors. Research techniques in animal ecology. Columbia University Press, New York, New York, USA.

Gilbert, J. H., J. L. Wright, D. J. Lauten, and J. R. Probst. 1997. Den and rest site characteristics of American marten and fisher in northern Wisconsin. Pages 135–145 *in* G. Proulx, H. N. Bryant, and P. M. Woodard, editors. *Martes*: taxonomy, ecology, techniques, and management. The Provincial Museum of Alberta, Edmonton, Alberta, Canada.

Gleason, H. A. 1926. The individualist concept of plant association. Torrey Botanical Club Bulletin 53:7–26.

Hall, L. S., P. R. Krausman, and M. L. Morrison. 1997. The habitat concept and a plea for standard terminology. Wildlife Society Bulletin 25:173–182.

Herrmann, M. 1994. Habitat use and spatial organization by the stone marten. Pages 122–136 *in* S. W. Buskirk, A. S. Harestad, M. G. Raphael, and R. A. Powell, editors. Martens, sables and fishers: biology and conservation. Cornell University Press, Ithaca, New York, USA.

Jones, J. J., and E. O. Garton. 1994. Selection of successional stages by fishers in north-central Idaho. Pages 377–388 *in* S. W. Buskirk, A. S. Harestad, M. G. Raphael, and R. A. Powell, editors. Martens, Sables and Fishers: Biology and Conservation. Cornell University Press, Ithaca, New York, USA.

Kareiva, P. 1990. Population dynamics in spatially complex environments: theory and data. Philosophical Transactions of the Royal Society of London 330:175–190.

Krohn, W. B., W. J. Zielinski, and R. B. Boone. 1997. Relations among fishers, snow, and martens in California: results from small-scale spatial comparisons. Pages 211–232 *in* G. Proulx, H. N. Bryant, and P. M. Woodard, editors. *Martes*: Taxonomy, Ecology, Techniques, and Management. The Provincial Museum of Alberta, Edmonton, Alberta, Canada.

Leopold, A. 1933. Game Management. Charles Scribner and Sons, New York, USA.

_____. 1949. A sand county almanac and sketches here and there. Oxford University Press, New York, USA.

Mitchell, M. S. 1997. Optimal home ranges and application to black bears. Dissertation, North Carolina State University, Raleigh, North Carolina, USA.

_____. and R. A. Powell. 2003. Linking fitness landscapes with the behavior and distribution of animals. Pages 93–124 *in* J. A. Bissonette and I. Storch, editors. Landscape ecology *and* resource management. Island Press, Washington, USA.

Morrison, M. L., B. G. Marcot, and R. W. Mannan. 1992. Wildlife-habitat relationships: concepts and applications. University of Wisconsin Press, Madison, Wisconsin, USA.

Peters, R. 1978. Communication, cognitive mapping, and strategy in wolves and hominids. Pages 95-108, *in* R. L. Hall and H. S. Sharp, editors. Wolf and man: evolution in parallel. Academic Press, New York, USA.

Powell, R. A. 1978. A comparison of fisher and weasel hunting behavior. Carnivore 1(1): 28-34.

_____. 1979. Ecological energetics and foraging strategies of the fisher (*Martes pennanti*). Journal of Animal Ecology 48:195-212.

_____. 1981. Fisher food requirements and hunting behavior. Pages 883-917 *in* J. A. Chapman and D. Pursley, editors. Proceedings of the First Worldwide Furbearer Conference. Worldwide Furbearer Conference, Inc., Baltimore, USA.

_____. 1993. The Fisher: Life History, Ecology and Behavior. Second edition. University of Minnesota Press. Minneapolis, Minnesota, USA.

_____. 1994a. Structure and spacing of *Martes* populations. Pages 101-121 *in* S. W. Buskirk, A. S. Harestad, M. G. Raphael, and R. A. Powell, editors. Martens, sables and fishers: biology and conservation. Cornell University Press, Ithaca, New York, USA.

_____. 1994b. Effects of scale on habitat selection and foraging behavior of fishers in winter. Journal of Mammalogy 75:349-356.

_____. 2000. Animal home ranges and territories and home range estimators. Pages 65–110 *in* L. Boitani and T. K. Fuller, editors. Research Techniques in Animal Ecology: Controversies and Consequences. Columbia University Press, New York, USA.

_____, J. W. Zimmerman, and D. E. Seaman. 1997a. Ecology and Behaviour of North American Black Bears: Home Ranges, Habitat and Social Organization. Chapman and Hall, London.

Powell, S. M., E. C. York, J. J. Scanlon, and T. K. Fuller. 1997b. Fisher maternal den sites in central New England. Pages 265–278 *in* G. Proulx, H. N. Bryant, and P. M. Woodard, editors. *Martes*: Taxonomy, Ecology, Techniques, and Management. The Provincial Museum of Alberta, Edmonton, Alberta, Canada.

Pyke, G. H. 1984. Optimal foraging theory: A critical review. Annual Review of Ecology and Systematics 15:523–575.

_____, H. R. Pulliam, and E. L. Charnov. 1977. Optimal foraging: A selective review of theory and tests. Quarterly Review of Biology 52:137–154.

Spencer, W. D. 1992. Space in the lives of vertebrates: On the ecology and psychology of space use. Dissertation. University of Arizona, Tucson, Arizona, USA.

Stearns, S. C. 1992. The evolution of life histories. Oxford University Press, Oxford, England.

Stephens, D. W., and J. R. Krebs. 1986. Foraging theory. Princeton University Press, Princeton, New Jersey, USA.

Sturtevant, B. R., and J. A. Bissonette. 1997. Stand structure and microtine abundance in Newfoundland: implications for marten. Pages 182–199 *in* G. Proulx, H. N. Bryant, and P. M. Woodard, editors. *Martes*: taxonomy, ecology, techniques, and management. The Provincial Museum of Alberta, Edmonton, Alberta, Canada.

Thomasma, L. E., T. Drummer, R. O. Peterson. 1991. Testing the habitat suitability index model for the fisher. Wildlife Society Bulletin 19:291-297.

_____, T. Drummer, and R. O. Peterson. 1994. Habitat selection by the fisher. Pages 316–325 *in* S. W. Buskirk, A. S. Harestad, M. G. Raphael and R. A. Powell, editors. Martens, sables and fishers: biology and conservation. Cornell University Press, Ithaca, New York, USA.

Thompson, I. D. and A. S. Harestad. 1994. Effects of logging on American martens, and models for habitat management. Pages 355–367 *in* S. W. Buskirk, A. S. Harestad, M. G. Raphael, and R. A. Powell, editors. Martens, sables and fishers: biology and conservation. Cornell University Press, Ithaca, New York, USA.

White, G. C. and R. A. Garrott. 1990. Analysis of wildlife radio-tracking data. Academic Press, New York, USA.

Wiens, J. A. 1989. Spatial scaling in ecology. Functional Ecology 3:385–397.

Wright, S. 1932. The roles of mutation, inbreeding, crossbreeding, and selection in evolution. Proceedings VI, International Congress of Genetics 1:356–366.

Wright, S. 1978. Evolution and the genetics of populations. Volume 4: Variability within and among natural populations. University of Chicago Press, Chicago, USA.

Zielinski, W. J., W. D. Spencer, and R. D. Barrett. 1983. Relationship between food habits and activity patterns of pine martens. Journal of Mammalogy 64:387–396.

Chapter 7

RELATIONSHIPS BETWEEN STONE MARTENS, GENETS AND CORK OAK WOODLANDS IN PORTUGAL

Margarida Santos-Reis, Maria João Santos, Sofia Lourenço, João Tiago Marques, Iris Pereira, and Bruno Pinto

Abstract: Although the stone marten (*Martes foina*) is widely distributed in Europe, little is known about its ecological and spatial requirements. Using radiotelemetry and food habits, we investigated the use of cork oak (*Quercus suber*) woodlands by an unharvested population of stone martens in SW Portugal. We also evaluated spatial ecology and food habits of the common genet, (*Genetta genetta*), and its niche overlap with stone martens. Home range size (minimum convex polygon) of martens (n = 5) and genets (n = 7) averaged 2.6 and 3.3 km^2, respectively. Variation in home range area during the year was associated with breeding activities. Tolerance among stone martens and genets was greater than between conspecifics of the same sex. Both stone martens and genets were nocturnal and used riparian vegetation and cultivated fields more than expected. However, genets also used oak woodland more than expected for foraging and resting. Stone martens and genets used similar rest sites. The seasonal variation in resource exploitation by stone martens (58 scats) and genets (75 scats) followed the same pattern. However, in summer and autumn, genets ate more crayfish (*Procambarus clarkii*) and mammals, while stone martens ate more fruits. Despite similarities in habitat utilization and food habits, core areas of both species were mutually exclusive, thus suggesting the existence of some present or past interspecific interaction. This study suggests that, except for resting, stone martens do not depend on cork oak trees and its products for existence. However, intensive destruction of understory cover and riparian corridors might alter use of cork oak trees by stone martens and their interspecific interactions with genets.

1. INTRODUCTION

Research on the genus *Martes* is greatly biased towards American martens (*Martes americana*), fishers (*Martes pennanti*), and pine martens (*Martes martes*) (e.g. Buskirk et al. 1994, Proulx et al. 1997, Griffith 2000) in spite of the wide distribution of other *Martes*, such as stone martens (*Martes foina*) (Proulx et al. 2003). One possible explanation is that stone marten is the single *Martes* species that does not behave as a habitat specialist, and is therefore assumed to be less vulnerable to habitat change. Another contributing factor is that stone martens are lightly harvested because of their lower pelt quality, and are apparently not endangered by human exploitation. In Portugal, harvesting or hunting stone martens is illegal (Law-Decret nº338/2001).

The current knowledge on European stone martens is scarce, dispersed over time and space, and mostly devoted to distribution and food habits (Clevenger 1994, Genovesi et al. 1996). Few studies have addressed population ecology (Lodé 1991, Lopez-Martin et al. 1991, Hermann 1994, Genovesi and Boitani 1997, Genovesi et al. 1997, Vadillo et al. 1997), but conclusions are still insufficient for a comprehensive understanding. Hence, investigating the basic characteristics and spatial requirements of natural populations of stone martens is still needed for successful management.

In Portugal, the stone marten is a common species with a wide distribution, inhabiting forests and human-altered environments. Its distribution is largely allopatric with the pine marten (*Martes martes*), a rare species with a fragmented and more restricted range (Santos-Reis 1983). However, until now, no study has specifically addressed the ecology of stone martens in Portugal. In 1997, we began a long-term project to analyze intraguild relations in a forest community of mesocarnivores. The selected study site was a cork oak (*Quercus suber*) woodland supporting a rich community of carnivores: red fox (*Vulpes vulpes*), least weasel (*Mustela nivalis*), western polecat (*Mustela putorius*), Eurasian badger (*Meles meles*), Eurasian otter (*Lutra lutra*), common genet (*Genetta genetta*), Egyptian mongoose (*Herpestes ichneumon*) and stone marten (one of the most abundant mesocarnivores) (Santos-Reis et al. 1999). Cork oak woodlands are the result of a complex process of co-evolution between natural ecosystems and humans, involving centuries of land use practices (Diáz et al. 1997, Blondel and Aronson 1999). Thirty-three percent of the worldwide cork oak range is found in Portugal and its survival is threatened due to forestry and agricultural activities (Pinto Correia 1993), reforestation with exotic species (e.g., *Eucalyptus* spp.), diseases associated with changes in edaphic conditions, and destructive management practices (e.g., understory removal).

Here, we investigated spatial ecology, habitat selection, and food use by stone martens in cork oak woodlands. We also evaluated niche overlap between stone martens and the common genet, a sympatric carnivore of similar size, equal arboreal aptitude, and similar energetic requirements (Livet and Roeder 1987). Our objectives for both species were to: (1) determine the size and distribution of home ranges and core areas; (2) quantify patterns of habitat selection; (3) examine temporal and spatial activity patterns, including resting and feeding, and (4) analyze interspecific interactions. Based on the available literature on coexistence principles and resources partitioning (e.g., Powell and Zielinski 1983, Maddock and Perrin 1993, Dayan and Simberloff 1996), spacing strategies of solitary carnivores (e.g., Powell 1979, Sandell 1989), and the biology and ecology of stone martens and genets (Livet and Roeder 1987, Libois and Waechter 1991, Clevenger 1994), we hypothesized that: (1) home ranges and core areas would differ by sex and/or season, and their overlap would be minimal among neighboring individuals but maximal among mated pairs; (2) habitat selection would not differ by sex, but would differ by species; (3) intrasexual bonds of mature individuals would be restricted to the mating season; and (4) activity patterns, rest site selection, and food preference would be species-specific rather than sex-specific.

2. STUDY AREA

This research was conducted near the south-western coast of Portugal in a 35 km^2 area (38° 05'N - 38° 08'N and 8° 29'W - 8° 38'W), located on the Grândola hills. The study area included the "Herdade da Ribeira Abaixo" (HRA), a 220-ha field station of the Environmental Biology Center (University of Lisbon), and surrounding private lands characterized by small holdings connected by dirt roads. Topography was gently to moderately rolling, with 0–15% slope and elevations varying from 150 to 270 m above sea level. Climate is Mediterranean under Atlantic influence with very dry and hot summers and cold and rainy winters. Mean annual temperature was 15.6 °C, and mean annual precipitation was 500 mm/year, occasionally reaching 800 mm (Correia and Santos-Reis 1999).

The landscape was dominated (70 to 80%) by cork oak woodland, and some areas had a developed understory of sage-leaved cistus (*Cistus salvifolius*). Riparian vegetation was composed of black poplar (*Populus nigra*), common alder (*Alnus glutinosa*), willow (*Salix atrocinerea*), ash (*Fraxinus angustifolia*), and blackberry (*Rubus ulmifolius*) thickets. The study area also included a few patches of exotic tree species (*Eucalyptus globulus* and *Pinus pinaster*), pastures, and small cultivated fields (orchards, vegetable-gardens and olive-yards) associated with scattered and mostly abandoned farm houses.

Human population density in the area was low, but anthropogenic influence on the landscape was high due to activities such as cork extraction in summer, livestock raising year round, and hunting (wild rabbits [*Oryctolagus cuniculus*], partridges [*Alectoris rufa*] and, more recently, wild boars [*Sus scrofa*]) from August to February. Although trapping for fur and/or predator control was illegal in the area, some poaching occurred (Santos-Reis, personal observation).

3. METHODS

3.1 Capture and Radio-tracking

Livetrapping, radiocollaring, and radio-tracking occurred in 1997 and 1998. We subdivided a 15-km^2 sub-area into 1-km^2 squares; in each square, we placed ≥2 box traps (Tomahawk Live Trap Co., Wisconsin, USA) in sites selected according to cover density and accessibility. To increase trapping success, we also used soft-catch foot-hold traps (#2 Victor® Long-Spring–Woodstream Corporation, Lititz, Pennsylvania, USA). We set sardine-baited traps for an average of 10 days/mo (March–August in 1997 and February–July in 1998), and visited them daily. Animals were immobilized with an intramuscular injection of 0.1 ml/kg HCl ketamine (Imalgene 1000, Rhône Mérieux, Lyon, France). We assessed weight, sex, condition (i.e., signs of pregnancy, lactation or estrus) and took body measurements. Relative age was estimated on the basis of tooth development and characteristics. Three age classes were considered: adults (full dentition with teeth showing some degree of wearing), juveniles (full dentition with teeth still growing and with sharp cusps), and cubs (with 1 or more milk teeth). Individuals were permanently identified with ear notches, and several adults were fitted with 35–40 g mortality-sensitive radio-transmitters (Telonics Inc., Mesa, Arizona, USA). We tested for differences in sex-ratio using a chi-square test.

Animals were located within a 2 km range using a RA-5A omnidirectional antenna (Telonics) fixed to the top of a 4-wheel-drive vehicle, and linked to a TR2 receiver with a TS1 scanner (Telonics). Accurate locations were obtained by walking with a hand-held RA-2A "H" antenna (Telonics) and a portable TRX – 1000S receiver (Wildlife Materials, Carbondale, Illinois, USA). We used ground triangulation only in places where access was difficult. To reduce error, azimuths were recorded within a 15 min interval, and we considered triangulation locations only for values of crossbearings between 60° and 120°. We investigated animal locations with 2 radio-tracking techniques: (1) systematic point locations, where the daily location was recorded for a period of 15 days/mo, and (2) sequential locations (focal runs), were recorded once per month on a 24 h basis, or within the activity period with at least 1 location/h.

3.2 Home Ranges

Homing locations were geo-referenced using a Global Positioning System (Garmin®, Model 75, Olathe, Kansas, USA) and triangulation locations were converted to UTM coordinates (Universal Transverse Mercator) using program TRACKER (version 1.1—Camponotus AB and Radio Location Systems 1994, Solna, Sweden). We analyzed seasonal (summer: 21 June–20 September and autumn: 21 September–20 December), and breeding patterns (mating: 15 June–14 August; diapause I: 15 August–14 October; and diapause II: 15 October–15 December) following Genovesi and Boitani's (1997) criteria. Genets do not have delayed implantation and an extended breeding period. On the basis of Aymerich (1982) and our capture results, we identified 2 periods: breeding (15 June to 14 August), and non-breeding (15 August to 15 December). Seasonal patterns are dependent on resource availability, and breeding patterns reflect different social behavior.

Home range, core areas, and overlap computations were performed with TRACKER software using the minimum convex polygon (MCP) (Mohr 1947 *in* White and Garrot 1990), and the adaptive kernel estimator (AK) (Worton 1989). To estimate a non-biased sample size, we determined at what point home range size reached an asymptote (Stickel 1954 *in* Harris et al. 1990). For home range calculations, we used 100% of the locations for the MCP (Genovesi and Boitani 1997), and 95% of the locations for the AK using the Epanechnickov adaptative model (Worton 1989). Core areas were defined using the 50% AK (Harris et al. 1990). We used a coefficient of variation of 1.8 and a 130 m grid size. We tested for autocorrelation using Schoener's index at the 0.25 significance level (Schoener 1981 *in* Swihart and Slade 1985). We tested for differences in home-range sizes using a Mann-Whitney *U*-test (Siegel and Castellan 1988). Data were combined across years when calculating average home range areas.

3.3 Habitat Selection

Five habitat types were considered according to land use pattern in the study area: cork oak woodland with pasture, cork oak woodland with shrubs, pasture, riparian vegetation, and cultivated fields (olive-yards, orchards, and vegetable-gardens). We used Chi-square goodness of fit test to evaluate whether use was distributed proportional to availability within each MCP home range (Sokal and Rohlf 1995). Whenever H_0 was rejected, the degree of selection was estimated using the *D* index that ranges from -1 (avoidance) to +1 (selection) (Jacobs 1974).

3.4 Activity Pattern

Homing and triangulation locations that referred to the onset and offset of activity or to the hourly diel intervals were expressed in terms of Greenwich Meridian Time. Values presented for a given hour (e.g., 21 h) represent the activity observed in the next diel period (e.g., 21:00–21:59).

Circadian time budgets were quantified using the focal runs when the contact with the individual was not lost for more than 2 h during the active period (Stahl 1986). Activity pattern was estimated using the percentage of active locations (Palomares and Delibes 1991a, Ciucci et al. 1997). Tests for differences among individuals were performed using a Kruskal-Wallis (*H*) test; correlations were obtained with Spearman's coefficient (r_s) (Zar 1984).

3.5 Diurnal Resting Sites

Homing locations allowed us to identify rest sites that were described in 1998 on the basis of structure (e.g., tree, bush, pile of cork). Re-use rates (no. locations in rest sites / total no. of rest sites) and intra-sexual and intra-specific partioning of rest sites (frequency of locations in a given rest site per individual of different sex or species) were also calculated .

3.6 Food Habits

In 1997, we collected fecal samples monthly from stone martens and genets along 20 km of line transects and from tree branches. Transects were selected along dirt roads that crossed the dominant land use type in proportion to their availability, and with a distance that maximized the number of home ranges that were intersected by transects. Scats were dried and sieved under water to separate food remains (e.g., hairs, feathers, insects, seeds). These were identified to the lowest possible taxon for each food category (e.g., species for mammals, order for birds, and family for insects), using identification keys (Debrot et al. 1982, Brom 1986), reference collections, or expert determination. Remains with low or no nutritive value (e.g., vegetable matter other than fruits or berries) were not included in the analysis because they were considered to be ingested incidentally with other food (Libois and Waechter 1991). For each item or food category, results were expressed as percent occurrence and percent biomass. Biomass was evaluated using published digestibility coefficients (e.g., Palomares and Delibes 1990) or by multiplying the minimum number of identified food items by their mean weight.

Niche breath was estimated using Levins standardized index (Ludwig and Reynolds 1988). For intraspecific comparisons, the obtained value was converted to the Shannon diversity index (H'_I) and tested with the Hutcheson *t* test (Zar 1984). We used Pianka's index to calculate niche overlap between seasons and species (Ludwig and Reynolds 1988).

According to available literature (e.g., Livet and Roeder 1987, Libois and Waechter 1991), we live-trapped in 3 main habitats (cork oak woodland with pastures, cork oak woodland with shrubs, and riparian vegetation) to examine the relationship between food habits and the abundance of small mammals, the most common prey for both species. In 1997, we established a trapline of 100 Sherman live-traps (H.B. Sherman Traps, Inc., Tallahassee, Florida, USA). In each habitat, traps were 10 m equidistant, and baited with a mixture of sardines in olive oil and oats. Trapping was conducted for 5 consecutive nights per season. A mark (fur-shaving) was used to identify each captured individual, and their sex and age were recorded before release. We used capture success and trapping effort to assess relative abundance (Pounds 1981).

All statistical tests were performed at the significance level of $\alpha = 0.05$, unless specified otherwise.

4. RESULTS

4.1 Capture Success

Twenty-four stone martens were captured in 2,614 trapnights (TN). Only 2 juveniles were captured; both were males. Sex ratio was not significantly different from 1:1 (13 M: 11 F; $\chi^2 = 0.001$, $P > 0.05$). Five adults (3 M, 2 F) were equipped with radiocollars and were radio-tracked an average of 6 months in summer and autumn. All individuals were monitored until radio failure, except for 1 adult male (Mf_{17}) that was found dead after 144 days of radio-tracking (Table 7.1).

During the same trapping period, 46 other carnivores were captured, including 20 genets. Sex distribution of genets was nearly even (7 F: 8 M, $\chi^2 = 0.001$, $P > 0.05$), and most individuals were adults ($n = 15$). Seven (4 M, 3 F) adult genets were radiocollared and monitored simultaneously with stone martens (Table 7.1). We lost the radio signal of 1 female (Gg_3) after one month of monitoring. One male (Gg_{12}) behaved abnormally after release, and was recaptured for veterinary treatment and removal of the collar. Another genet male (Gg_2) was monitored from April to December 1997, recaptured and re-equipped in July 1998, but the radio failed soon after release.

Table 7.1. Number of telemetry fixes, time to independence and associated data for radiocollared stone martens and genets in a cork oak woodland of the Grândola Hills in southwestern Portugal, 1997–1998.

Individuals[a]	Sex	Age	Tracking period	Focal runs[b]	Fixes[c]	Independence (h)[d]
Mf_1	F	Adult	05/24/97–10/28/97	4	60	5:30
Mf_2	M	Adult	07/22/97–11/24/97	6	110	5:00
Mf_{17}	M	Adult	06/07/98–10/12/98	3	70	5:00
Mf_{18}	M	Adult	06/09/98–11/15/98	8	170	4:30
Mf_{19}	F	Adult	06/11/98–11/15/98	7	170	4:30
Gg_1	M	Adult	04/24/97–12/14/97	8	110	6:30
Gg_2	M	Juvenile	05/02/97–12/17/97	9	220	6:30
Gg_{2a}	M	Adult	07/12/98			
Gg_3	F	Adult	05/05/97–05/31/97			
Gg_{12}	M	Adult	05/03/98			
Gg_{15}	F	Adult	06/02/98–12/01/98	6	180	5:30
Gg_{16}	M	Adult	06/05/98–11/15/98	7	150	5:00
Gg_{20}	F	Adult	07/10/98–11/15/98	8	180	5:00

[a]Mf = Stone marten; Gg = Common genet; a = recapture of radiocollared animal.
[b]Number of sequential locations (see methods section).
[c]Number of fixes to attain the asymptote. (Total number of fixes see Table 7.2.)
[d]Time period for the temporal independence of sequential fixes.

4.2 Home Ranges

4.2.1 Stone Martens

Home range areas of stone martens averaged 2.6 km^2, (100% MCP); estimates were similar (2.5 km^2) to those obtained using the 95% AK technique (Table 7.2). The large MCP value of female Mf_{19} resulted from an expansion of its home range following the death of a neighbor Mf_{17}. Core areas were on average one-fifth smaller (0.5 km^2) than home-ranges, and were mutually exclusive (Fig. 7.1). Each individual (except Mf_1) had ≥2 core areas; 1 was used mostly for foraging and the others for diurnal resting.

Spatial segregation was suggested between 2 males (Mf_{17} and Mf_{18}), whose home ranges overlapped only slightly (Fig. 7.2). The degree of overlap of males with females ranged from 13% (Mf_2/Mf_1 and Mf_{18}/Mf_{19}) to 33% (Mf_{17}/Mf_{19}). Because all monitored individuals were sexually mature, we assumed that Mf_{17} and Mf_{19} constituted a mating pair. This assumption was supported by a greater extent of home-range overlap during the mating season (0.48 km^2) than during the following diapause (0.14 km^2). During the mating period, the overlap included part of the female's core area and diurnal rest sites. The home range of

Table 7.2. Home range size and seasonal variation of stone martens and genets in a cork oak woodland of the Grândola Hills in southwestern Portugal, 1997–1998.

		Home Range					Seasonal Home Ranges					
		MCP (km^2)		Kernel (km^2)			Spring		Summer		Autumn	
Individual[a]	Sex	100%	n^b	95%	50%	n^b	MCP	n^b	MCP	n^b	MCP	n^b
Mf_1	F	1.65	99	2.37	0.42	17			0.99	65	0.35	32
Mf_2	M	2.48	180	1.80	0.26	28			1.03	101	2.24	79
Mf_{17}	M	2.18	134	2.21	0.92	67			2.15	81	0.95	41
Mf_{18}	M	3.67	210	3.53	1.00	83			2.20	105	2.01	109
Mf_{19}	F	2.87	239	2.58	0.11	91			1.29	113	2.41	124
Gg_1	M	3.46	283	3.03	0.59	43	0.80	63	1.89	125	3.04	96
Gg_2	M	1.90	319	2.60	0.39	36	0.86	94	1.39	109	1.38	107
Gg_{15}	F	3.69	216	3.49	0.19	87			2.26	89	3.21	102
Gg_{16}	M	4.67	184	4.52	0.42	83			4.31	90	3.57	92
Gg_{20}	F	2.72	230	3.42	0.09	80			0.85	102	2.56	128

[a]Mf = Stone marten; Gg = Common genet
[b]Number of fixes.

Figure 7.1. Home ranges and core areas (95% and 50% adaptive kernel) of stone martens (Mf) in a cork oak woodland of the Grândola Hills in southwestern Portugal (numbers correspond to UTM coordinates).

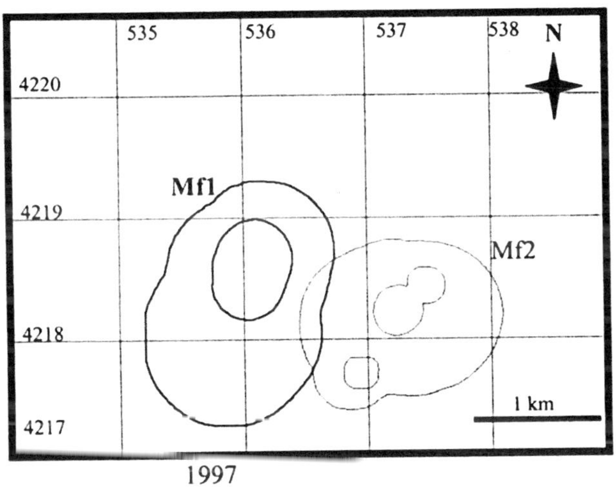

male Mf_{18} also overlapped with that of female Mf_{19}, and both individuals occasionally shared a diurnal rest site.

Seasonal variation in average MCP home ranges were not significant, either between seasons (summer = 1.53 km^2, autumn = 1.59 km^2) (Table 7.2) or breeding periods (mating = 1.14 km^2, diapause I = 1.03 km^2, diapause II = 1.44 km^2) (Table 7.3).

Figure 7.2. Home ranges and core areas (95% and 50% adaptive kernel) of stone martens (Mf) in a cork oak woodland of the Grândola Hills in southwestern Portugal (numbers correspond to UTM coordinates).

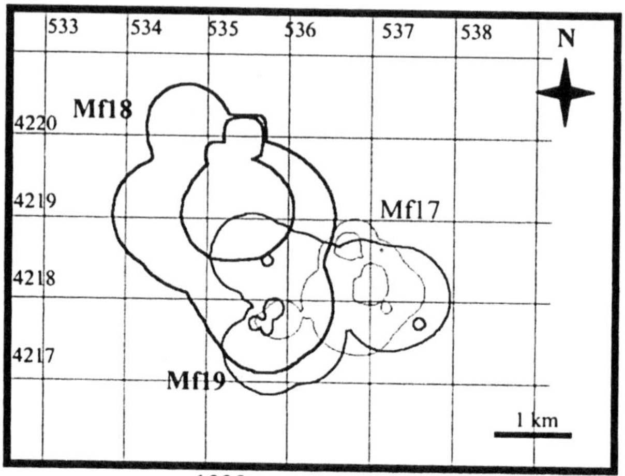

4.2.2 Genets

Mean home range areas of genets were larger (MCP = 3.29 km^2, 95% AK = 3.41 km^2) than for stone martens but core areas of genets averaged only 0.34 km^2 (Table 7.2). One-third of a male genet's (Gg_1) home-range overlapped with approximately half of another male's (Gg_2) home range, but no overlap was observed between females Gg_{15} and Gg_{20}. Overlap between the home ranges of male Gg_{16} and female Gg_{15} was high (94% of the female range and 75% of that of the male), suggesting a mated pair (Fig. 7.3).

Average home range areas (MCP) of genets were not significantly different between seasons (summer = 2.14 km^2, autumn = 2.75 km^2) (Table 7.2). However, the average home range during the non-breeding period (2.56 km^2) was 56% larger than during the mating season (1.44 km^2) (Table 7.3).

Core areas did not overlap among individuals, except for male Gg_{16} and female Gg_{15}. The male had 4 core areas inside its range, and the female had 5 (Fig. 7.3). The main core areas of both individuals overlapped >50%; the shared areas were used for both foraging and resting. The male and the female shared the same rest site for at least 3 consecutive days, from 17 October to 6 November.

Interspecific overlap was extensive among MCP ranges, but core areas were usually exclusive between species. No statistical differences were observed in home range area between species ($Z = 1.149$, $P = 0.251$) or among sexes ($Z = 0.213$, $P = 0.831$).

Table 7.3. Home range variation according to breeding season of stone martens and genets in a cork oak woodland of the Grândola Hills in southwestern Portugal, 1997–1998.

	Stone Marten Breeding Season							Genet Breeding Season			
	Mating		Diapause I		Diapause II			Breeding Season		Non-breeding Season	
Individuals	MCP[a]	n[b]	MCP[a]	n[b]	MCP[a]	n[b]	Individuals	MCP[a]	n[b]	MCP[a]	n[b]
Mf_1	0.300	19	0.714	46	0.349	31	Gg1	1.007	56	3.108	174
Mf_2	0.564	46	0.878	55	2.239	79	Gg2	1.220	36	1.637	194
Mf_{17}	2.033	48	1.237	75			Gg15	2.370	76	3.232	70
Mf_{18}	2.020	70	1.060	88	1.347	50	Gg16	4.118	61	3.150	63
Mf_{19}	0.795	53	1.261	82	1.817	86	Gg20	0.290	32	1.654	108

[a]In km^2.
[b]Number of fixes.

Figure 7.3. Home ranges and core areas (95% and 50% adaptive kernel) of genets (Gg) in a cork oak woodland of the Grândola Hills in southwestern Portugal (the arrow indicates a shared core area).

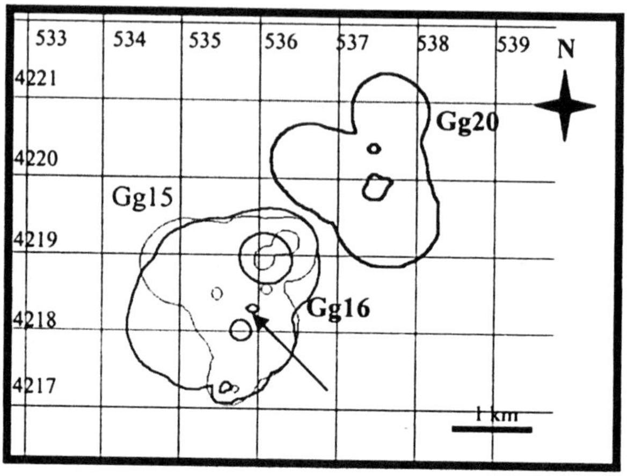

1998

4.3 Habitat Selection of Stone Martens and Genets

Stone martens used riparian vegetation and cultivated fields greater than expected within their home ranges (Table 7.4). These habitats were used most frequently by foraging animals during resting bouts both at night and during the day. However, there were individual differences by type of activity (Table 7.5). Some individuals specialized on one habitat for all activity types (e.g., Mf_1, Mf_2), while others showed a greater use of different habitats for foraging versus resting (e.g., Mf_{17}, Mf_{18}, Mf_{19}). Oakland, either with or without shrubs, was consistently used disproportionately less than its availability (Table 7.4). We observed a similar pattern for genets. They also used riparian vegetation and cultivated fields greater than expected. However, some genets also used oak woodlands more than expected for foraging or resting (Tables 7.4).

4.4 Activity

4.4.1 Stone Martens

Stone martens were exclusively nocturnal. On average, activity began 49 min (SD = 59.5 min, n = 30) after sunset and ended 41 min (SD = 40.8 min, n = 30) before sunrise. Some variability was observed among individuals, with male Mf_{18} showing the most striking differences from the norm (Table 7.5); his activity started after the other martens and ended much sooner. This individual

Table 7.4. Proportion of habitats[a], chi-square value and *P*-value within the MCP home range of stone martens and genets in a cork oak woodland of the Grândola Hills in southwestern Portugal, 1997–1998 (# Loc = number of locations).

	WS (# Loc)	WP (# Loc)	Rp (# Loc)	CF (# Loc)	P (# Loc)	Chi-square	*P*-value
Mf_1[b]	0.560 (30)	0.360 (8)	0.020 (55)	0.060 (5)	*******	1436.86	<0.001
Mf_2	0.778 (121)	0.145 (2)	0.028 (55)	0.048 (2)	*******	530.18	<0.001
Mf_{17}	0.570 (71)	0.360 (7)	0.050 (29)	0.020 (14)	0,000 (0)	201.60	<0.001
Mf_{18}	0.180 (11)	0.580 (146)	0.070 (13)	0.050 (71)	0.120 (47)	264.55	<0.001
Mf_{19}	0.450 (126)	0.460 (51)	0.070 (100)	0.020 (13)	0,000 (2)	429.12	<0.001
Gg_1	0.501 (130)	0.362 (22)	0.058 (119)	0.079 (11)	*******	681.10	<0.001
Gg_2	0.736 (89)	0.131 (62)	0.074 (143)	0.059 (12)	*******	722.39	<0.001
Gg_{15}	0.500 (88)	0.420 (41)	0.060 (61)	0.000 (20)	0.020 (2)	347.00	<0.001
Gg_{16}	0.510 (45)	0.400 (89)	0.050 (21)	0.020 (21)	0.020 (23)	115.00	<0.001
Gg_{20}	0.560 (97)	0.100 (52)	0.080 (105)	0.150 (26)	0.110 (1)	403.00	<0.001

[a]WS = cork oak woodland with shrubland, WP = cork oak woodland with pasture, Rp = riparian vegetation, CF = cultivated fields, P = pastures.
[b]Mf = stone marten, Gg = common genet.

was the most synanthropic (depending heavily on humans and their immediate surroundings; Delibes 1983) and occupied a home range that included a small village.

During summer and autumn, stone martens spent an average of 60% of their time resting during daytime. Locomotory and foraging activities accounted for about 30% of the circadian period, with short nightly resting bouts (<1 hr) accounting for less than 10% (Table 7.5). From summer to autumn, with decreasing daylight hours, individuals became more active (31.5% *vs.* 35.8%), increased night resting (4.0% *vs.* 9.1%), and reduced day resting (64.8% *vs.* 55.1%). The amount of activity did not differ significantly among individuals ($H = 5.761$, $P = 0.22$) and we observed no correlation among seasons ($P > 0.05$).

4.4.2 Genets

Genets were also strictly nocturnal, but they were more active than stone martens. On average, they left rest sites 25 min (SD = 37 min, $n = 43$) after sunset and ended 51 min (SD = 52 min, $n = 41$) before sunrise. Differences were observed at the individual level; one animal (Gg_1) typically started activity just before sunset (1–25 min, $n = 10$) and another (Gg_2) soon after sunset (8–61 min, $n = 9$).

The pattern of circadian activity of genets was generally similar to stone martens (Table 7.5). More than half of the 24 h period was spent in a continuous period of resting in a safe refuge during daytime, and almost all nighttime hours were spent moving, marking, and foraging. Genets were active longer

Table 7.5. Circadian activity pattern of stone martens and genets in a cork oak woodland of the Grândola Hills in southwestern Portugal, 1997–1998.

Individual[a]	Activity (%)						Nocturnal inactivity (%)						Diurnal inactivity (%)					
	Summer			Autumn			Summer			Autumn			Summer			Autumn		
	$\bar{x}$	SD	n[b]	$\bar{x}$	SD	n[b]	$\bar{x}$	SD	n[b]	$\bar{x}$	SD	n[b]	$\bar{x}$	SD	n[b]	$\bar{x}$	SD	n[b]
Mf_1	25.0	4.9	2	40.0		1	5.0	3.5	2	8.0		1	71.0	8.4	2	52.0		1
Mf_2	32.3	2.2	2	39.3	1.7	2	1.4	1.2	2	3.5	1.9	2	66.4	3.2	2	57.3	0.2	2
Mf_{17}	36.3	3.2	3	39.2		1	1.4	0.6	3	6.3		1	62.3	3.8	3	54.5		1
Mf_{18}	27.6	4.1	4	19.6	5.5	4	8.1	4.0	4	18.5	6.1	4	64.3	4.0	4	61.9	6.9	4
Mf_{19}	36.2	5.1	4	40.7	2.6	4	3.9	2.0	4	9.4	3.1	4	59.9	4.7	4	50.0	2.0	4
Gg_1	39.3	3.9	3	39.1	6.1	3	0.0	0.0	3	4.6	6.2	3	60.8	4.0	3	56.2	3.1	3
Gg_2	34.7	4.2	4	41.6	5.7	3	2.1	4.3	4	2.2	3.9	3	63.1	0.9	4	56.2	2.7	3
Gg_{15}	29.8	4.7	3	38.8	3.3	3	7.6	3.7	3	10.1	3.3	3	62.9	1.9	3	51.2	1.3	3
Gg_{16}	31.7	3.4	4	34.8	8.0	3	4.9	1.0	4	7.6	4.9	3	63.3	3.8	4	57.6	5.0	3
Gg_{20}	33.3	5.5	4	41.0	2.8	4	6.0	2.6	4	7.6	2.5	4	60.7	5.0	4	51.4	1.6	4

[a] Mf = stone marten, Gg = common genet.
[b]Number of focal runs.

Table 7.6. Number of locations (# loc)[a], different diurnal resting sites (Nn)[b], and re-use rates of stone martens and genets in a cork oak woodland of the Grândola Hills in southwestern Portugal, 1997–1998.

Individual[c]	Tree # loc (Nn)	Shrub # loc (Nn)	Other # loc (Nn)	Total # loc (Nn)	Re-use rate
Summer					
Mf_{17}	15 (10)	19 (12)	5 (1)	39 (23)	1.70
Mf_{18}	7 (5)	4 (4)	25 (1)	36 (10)	3.60
Mf_{19}	8 (5)	31 (14)	0 (0)	39 (19)	2.05
Mf_{tot}	30 (20)	54 (30)	30 (2)	114 (52)	
Total (%)	26.3(38.5)	47.4(57.7)	26.3(3.8)	100	
Gg_{15}	28 (21)	11 (5)	2 (2)	41 (28)	1.46
Gg_{16}	34 (20)	3 (3)	0 (0)	37 (23)	1.61
Gg_{20}	9 (2)	22 (5)	0 (0)	31 (7)	4.43
Gg_{tot}	71 (43)	36 (13)	2 (2)	109 (58)	
Total (%)	65.1 (74.1)	33.0 (22.5)	1.9 (3.4)	100	
Autumn					
Mf_{17}	12 (7)	2 (2)	1 (1)	15 (10)	1.50
Mf_{18}	0 (0)	0 (0)	35 (3)	35 (3)	11.60
Mf_{19}	10 (6)	22 (8)	3 (1)	35 (15)	2.30
Mf_{tot}	22 (13)	24 (10)	39 (5)	85 (28)	
Total (%)	25.9(46.4)	28.2(35.7)	45.9(17.9)	100	
Gg_{15}	7 (7)	28 (8)	1 (1)	36 (16)	2.25
Gg_{16}	25 (21)	9 (1)	3 (3)	37 (25)	1.48
Gg_{20}	7 (5)	29 (13)	0 (0)	36 (18)	2.00
Gg_{tot}	39 (33)	66 (22)	4 (4)	109 (59)	
Total (%)	35.8 (55.9)	60.6 (37.3)	3.6 (6.8)	100	
Total					
Mf_{17}	27 (14)	21 (13)	6 (1)	54 (28)	1.93
Mf_{18}	7 (5)	4 (4)	60 (3)	71 (12)	5.92
Mf_{19}	18 (11)	53 (19)	3 (1)	74 (31)	2.39
Mf_{tot}	52 (30)	78 (36)	69 (5)	199 (71)	
Total (%)	26.1(42.3)	39.2(50.7)	34.7(7.0)	100	
Gg_{15}	35 (26)	39 (10)	3 (2)	77 (38)	2.03
Gg_{16}	59 (37)	12 (3)	3 (3)	74 (43)	1.72
Gg_{20}	16 (7)	51 (16)	0 (0)	67 (23)	2.91
Gg_{tot}	110 (71)	102 (29)	6 (5)	218 (104)	
Total (%)	50.5 (68.3)	46.8 (27.9)	2.7 (4.8)	100	

[a]# loc = number of locations in a class of rest site.
[b]Nn = number of different rest sites within a class.
[c]Mf = stone martens, Gg = common genets.
NOTE: Because of the intra and inter-specific partitioning of rest sites, the sum of the rest sites used by each species is slightly higher than those values referred to in the text.

(summer: 33.8%, autumn: 39.1%) than were stone martens (summer: 31.5%, autumn: 35.8%). Activity of genets increased during autumn (Table 7.5).

Genets did not exhibit significant individual variability in their activity index ($H = 3.59$, $P = 0.46$), and no correlation in activity patterns among seasons was observed ($P > 0.05$).

4.5 Diurnal Rest Sites

4.5.1 Stone Martens

The 3 stone martens monitored from June to November 1998 were located 199 times while resting during the day: 39.2% of the time within shrubs, 26.1% in tree cavities, and 34.7% in other structures (Table 7.6). There was substantial individual variability. Male Mf_{18}, with his core area located in a village, was located 86.9% of the time in human-made structures near or in the village during summer and autumn. The 2 individuals that used natural features of the landscape had 57.8% of rest sites in shrubs, 35.2% in trees, and 7% in other structures. The preference for shrubs was consistent during both seasons (Table 7.6).

Martens were located in 63 different rest sites. With the exception of Mf_{18} who used only 12 sites, the number of rest sites per individual increased with the number of locations (Table 7.6). Of the rest sites used by stone martens, 44.4% (28 of 63) were in oak trees, 47.6% (30 of 63) were within shrubs, and only 8% (5 of 63) were in other structures. Stone martens used cavities in old trees, sometimes with the entrance at ground level; tree branches were never selected for diurnal resting. Riparian shelterbelts were frequently used. These were thick patches of bushy vegetation, varying in length from 12 to 500 m and composed of blackberry shrubs and, less commonly, by a heterogeneous mixture of blackberries, strawberry trees (*Arbutus unedo*), heathers (*Erica* spp.), and creeping plants around poplars, alders, willows, or ashes.

Stone martens used most rest sites only once (trees = 62.1%, shrubs = 51.7%), and the average number of rest sites per individual was 23.7. Re-use rates were low. The most notable exception was male Mf_{18}, especially in autumn when he used only 3 different rest sites (Table 7.6). The maximum number of times that a rest site was used by stone martens was 17 for the same bush, and 9 for the same tree. No rest site was ever used by 2 individuals at the same time, and the proportion of allopatrically shared sites was extremely low ($n = 8$). All sites shared allopatrically involved female Mf_{19}. Six times female Mf_{19} shared a rest site with male Mf_{17}, and 2 times she shared with male Mf_{18}.

4.5.2 Genets

The 3 genets monitored from June to November 1998 were located 218 times while resting during the day. Locations occurred in tree cavities (50.5%), shrubs (46.8%), and in other structures (2.8%) (Table 7.6). In summer, genets frequently used trees (65.1%); in autumn, they were commonly found sleeping under shrubs (61.0%).

Genets used a greater number of rest sites than stone martens. We identified 91 sites, of which 64% were in trees, 31% in shrubs; only 5% were in other structures (Table 7.6). Characteristics of selected trees and shrubs were identical to those described for stone martens.

The number of different rest sites increased with a greater number of locations suggesting a time interaction. We estimated an average of 34.7 sites/individual. Re-use rates of the same sites were lower for genets than for stone martens (Table 7.6). The majority of sites were only used once by any individual (trees = 62.1%, shrubs = 46.4%); however, the maximum number of times that the same rest site was used by several animals was 15 times for the same bush and 13 times for the same tree cavity. The highest re-use rate was by female Gg_{20} during summer, when only 7 nearby rest sites were used ($\bar{x}$ distance = 138 m; SD = 114 m).

No pair of individuals was ever found sleeping in the same rest site, but 13 sites (10 trees, 2 shrubs and 1 tree stump) were shared allopatrically between male Gg_{16} and female Gg_{15}. One of the shrubs, the most common rest site of male Gg_{16} during autumn, was shared allopatrically 14 times with female Gg_{15} between 9 October and 15 November, on alternate days. Six of 148 sites (3 trees and 3 shrubs) were shared between the 3 stone martens and female genet Gg_{15}.

4.6 Food Habits

4.6.1 Stone Martens

Diet was described using 58 scats collected from January to November 1997. Insects were the staple food representing 81.8% of prey occurrences. Fruits and berries represented 12.3%, and all other food items (mammals, birds and other invertebrates) represented ≤2%. When we converted frequency of occurrence to biomass, fruits became the most important (57.9%), followed by mammals (21.6%), and insects (12.8%). Birds increased in importance (4.3%), and other invertebrates continued to represent occasional prey (<2%). Pears (*Pyrus* sp.) accounted for almost all the fruit intake by stone martens. The wood mouse (*Apodemus sylvaticus*) and the Algerian mouse (*Mus spretus*) were the most frequent mammals eaten. Insects eaten were mostly beetles and crickets; birds were all passerines. Acorns were never detected in scats.

Based on percent occurrence, insects were the most frequently consumed food year round, representing 93% of total food items consumed during winter and spring, and decreasing to 72.4% of food items consumed during summer and autumn (Fig. 7.4). Mammals were the second most important prey during colder months (4.1%). In warmer months, fruits accounted for 22.6% of the occurrences. Biomass values for insects declined to 30.6% in winter and spring and to 11.3% in summer and autumn. Mammals (33.6% biomass) and birds (34.3% biomass) were major prey items from January to June, but were replaced by fruits (83.6%) during summer-autumn. Other invertebrates were not prevalent in the diet of stone martens.

4.6.2 Genets

Diet of the common genet was described using 75 scats. Based on percent occurrence, insects (75.5%) were the most frequent item, followed by fruits (11.1%), mammals (3.7%), other invertebrates (2.6% to 3.4%), and birds (0.8%). On the basis of biomass, insects decreased in importance (11.4%). Three resources were of similar importance: mammals (29.9%), fruits (24.6%), and crayfish (*Procambarus clarkii*) (22.3%). Birds accounted for 9.3% of the biomass, and invertebrates <2%.

Compared to stone martens, genets consumed a greater number of greater-white-toothed-shrews (*Crocidura russula*), a wider variety of fruits and seeds (figs, *Ficus carica*; olives, *Olea europaea*; wine grapes, *Vitis vinifera*; blackberries, *Rubus* spp.; pears), and crayfish. Acorns were only detected in 2 genet scats.

The seasonal variation in resource exploitation by genets followed the pattern reported for stone martens, and major differences were apparent only in biomass (Fig. 7.4). In summer and autumn, genets ate more crayfish (29.6%) and mammals (16.7%), while stone martens ate more fruits.

4.7 Trophic Niches

Niche breath values ($B_{Mf} = 1.949$, $B_{Gg} = 2.533$) indicated that the diet of stone martens and genets were not significantly different in diversity ($t_H = 8.1$, $P > 0.05$). Further, niche overlap was almost equal to 1 (winter and spring = 0.998, summer and autumn = 0.991) when using percent occurrence of food items. These values were likely biased because of the greater number of insect species. Although still high, niche breath values were lower when calculated on the basis of biomass values, and lower in the dry season (winter–spring = 0.795, summer–autumn = 0.727).

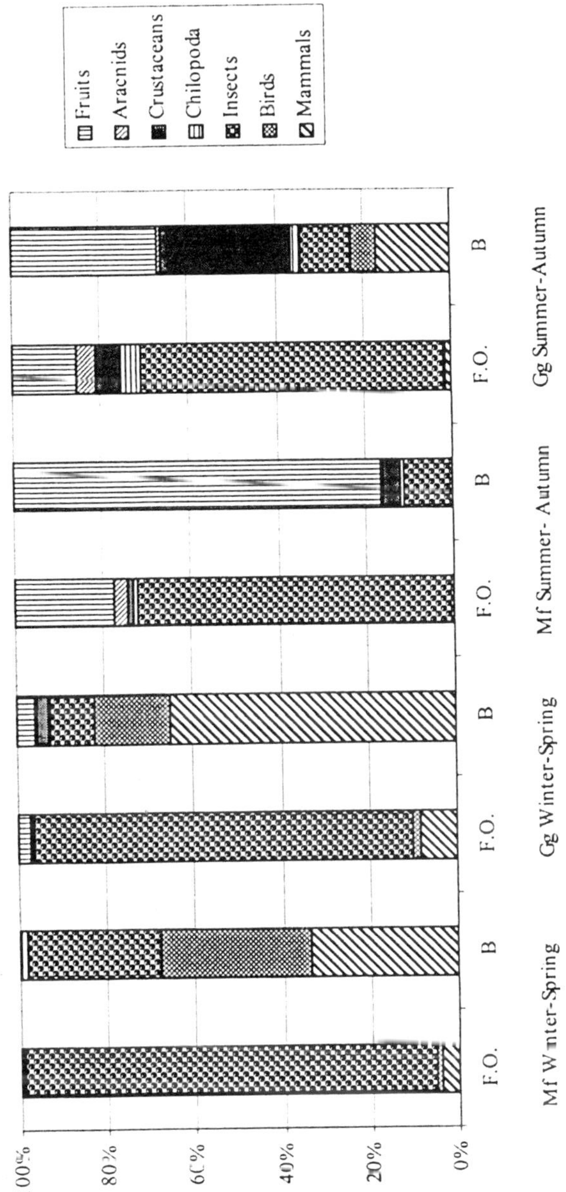

Figure 7.4. Seasonal variation of the diet of stone martens (Mf) and genets (Gg) in a cork oak woodland of the Grândola Hills in southwestern Portugal, 1997–1998.

4.8 Small Mammal Abundance

Algerian house mice (*Mus spretus*) and black rats (*Rattus rattus*) were the most and the least frequently captured small mammals, respectively (Table 7.7). Species abundance varied seasonally and by habitat type (Table 7.8). Riparian vegetation showed the highest diversity and abundance of small mammals, followed closely by the woodland with shrubs. Pastures under trees still contained a significant number of Algerian mice, but the other species were rare (Table 7.8).

Table 7.7. Small mammal abundance in a cork oak woodland of the Grândola Hills in southwestern Portugal, 1997–1998.

	Species (n / 100 trap-nigths)			
Season	*Mus spretus* (n = 405)	*Apodemus sylvaticus* (n = 60)	*Crocidura russula* (n = 123)	*Rattus rattus* (n = 1)
Spring	3.9	1.6	2.7	0.1
Summer	5.5	0.5	1.3	0.0
Autumn	8.8	1.3	3.9	0.0
Total	6.0	1.1	2.5	0.0

Table 7.8. Number of captures of small mammals in three different habitats[a] of a cork oak woodland of the Grândola Hills in southwestern Portugal, 1997–1998.

	Species (n 1st captures)											
	Mus spretus			*Apodemus sylvaticus*			*Crocidura russula*			*Rattus rattus*		
Season	WS	WP	Rp	WS	WP	Rp	WS	WP	Rp	WS	WP	Rp
Spring	24	12	12	6	0	12	7	1	23	0	0	1
Summer	21	8	47	1	2	4	8	0	9	0	0	0
Autumn	45	39	28	0	0	15	7	0	39	0	0	0
Total	90	59	87	7	2	31	22	1	71	0	0	1

[a]WS = cork oak woodland with shrubs, WP = cork oak woodland with pasture, Rp = riparian vegetation.

5. DISCUSSION

Major findings of our study can be briefly summarized as: (1) variation in home range area of martens and genets during the year is primarily associated with breeding activities; (2) core areas are often mutually exclusive, even for individuals suspected of forming a pair bond; (3) tolerance among stone martens and genets is higher than between conspecifics of the same sex, but mutual avoidance is suggested by the exclusiveness of core areas; and (4) non-oak habitats are important as a food source and/or refuge for martens and genets on our study site.

Home ranges of genets were larger than for stone martens, as expected due to their slightly bigger size. Male ranges in both species were slightly larger than those of females, a fact explained by sexual dimorphism in the body size and by the higher energetic requirements of the larger sex (Harestad and Bunnell 1979). Home ranges of adult male and female stone martens were larger than those reported by Hermann (1994) in rural villages of southwest Germany, where males and females used areas <1 km^2. However, when compared to home ranges of animals inhabiting wooded areas (e.g., Skirnisson 1986 *in* Sandell 1989, Genovesi et al 1997), they were of similar size. Sexual dimorphism in home range size was not significant. Stone martens generally (exception Mf_{18}) avoided villages, which can be explained, at least in part, by the availability and predictability of food resources in orchards that are so uncommon in forested habitats, and by a higher susceptibility to human perturbation.

Spacing patterns of stone martens inhabiting cork oak woodlands were consistent with those found in other areas of the species range (Hermann 1994, Genovesi and Boitani 1997, Genovesi et al. 1997) and seem to fit the predictions of the intrasexual territoriality model, where individuals maintain territories only with respect to members of the same sex (Powell 1979). The maintenance of territories during the mating season was not consistent with Sandell's model (1989) for solitary carnivores, where males adopt a roaming strategy; however evidence suggests that access to females plays a role in the spacing pattern of mature males.

On the basis of our limited data, it appears that core areas are not shared among individuals, thus supporting the conclusion of Goodenough et al. (1993) that home-range overlap is less likely to occur in areas of maximum activity. This behavior, strongly contrasts with the findings of Genovesi and Boitani (1997), and could be interpreted as a strategy of resource defense. Maher and Lott (1995) postulated that coexistence is possible by maintaining exclusive areas that provide basic resources.

Trophic resources seemed to play an important role in the spatial ecology of stone martens. The staple food for stone martens in the study area was fruit. Clevenger (1994) reported that in 7 of 14 studies conducted in Europe, wild and cultivated fruits were the primary food item, followed by mammals. Similar results were obtained in 2 other studies conducted in northeastern Spain (Ruiz-Olmo and Palazon 1993) and Italy (Genovesi et al. 1996). On the basis of frequency of occurrence, insects were an important food item in this study, as was reported in other Mediterranean habitats in Spain and Italy (Alegre et al. 1991, Clevenger 1994). However, frequency of occurrence may overestimate the importance of small food items (Reynolds and Aebischer 1991).

Pears were mostly found in the small orchards dispersed across the study area, and they represented more than 90% of fruit occurrences in scats. Cultivated fields, where pear tree patches are found, were highly used by stone martens when foraging. Also, most core areas included 1 orchard. High use of this limited resource as well as the seasonality of its production suggests that fruit availability might be a limiting resource influencing the spacing of the population of stone martens.

Insects and fruits were also important foods for genets. This species differed mainly from the stone marten by the consumption of crustaceans during the summer and autumn. These represent the American species that was introduced to southern Portugal at the end of the 1970s, and has been spreading northwards representing an alternative food resource for many carnivores (Correia 1995), such as mink and otter (e.g., Beja 1996). Predation on crayfishes by genets was already reported in Spain by Palomares and Delibes (1991b) and by Ruiz-Olmo and López-Martin (1993).

In solitary carnivores such as stone martens, spacing patterns are shaped not only by conspecifics, but also by the interactions with other sympatric species with similar ecological requirements. This behavior has been documented for closely related species such as mustelids of the genus *Martes* and *Mustela* (Powell and Zielinski 1983). Within the guild of carnivores inhabiting the cork oak woodlands of southern Iberia, in eco-morphological terms, genets are the species most similar to stone martens and, therefore, are the strongest potential competitors. Indeed, we found that genets have similar home range sizes, intraspecific spacing patterns, activity patterns, habitat preferences, rest sites, and food resources. However, similar population densities (based on the number of captures per trapping effort), major overlap of home ranges, and the partitioning of rest sites do not suggest that resources were limiting. While both species are arboreal, they shared common resources. Rest sites were numerous and readily available, and were not a limiting resource. Genets seemed to prefer trees to shrubs, while stone martens used both equally. In spite of the above considerations, core areas of both species were mutually exclusive. This suggests the existence of spatial partitioning that could be interpreted as a strategy to minimize the potential for competition if and when resources might become limiting. However, more data is required to evaluate niche partitioning and potential competition between stone martens and genets.

This study suggests that except for resting, stone martens do not strictly depend on cork oak trees and its products for existence, but are much more dependent on other habitats in close association with oak woodlands.

6. MANAGEMENT IMPLICATIONS

The use of cork oak woodlands by stone martens is strictly influenced by management. The abundance and survival of stone martens clearly depends on other features of the landscape that guarantee food (cultivated fields), security (riparian corridors), and reduced interspecific competition. The apparent equilibrium between stone martens and other carnivores such as genets may be unstable through time. The desertification of rural areas in Portugal will result in a decrease in the number of cultivated fields, and possibly decreased productivity in the remainder. Modernization of agriculture, clear-cutting of shrubs to allow access for cork extraction, and extensive destruction of understory cover and riparian corridors, are all negative factors. As a consequence, we expect an increased use of cork oak trees by stone martens, and possibly greater competition with genets and other carnivores. The destruction of riparian vegetation might also have a negative impact on hydrology, and we predict a decline in the availability of crayfish for genets, and an increase in competition for other food resources.

Because the stone marten is not a threatened species (SNPRCN 1990) and lacks legal protection, landscapes might not be managed to ensure its survival. However, stone martens should be viewed as one of several wildlife values associated with cork oak woodlands, and its recreational, aesthetical, and ecological values should be taken into account in the same way as is the current economic output of this wooded habitat. The new markets for environmental goods and services offer good perspectives and may help slow the decline in cork oak woodlands in Portugal.

Management for stone marten conservation should consider the importance of a heterogeneous landscape where mature oak trees with understory cover should be maintained together with small patches of fruit trees. Extensive clear cutting of shrubs, including stream banks, may result in declines in stone marten populations. Also, because dead trees are important resting sites, non-selective clear cutting could be detrimental to the species.

7. ACKNOWLEDGMENTS

Funding was provided by the "Fundação para a Ciência e Tecnologia" (Project PRAXIS XXI/PCNA/C/BIA/105/96). Logistical support was provided by the field station of the Centre of Environmental Biology, a research unit of the Science Faculty of the Lisbon University. We thank Clara Espírito Santo, Luis Miguel Rosalino, Marina Rodrigues, Mário Mota, and Sónia Domingos for their field assistance. A preliminary draft of this paper was much improved by the comments of J. A. Bissonette, S. W. Buskirk and G. Proulx.

8. LITERATURE CITED

Alegre, J., A. Hernández, F. J. Purroy, J. M. Salgado, and B. Fuertes. 1991. Dieta otoño-invernal de la garduña, *Martes foina* (Erxleben, 1777), en un habitat rural de Leon (España). Ecología 5:265–273.

Aymerich, M. 1982. Contribution à l'étude de la biologie de la genette (*Genetta genetta* L.) en Espagne. Mammalia 46:389–393.

Beja, P. R. 1996. An analysis of otter *Lutra lutra* predation on introduced American crayfish *Procambarus clarkii* in Iberian streams. Journal of Applied Ecology 33:1156–1170.

Blondel, J., and J. Aronson. 1999. Biology and wildlife of the Mediterranean region. Oxford University Press, Oxford, UK.

Brom, T. G. 1996. Microscopic identification of feather-remains after collisions between birds and aircraft. Bijdragen tot de Dierkunde 56:181–204.

Buskirk, S. W. 1994. Introduction to the genus *Martes*. Pages 1–10 *in* S. W. Buskirk, A. S. Harestad, M. G. Raphael, and R. A. Powell, editors. Martens, sables and fishers: biology and conservation. Cornell University Press, Ithaca, New York, USA.

_____, A. S. Harestad, M. G. Raphael, and R. A. Powell, editors. 1994. Martens, sables and fishers: biology and conservation. Cornell University Press, Ithaca, New York, USA.

Ciucci, P., L. Boitani, F. Francisci, and G. Andreoli. 1997. Home range, activity and movements of a wolf pack in central Italy. Journal of Zoology 243:803–819.

Clevenger, A. P. 1994. Feeding ecology of Eurasian pine martens and stone martens in Europe. Pages 326–340 *in* S. W. Buskirk, A. S. Harestad, M. G. Raphael, and R.A. Powell, editors. Martens, sables and fishers: biology and conservation. Cornell University Press, Ithaca, New York, USA.

Correia, A. I., and M. Santos-Reis. 1999. Área de estudo. Pages 5–8 *in* M. Santos-Reis, and A. I. Correia, editors. Caracterização da flora e fauna do montado da Herdade da Ribeira Abaixo (Grândola, Baixo Alentejo). Centro de Biologia Ambiental, Lisboa, Portugal.

Correia, P. A. 1995. Biologia populacional e interacções tróficas de *Procambarus clarkii* (Crustacea, Decapoda, Cambaridae) no sistema hidrográfico do rio Tejo (Ribatejo, Portugal). Dissertation, Lisbon University, Portugal.

Dayan, T., and D. Simberloff. 1996. Patterns of size separation in carnivore communities. Pages 243–267 *in* J. L. Gittleman, editor. Carnivore behavior, ecology and evolution. Volume 2. Cornell University Press, Ithaca, New York, USA.

Debrot, S., G. Fivaz, C. Mermod, and J. M. Weber. 1982. Atlas des poils de mammifères d'Europe. Institut de Zoologie, Université de Neuchâtel, Neuchâtel, Switzerland.

Delibes, M. 1983. Interspecific competition and the habitat of the stone marten *Martes foina* (Erxleben 1777) in Europe. Acta Zoologica Fennica 174:229–231.

Díaz, M., P. Campos, and F. Pulido. 1997. The Spanish dehesas: a diversity in land-use and wildlife. Pages 178–209 *in* D. J. Pain, and M. W. Pienkowski, editors. Farming and birds in Europe: the common agriculture policy and its implications for bird conservation. Academic Press, London, UK.

Genovesi, P., and L. Boitani. 1997. Social ecology of the stone marten in central Italy. Pages 110–120 *in* G. Proulx, H. N. Bryant, and P. M. Woodard, editors. *Martes*: taxonomy, ecology, techniques and management. Provincial Museum of Alberta, Edmonton, Alberta, Canada.

_____, M. Secchi, and L. Boitani. 1996. Diet of stone martens: an example of ecological flexibility. Journal of Zoology 238:545–555.

_____, I. Sinibaldi, and L. Boitani. 1997. Spacing patterns and territoriality of the stone marten. Canadian Journal of Zoology 75:1966–1971.

Goodenough, J., B. Mcguire, and R. Wallace. 1993. Perspectives on animal behavior. John Wiley & Sons, Inc., New York, USA. 764 pages.

Griffith, H. I. 2000. Mustelids in a modern world. Management and conservation aspects of small carnivore:human interactions. Backhuys Publishers, Leiden, The Netherlands. 342 pages.

Harestad, A. S., and F. L. Bunnell. 1979. Home range and body weight— a reevaluation. Ecology 60:389–402.

Harris, S., W. J. Cresswell, P. G. Forde, W. J. Trewhella, T. Woolard, and S. Wray. 1990. Home range analysis using radio-tracking data—a review of problems and techniques particularly as applied to the study of mammals. Mammal Review 20:97–123.

Herrmann, M. 1994. Habitat use and spatial organization by the stone marten. Pages 122–136 *in* S. W. Buskirk, A. S. Harestad, M. G. Raphael, and R. A. Powell, editors. Martens, sables and fishers: biology and conservation. Cornell University Press, Ithaca, New York, USA. 484 pages.

Jacobs, J. 1974. Quantitative measurement of food selection. Oecologia 14: 413–417.

Libois, R., and A. Waechter. 1991. La fouine (*Martes foina* Erxleben, 1777). Encyclopédie des carnivores de France. Société Française pour l'Étude et la Protection des Mammifères 10:1–53.

Livet, F., and J-J. Roeder. 1987. La genette (*Genetta genetta* Linnaeus, 1758). Encyclopédie des carnivores de France. Société Française pour l'Étude et la Protection des Mammifères 16:1–33.

Lodé, T. 1991. Exploitation des milieux et organisation de l'espace chez deux mustelidés Européens: la fouine et le putois. Vie et Milieu 41:29–38.

Lopez-Martin, J. M., J. Ruiz-Olmo, and S. Cahill. 1991. Autumn home range and activity of a stone marten (*Martes foina* Erxleben, 1777) in northeastern Spain. Acta Theriologica 36:134–138.

Ludwig, J. A., and J. F. Reynolds. 1988. Statistical ecology. A primer on methods and computing. John Wiley & Sons, Inc, New York, USA. 329 pages.

Maddock, A. H., and M. R. Perrin. 1993. Spatial and temporal ecology of an assemblage of viverrids in Natal, South Africa. Journal of Zoology 229:277–287.

Maher, C., and D. Lott. 1995. Definitions of territoriality used in the study of variation in vertebrate spacing systems. Animal Behaviour 49:1581–1597.

Palomares, F., and M. Delibes. 1990. Factores de transformación para el cálculo de la biomasa consumida por gineta (*Genetta genetta*) y meloncillo (*Herpestes ichneumon*) (Carnivora, Mammalia). Miscellánea Zoológica 14:233–236.

_____, and M. Delibes. 1991a. Assessing three methods to estimate daily activity patterns in radio tracked mongooses. Journal of Wildlife Management 55:698–700.

_____, and M. Delibes. 1991b. Alimentacion del melloncillo *Herpestes ichneumon* y de la gineta *Genetta genetta* en la Reserva Biologica de Doñana, S.O. de la Peninsula Iberica. Doñana Acta Vertebrata 18:5–20.

Pinto-Correia, T. 1993. Threatened landscape in Alentejo, Portugal: the montado and other agro-silvo-pastoral systems. Landscape and Urban Planning 24:43–48.

Pounds, C. J. 1981. Niche overlap in sympatric populations of stoats (*Mustela erminea*) and weasels (*Mustela nivalis*) in north-east Scotland. Dissertation, Aberdeen University, UK.
Powell, R. A. 1979. Mustelid spacing patterns: variations on a theme by *Mustela*. Zeitschrift fur Tierpsychologie 50:153–165.
_____, and W. J. Zielinski. 1983. Competition and coexistence in mustelid communities. Acta Zoologica Fennica 174:223–227.
Proulx, G., K. B. Aubry, J. Birks, S. W. Buskirk, C. Fortin, H. C. Frost, W. B. Krohn, L. Mayo, V. Monakov, D. Payer, M. Saeki, M. Santos-Reis, R. Weir, and W. J. Zielinski. 2004. World distribution and status of the genus *Martes*. Pages 21–76 *in* D. J. Harrison, A. K. Fuller, and G. Proulx, editors. Martens and fishers *(Martes)* in human altered environments: An international perspective. Kluwer Academic Publishers, Boston, Massachusetts, USA.
_____, H. N. Bryant, and P. M. Woodard, editors. 1997. *Martes*: Taxonomy, ecology, techniques and management. The Provincial Museum of Alberta, Edmonton.
Reynolds, J. C., and N. J. Aebischer. 1991. Comparison and quantification of carnivore diet by faecal analysis: a critique, with recommendations, based on a study of the fox *Vulpes vulpes*. Mammal Review 21:97–122.
Ruiz-Olmo, J., and J. M. López-Martin. 1993. Note on the diet of the common genet (*Genetta genetta* L.) in the mediterranean riparian habitats of N.E. Spain. Mammalia 57:607–610.
_____, and S. Palazon. 1993. Diet of the stone marten (*Martes foina* Erxleben, 1777) in northeastern Spain. Doñana Acta Vertebrata 20:59–67.
Sandell, M. 1989. The mating tactics and spacing patterns of solitary carnivores. Pages 164–182 *in* J. L. Gittleman, editor. Carnivore behaviour, ecology and evolution. Volume 1. Cornell University Press, Ithaca, New York, USA.
Santos-Reis, M. 1983. Status and distribution of the Portuguese mustelids. Acta Zoologica Fennica 174:213–216.
_____, L. M. Rosalino, and M. Rodrigues. 1999. Lagomorfos, carnívoros e artiodáctilos. Pages 249–262 *in* M. Santos-Reis, and A.I. Correia, editors. Caracterização da flora e fauna do montado da Herdade da Ribeira Abaixo (Grândola, Baixo Alentejo). Centro de Biologia Ambiental, Lisboa, Portugal.
Siegel, S., and N. J. Castellan, Jr. 1988. Nonparametric statistics for the behavioral sciences. Second edition. McGraw-Hill, Inc., New York, USA.
SNPRCN (Serviço Nacional de Parques, Reservas e Conservação da Natureza). 1990. Livro Vermelho dos Vertebrados Terrestres de Portugal. Vol. I – Mamíferos, aves, répteis e anfíbios. Ministério do Ambiente e Defesa do Consumidor, Lisboa, Portugal.
Sokal, R., and F. Rohlf. 1995. Biometry: the principles and practice of statistics in biological research. Third edition. W. H. Freeman and Company, New York, USA. .
Stahl, P. 1986. Le chat forestier d'Europe (*Felis silvestris* Schreber, 1777). Exploitation des resources et organization spatial. Dissertation, Nancy University, France.
Swihart, R., and N. Slade. 1985. Testing for independence of observations in animal movements. Ecology 66:1176–1184.
Vadillo, J. M., J. Reija, and C. Vilà. 1997. Distribución y selección de habitat de la garduña (*Martes foina* Erxleben, 1777) en Vizcaya y Sierra Salvada (Burgos). Doñana Acta Vertebrata 22:39–49.
White, G. C., and R. A. Garrott. 1990. Analysis of wildlife radio-tracking data. Academic Press, San Diego, California, USA.
Worton, B. J. 1989. Kernel methods for estimating the utilization distribution in home-range studies. Ecology 70:164–168.
Zar, J. H. 1984. Statistical analysis. Second edition. Prentice Hall, New Jersey, USA.

Chapter 8

RELATIONSHIPS BETWEEN FOREST STRUCTURE AND HABITAT USE BY AMERICAN MARTENS IN MAINE, USA

David Payer and Daniel Harrison

Abstract: Regional differences in stand-scale habitat selection by American martens (*Martes americana*) suggest that attributes other than stand age and dominant overstory-species composition are responsible for observed habitat associations. Although several investigators have suggested that martens require complex forest structure, few studies have attempted to quantify the relationship between structural attributes and patterns of habitat occupancy by martens. In a forest reserve in northern Maine, USA, we compared characteristics of coarse woody debris (CWD), understory vegetation, and overstory vegetation between 16-ha areas with high use versus low use by 57 (34 M, 23 F) resident, nonjuvenile (≥1 yr) martens. High-use areas had relatively greater volumes of CWD ($P = 0.07$), primarily associated with higher volume ($P = 0.03$) and density ($P = 0.08$) of root masses, and lower densities of live trees ($P = 0.09$); those areas were typically mixed coniferous-deciduous stands with substantial mortality of balsam fir trees from a spruce-budworm *(Choristoneura fumiferana)* epidemic in the early 1980s. Structural differences between high-use and low-use areas were small, however, and a logistic regression model of use intensity based on these characteristics did not reliably differentiate patterns of spatial use. We conclude that forest structure occurred at or above minimum thresholds required by martens throughout our study area, which had high marten densities and was dominated by mature, well-stocked forests. We present estimates of forest structure that may be used as conservative minimum structural thresholds for martens.

1. INTRODUCTION

Habitat associations of American martens have been well studied at the scale of the forest stand, and studies have typically related patterns of use to forest overstory characteristics such as stand age and relative composition of dominant species. At this scale, studies conducted in different regions of North America have yielded inconsistent results. For example, martens selected late-successional, conifer-dominated forests in the northwestern USA (Koehler et al. 1990), the southern Rocky Mountains of the western USA (Koehler and Hornocker 1977, Buskirk et al. 1989), and boreal regions of central Canada (Thompson and Curran 1995). Deciduous forests (Buskirk and Ruggiero 1994) and mid-successional second-growth forests (Thompson and Curran 1995) in these regions were generally avoided. Martens in the boreal forests of Quebec, Canada (Potvin et al. 2000) and the transitional forests of Maine, USA did not select against deciduous stands (Katnik 1992, Chapin et al. 1997a, Payer 1999). In Maine, stands with extensive spruce-budworm-caused mortality had the highest selection index of all stand types (Chapin et al. 1997a). These stands were characterized by extensive mortality of mature conifers, reduced overstory canopy closure, abundant coarse woody debris (i.e., downed logs, snags, stumps, and exposed root masses; CWD), and vigorous early-successional regeneration (Payer and Harrison 2000).

Regional differences in stand-scale habitat associations suggest that stand-age and tree-species composition are surrogates for within-stand habitat components that are required by martens. Marten habitat requirements may best be described in relation to complex physical structure provided by CWD, shrubs, low-hanging branches, and multistoried overhead cover (Buskirk and Powell 1994). These features provide protection from predators (Hargis and McCullough 1984, Hodgman et al. 1997), denning and thermoneutral resting sites (Buskirk et al. 1989, Chapin et al. 1997b, Ruggiero et al. 1998, Bull and Heater 2000), and access to prey (Sherburne and Bissonette 1994, Thompson and Curran 1995).

The relationship between patterns of spatial use and forest structural characteristics may be a unifying principle for understanding habitat associations throughout the range of the American marten. Further, quantification of specific structural components in relation to spatial distribution of use by martens within a landscape would be valuable for developing silvicultural guidelines in intensively managed forests. Therefore, our objectives were to provide baseline measurements of forest structure within a reserve that was closed to logging and trapping, and supported a dense population of martens; and to identify and quantify structural characteristics associated with areas of high versus low use intensity by martens.

2. STUDY AREA

Our study area was located in Piscataquis County, Maine, USA, within the west-central portion of Baxter State Park (BSP). The 55-km^2 area was managed as wilderness without timber harvesting (>35 yr) or trapping (>50 yr). Prior to protection, some large-diameter red spruce (*Picea rubens*) and eastern white pine (*Pinus strobus*) was selectively harvested. The study area was surrounded by a >3-km buffer of similarly managed land except to the west, where the area was contiguous with an intensively managed industrial forest. Elevation ranged from 290 to 735 m. Mean maximum July temperature was 25°C and mean minimum January temperature was -17°C (McMahon 1990). Average annual snowfall was approximately 3 m. Access was provided by a single-lane road, yielding a road density of 0.6 km/km^2. From 1991 to 1997, marten density was 0.63 martens/km^2 and marten home ranges occupied >70% of the study area, suggesting that the population was at or near carrying capacity (Phillips 1994, Payer 1999).

The landscape was characterized by mature (70–100-yr) deciduous (>75% deciduous overstory), coniferous (>75% coniferous overstory), and mixed coniferous-deciduous (25–75% coniferous overstory) forests, which comprised 33%, 15% and 24% of the study area, respectively. Mature coniferous stands were dominated by red spruce and balsam fir (*Abies balsamea*), and also included white pine, eastern hemlock (*Tsuga canadensis*), black spruce (*Picea mariana*), northern white cedar (*Thuja occidentalis*), and larch (*Larix laricina*). Common tree species in mature deciduous stands included sugar maple (*Acer saccharum*), red maple (*A. rubrum*), paper birch (*Betula papyrifera*), yellow birch (*B. allegheniensis*), and American beech (*Fagus grandifolia*). Mature stands were interspersed with stands regenerating following a spruce-budworm epidemic, which occurred from 1974 to 1984 (Irland et al. 1988). Regenerating stands comprised 26% of the area. The spruce-budworm epidemic caused extensive mortality of mature coniferous trees, especially balsam fir. Affected stands had abundant CWD and <50% canopy closure of mature trees. Many of these stands also had a dense layer of understory vegetation, including regenerating balsam fir, maples, paper birch, American mountain-ash (*Sorbus americana*), and raspberry (*Rubus* spp.) (Payer and Harrison 2000).

3. METHODS

3.1 Study Design

We used radiolocation data collected from May to October 1991–1995 from 57 (34 M, 23 F) resident (≥25 locations with error polygons <25 ha over ≥90 days), nonjuvenile (≥1 yr) martens. We obtained radiolocations from fixed-wing aircraft (53%) and ground-based telemetry (47%). Mean distance between estimated and true locations for 40 test transmitters located from aircraft was 93.3 m (*SD* = 46.0 m), yielding an estimated error for aerial telemetry of 2.7 ha (i.e., area of circle with radius = 93.3 m). For radiolocations obtained from the ground, we calculated error polygons of radiolocations with the program TRIANG (White and Garrott 1984) using observer-specific angular errors (Payer 1999). Thirteen martens were monitored >1 yr. We included each marten-year separately in subsequent analyses, yielding a total of 79 (54 M, 25 F) marten-years. Capture, handling, and telemetry protocols were described by Payer et al. (2004), and were approved by the Institutional Animal Care and Use Committee at the University of Maine, Orono.

We calculated 95% minimum-convex-polygon (MCP) home ranges for each marten using all telemetry locations with error polygon <25 ha. We defined the boundary of the study area as a minimum-area concave polygon surrounding the pooled locations within home ranges of all martens. To increase the mean accuracy of locations used to determine site-specific use intensity, we deleted marten locations with error polygons >10 ha from further analysis. This procedure yielded a minimum of 24 locations per marten-year, with mean error polygon of 2.8 ha (SD = 1.7 ha). For martens with >24 locations/yr, we used random subsets of 24 locations so that each marten contributed equally to subsequent analyses.

We overlayed a grid of 16-ha (400 m × 400 m) cells and the selected marten locations with error polygon <10 ha on a map of the study area using a vector-based geographic information system (PC ARC/INFO 3.4.1, Environmental Systems Research Institute, Redlands, CA, USA). We excluded cells if >50% of the cell area was >400 m from the access road because marten locations obtained from ground-based telemetry were biased, with fewer locations obtained at distances >400 m from the road (Chapin et al. 1997a). Forest-type composition did not differ between areas <400 m and >400 m from the road (Chapin et al. 1997a). Therefore, our screening process avoided rather than introduced a sampling bias.

Among selected marten locations, the ratio of grid-cell size to mean error-polygon size was 5.7:1. Further, for 95% of the selected locations, the size of the grid cells was >2.4× the size of the error polygon. Therefore, our relatively small telemetry errors did not cause significant bias for testing the relationship between habitat use and structural characteristics within grid cells (Nams 1988).

Based on our GIS overlay, all grid cells were used by resident martens. We assigned each cell to 1 of 3 marten use-intensity categories (high, medium, or low use) based on whether the number of locations within the cell fell within the upper, middle, or lower quantile of observed use. Low-use cells contained 2–7 marten locations, medium-use cells contained 8–13 locations, and high-use cells contained 14–41 locations. There were 27 cells each in the low-use and medium-use categories, and 28 cells in the high-use category.

3.2 Habitat Sampling

We sampled habitat characteristics in randomly selected high-use ($n = 18$) and low-use ($n = 18$) cells during June–August 1996. We did not sample medium-use cells, choosing instead to focus our efforts on the upper and lower quantiles of use intensity. We randomly selected 8 sampling points within each cell, and estimated 18 habitat characteristics (Table 8.1) at each point. We established a 0.04-ha circular plot (radius = 11.3 m) centered on each sampling point, and recorded the number, height, and basal diameter of snags (≥7.6 cm basal diameter, ≥2.0 m tall) within each plot. We used the formula for volume of a cone to calculate the volume of snags (Spies et al. 1988). We also recorded end diameters and length of exposed root masses (minimum diameter ≥7.6 cm), and height and mid-point diameter of stumps (≥7.6 cm mid-point diameter, <2.0 m tall) within each 0.04-ha plot. We estimated the percent of each root mass that was exposed (i.e., above ground) to the nearest 5%. We calculated the volume of root masses as the frustum of a cone, adjusted for percent exposed, and the volume of stumps as a cylinder (Corn and Raphael 1992). We included snags and stumps that intercepted the plot boundary if ≥50% of their basal diameter was within the plot. Root masses were included if ≥50% of their overall volume was within the plot.

We estimated volume of downed logs using the planar intersection method (Brown 1971), as adapted by the U.S. Forest Service for the fourth forest inventory of Maine (U.S. Department of Agriculture 1995). We measured large-end diameter, small-end diameter (≥7.6 cm), and length of logs that intersected a randomly oriented 22.6-m transect centered on each sampling point. We included logs in decay classes 1–3 (sound to moderately rotten [U.S. Department of Agriculture 1995]) that were ≤2 m from ground level with diameter ≥7.6 cm at the point of intersection. Logs in decay class 3 had sloughing or detached bark, were soft in rotten areas, and had branch stubs that pulled out easily. We excluded logs in decay class 4 (advanced decomposition), which generally lacked bark and branch stubs, and had extensive rotten areas with a "doughy" consistency (U.S. Department of Agriculture 1995). We calculated the volume of each log as a frustum of a cone, and estimated the volume of logs per hectare using the method of de Vries (1986:258–261). We measured depth of litter (i.e., leaves, twigs, bark, and fruits) at 5-m intervals along each log transect, and calculated mean depth per transect.

We estimated understory stem density by counting the number of deciduous and coniferous stems (<2.0 m tall) at ground level within 2 non-overlapping, randomly oriented 11.3-m × 1.0-m strip quadrats originating at the sampling point. We included stems on the quadrat boundary if ≥50% of the basal diameter was within the quadrat. We used a cover pole (Griffith and Youtie

1988) viewed from 11.3 m due east and west to estimate understory foliage density at <0.5 m and 0.5–2.0 m above ground level. At each sampling point we also determined basal area of live coniferous and deciduous trees (m^2/ha) with a wedge prism, and estimated density and height of live trees (≥7.6 cm dbh, ≥2.0 m tall) using the point-centered-quarter technique (Cottam and Curtis 1956). Finally, we estimated overhead cover with a spherical densiometer held 1.0 m above ground level. We averaged densiometer readings from the 4 cardinal directions at each sampling point.

3.3 Statistical Analysis

The grid cell was the experimental unit for analysis of the relationship between marten-use intensity and habitat characteristics. Therefore, we averaged values for habitat characteristics from the 8 sampling points within each cell and used cell means in subsequent analyses. We used univariate Mann-Whitney *U* tests to compare habitat characteristics of cells with high marten use to those with low marten use. We examined pairwise Spearman correlation coefficients (r_s) for the subset of variables with $P \leq 0.10$. If $|r_s| > 0.80$, we eliminated the variable that was least significant in the univariate test. We used logistic regression with retained variables to build explanatory models for differentiating low-use versus high-use cells. We built models in a forward, stepwise fashion, at each step adding the variable that provided the greatest increase in McFadden's r^2 (a statistic that reflected how well the model fit the data) (Hensher and Johnson 1981). Using a *G* statistic (twice the difference in log likelihoods), each model was compared with the model immediately preceding it to assess if the additional variable contributed significantly ($P \leq 0.05$) to model fit. Variables selected by this stepwise procedure were retained in the final model if the 95% confidence interval for their odds ratio did not include 1.0. Significance of the final model was assessed with a *G* statistic that compared the model to a constant-only model. Model performance was evaluated with McFadden's ρ^2 and a model prediction success table that summarized how observations from each level of the dependent variable were allocated to predicted outcomes.

4. RESULTS

The median volume of CWD estimated at 288 sampling points was 101 m^3/ha (interquartile range [IR]: 52–170 m^3/ha). Downed logs accounted for 63% of the total CWD volume, and median log density was 920 logs/ha (IR: 607–1,280 logs/ha). Exposed root masses were the next greatest contributor to

CWD volume and comprised 20% of the total; median density was 63 root masses/ha (IR: 44–104 root masses/ha). Snags were also common, comprising 11% of total CWD volume (median density [IR] = 163 snags/ha [130 – 223 snags/ha]). Stumps comprised 6% of the estimated volume of CWD (median density [IR] = 292 stumps/ha [243–364 stumps/ha]).

Most of the study area was densely forested, and the median density of live deciduous and coniferous trees was 596 trees/ha (IR: 468–718 trees/ha) with basal area of 17 m^2/ha (IR: 12–24 m^2/ha) for trees $\geq$7.6 cm dbh. Understory vegetation was plentiful, including 15,300 woody stems/ha (IR: 8,850–23,000 stems/ha). Median near-ground foliage density was 85% (IR: 68–98%). Generally, a dense growth of tall shrubs and trees resulted in a closed overhead canopy; median overhead cover was 98%.

There were few differences in forest structural characteristics between cells that received high use versus low use by martens (Table 8.1). High-use cells had greater volume ($\chi^2 = 4.63$, 1 df, $P = 0.03$) and density ($\chi^2 = 2.98$, 1 df, $P = 0.08$) of exposed root masses than low-use cells. Root-mass volume and density were highly correlated ($r_s = 0.90$). Total CWD volume was greater ($\chi^2 = 3.25$, 1 df, $P = 0.07$) in high-use cells (median 129.8 m^3/ha) than in low-use cells (median 99.9 m^3/ha). Further, high-use cells had lower density of live trees ($\chi^2 = 2.92$, 1 df, $P = 0.09$) than low-use cells. Root-mass volume and live-tree density were negatively correlated ($r_s = -0.61$). None of the other 15 habitat characteristics we measured differed between high-use and low-use cells ($P \geq 0.18$) (Table 8.1).

We performed a stepwise logistic regression with root-mass volume and tree density as potential explanatory variables. Our final model for differentiating low versus high use-intensity cells was significant ($G = 5.40$, 1 df, $P = 0.02$) and included only root-mass volume (m^3/ha), which had a positive association with a cell belonging to the high-use category (odds ratio = 1.05). The logistic model fit the data poorly ($\rho^2 = 0.108$), however, and correctly predicted group membership for only 57% of the 36 cells. Based on the habitat characteristics we measured, we did not identify structural variables that reliably differentiated areas of low versus high use by martens.

5. DISCUSSION

We detected few differences in forest-structural characteristics between areas with high and low use intensity by martens. Our observation that high-use areas had greater volume of CWD, which was primarily associated with greater volume and density of root masses, and lower density of live trees suggests that martens may have selected stands with substantial wind throw or

Table 8.1. Median (interquartile range) values of habitat characteristics in 16-ha cells receiving high use or low use by American martens in a forest reserve in Maine. Habitat characteristics were measured in 288 0.04-ha plots during June–August 1996.

Habitat variable	P[a]	High marten use (n = 18 cells)	Low marten use (n = 18 cells)
Volume of snags (m³/ha)	0.92	10.5 (8.3–16.9)	11.8 (8.6–14.1)
Snag density (no./ha)	0.58	144 (129–216)	170 (140–230)
Volume of stumps (m³/ha)	0.20	7.3 (5.9–8.3)	6.6 (4.4–7.5)
Stump density (no./ha)	0.24	319 (267–364)	267 (201–355)
Volume of exposed root masses (m³/ha)	0.03	25.2 (18.9–41.9)	11.8 (5.4–23.1)
Root-mass density (no./ha)	0.08	75 (45–119)	55 (39–85)
Volume of downed logs (m³/ha)	0.18	76.4 (63.0–117.7)	70.8 (44.2–81.0)
Downed-log density (no./ha)	0.24	975 (746–1,297)	834 (484–1,253)
Basal area of live deciduous trees (m²/ha)	0.40	10.0 (4.5–12.3)	9.9 (7.5–12.8)
Basal area of live coniferous trees (m²/ha)	0.51	7.3 (6.5–12.4)	7.4 (5.5–9.8)
Density of live trees (no./ha)	0.09	578 (407–654)	636 (505–740)
Tree height (m)	0.39	8.7 (8.4–9.4)	9.1 (8.6–9.7)
Overhead cover (%)	0.37	97.0 (91.0–99.2)	98.0 (93.8–99.4)
Litter depth (cm)	0.92	2.6 (2.1–3.4)	2.5 (2.3–3.1)
Understory deciduous stem density (no./ha)	0.47	6,750 (5,050–7,830)	5,700 (4,950–7,850)
Understory coniferous stem density (no./ha)	0.92	9,260 (6,570–16,400)	11,500 (6,290–15,100)
Foliage density, <0.5 m (%)	0.27	96.3 (93.7–98.4)	93.7 (89.4–97.5)
Foliage density, 0.5–2.0 m (%)	0.62	77.5 (67.3–85.9)	76.3 (65.6–83.6)

[a]Mann-Whitney *U* test

tree mortality. These conditions are associated with spruce-budworm defoliation, which was the predominant source of disturbance in our study area (Payer and Harrison 2000). Consistent with this observation, martens have been reported to select defoliated stands 15–20 years following a spruce-budworm epidemic, or to use such stands in proportion to their availability, during both winter and summer (Chapin et al. 1997a, Payer 1999, Potvin et al. 2000). The relationship between forest structure and marten-use intensity was weak, however, as evidenced by the low explanatory power of our logistic regression model. The weakness of this relationship, combined with the observation that martens occurred at high density and occupied most of the available landscape, suggests that the structural characteristics we measured were not limiting for martens within our forest-reserve study area. Further, the distribution of these features did not significantly influence patterns of habitat use at the scale of 16-ha grid cells. These results suggest that although structure is undoubtedly important for martens, it is not always the limiting factor determining patterns of habitat use. Where complex structure is abundant, martens may occupy a wide range of successional stages and stand types including both coniferous and deciduous tree species.

In a companion study of marten distribution in relation to forest structure within an adjacent industrial forest, areas used by martens were distinguished from unused areas primarily by characteristics related to stand maturity (Payer and Harrison 2003). The industrial forest was intensively managed for forest products (primarily pulpwood), and unused areas were generally young, regenerating stands that had been clearcut between 1974 and 1982. Relative to those clearcut areas, used areas had higher basal area and density of live trees, taller trees, greater snag volume, and denser overhead canopy. In contrast to that study, we did not observe any 16-ha cells that were unused by martens within the forest reserve. Budworm-defoliated stands had greater volumes of snags, downed logs, and root masses, and included taller trees and higher basal areas of live trees than regenerating clearcuts (Payer and Harrison 2000). The defoliated stands provided suitable marten habitat because vertical structure provided by large snags and surviving live trees (red spruce or deciduous species), in combination with plentiful downed woody debris and understory vegetation, offset the reduced live-tree basal area compared to intact mature stands.

In second-growth boreal forest in Ontario, Canada, 100-m^2 quadrats used by martens had more downed logs and snags, taller trees, and greater canopy closure than unused quadrats (Bowman and Robitaille 1997). These characteristics were positively associated with the proportion of spruce and fir in the overstory, suggesting a preference for coniferous stands dominated by spruce and fir over other stand types. Although we failed to document similar structural differences between high-use and low-use cells, our study was conducted at a broader spatial scale. Our overall estimates for overhead canopy closure and density of logs and snags exceeded those reported by Bowman and Robitaille (1997) for used quadrats, further suggesting that structure was not limiting on our study area. Chapin et al. (1997b) documented a greater density of logs and snags in coniferous stands versus deciduous or mixed coniferous-deciduous stands in our study area, although martens did not select coniferous stands over other stand types (Katnik 1992, Chapin et al. 1997a, Payer 1999). Therefore, we conclude that requirements for forest structure were met or exceeded within all stand types in our study area. The forests where we studied martens, which are transitional between northern hardwood and boreal forest types (Seymour and Hunter 1992), appear to be more structurally complex than most boreal forests. High levels of structural complexity may explain the more generalized habitat use observed at the scale of the forest stand for martens in the transitional forests of Maine, USA (Katnik 1992, Chapin et al. 1997a, Payer 1999) and Quebec, Canada (Potvin et al. 2000).

Abundance and availability of prey may be an important determinant of habitat selection by martens (Thompson and Colgan 1987, Buskirk and Powell 1994, Thompson and Curran 1995). Voles are the most common item in diets of American martens (Martin 1994), and southern red-backed voles (*Clethrionomys gapperi*) are a staple food in many areas (Buskirk and Ruggiero 1994), including northern Maine (Lachowski 1997). Red-backed voles are often associated with coniferous forests (Monthey and Soutiere 1985, Nordyke and Buskirk 1991), although they may also be abundant in other forest types (Kirkland 1990, DeGraaf et al. 1991, Fuller et al. 2004). In general, red-backed voles are associated with downed woody debris (Miller and Getz 1977, Hayes and Cross 1987) and dense understory vegetation (Nordyke and Buskirk 1991). On a study area adjacent to ours, Fuller et al. (2004) concluded that abundance of red-backed voles was greatest in mature stands, especially those with a significant deciduous component. At the scale of the microsite, Lachowski (1997) found that red-backed vole occurrence was positively associated with availability of downed logs and overhead canopy closure, and he concluded that sufficient structure was available in all mature forest-stand types and in budworm-killed stands to meet habitat requirements of voles. Therefore, both martens and their primary prey exhibited similar generalist patterns of stand-scale habitat use among mature, mixed, deciduous and coniferous stand types, and budworm-killed stands. These patterns resulted from abundant CWD, horizontal structure, and vertical structure in all forest types except regenerating clearcuts.

Factors other than the structural characteristics that we measured may have influenced patterns of marten occurrence. High marten density likely caused competition for space (Phillips et al. 1998), and may have resulted in more generalized population-level habitat use (e.g., Fretwell and Lucas 1970). High density was also associated with smaller home ranges (Phillips 1994) and greater intrasexual home-range overlap among males (Payer et al., Chapter 4, this volume), as well as increased incidence of range abandonment by females (Phillips et al. 1998). Despite apparently greater intraspecific competition in the reserve, reproductive success of individual females, as measured by the proportion of adults ($\geq$2 yr) lactating in early summer, did not differ between the reserve and the industrial forest (Phillips 1994, Payer 1999). Therefore, increased competition may constrain habitat choices by martens in high-density populations, but reproductive performance of females may be maintained if their territories contain adequate structure and prey.

Our study was based on data collected during 1 May–31 October. During the remainder of the year, cold periods with snow accumulation occur, and near-ground forest structure becomes important for thermoregulation (Buskirk et al. 1989) and subnivean access (Sherburne and Bissonette 1994, Thompson and Curran 1995). We observed a high degree of home-range fidelity between consecutive seasons (Phillips et al. 1998, Payer et al. 2004), however, and patterns of landscape-scale and stand-scale habitat selection were similar between summer and winter (Chapin et al. 1997a, Payer 1999). Therefore, available evidence suggests that broad-scale patterns of spatial use did not change seasonally, and that CWD and other components of near-ground structure were plentiful and well distributed. Although martens may have used specific near-ground features more intensively during winter, the availability of these features did not appear to affect habitat use at the scale of 16-ha forested cells within the home ranges of resident adult martens.

6. CONCLUSIONS AND MANAGEMENT IMPLICATIONS

Mature second-growth forests and stands defoliated 10–20 years previously by eastern spruce budworms provided suitable habitat for martens in our study area, and martens occupied most of the available landscape (Chapin et al. 1997a, Payer 1999). Although areas with high use by martens had larger volumes of CWD (primarily associated with higher volume and density of root masses) and lower densities of live trees than low-use areas, differences were small. Further, observed differences did not reliably differentiate high-use from low-use areas. We conclude that availability of forest structure exceeded minimum thresholds required by martens throughout our study area, regardless of forest type.

Our overall estimates for forest structural characteristics compared favorably to areas occupied by martens in an intensively managed industrial forest (Payer and Harrison 2003), and may be used as conservative minimum structural thresholds for martens. In our region, suitable marten habitat had approximately 100 m^3/ha CWD (interquartile range [IR]: 52–170 m^3/ha), including 900 sound to moderately rotten downed logs/ha with minimum diameter 7.6 cm. Root masses (63 per ha), standing dead trees (160 per ha), and stumps (290 per ha) also contributed to available CWD. Other characteristics of suitable habitat included basal area ≥18 m^2/ha for trees ≥7.6 cm dbh, ≥75% overhead canopy, and dense understory vegetation (approximately 15,000 woody stems/ha).

We recommend that further research be conducted to specifically address the influence of forest structure on martens in managed forests. Monitoring the effects of experimental manipulation of structure on habitat occupancy and fitness of martens will increase understanding of marten-habitat relationships, and will allow refinement of minimum-threshold estimates for structural characteristics. Effects of alternative silvicultural practices (e.g., clearcutting and various forms of partial harvesting) on forest structure required by martens should also be investigated. This knowledge, combined with landscape-scale approaches to marten-habitat management (e.g., Chapin et al. 1998, Hargis et al. 1999, Potvin et al. 2000), may facilitate development of harvest prescriptions that minimize adverse effects on marten populations.

7. ACKNOWLEDGMENTS

This project was funded by the National Council of the Paper Industry for Air and Stream Improvement, the Maine Department of Inland Fisheries and Wildlife, Federal Aid in Wildlife Restoration Project W-82-R-II-368, the Maine Forest Service, and the Maine Agricultural and Forest Experiment Station. We received logistical support from the Baxter State Park Authority and the Cooperative Forestry Research Unit, University of Maine. M. Estabrook, T. Hodgman, R. Kelshaw, S. McLellan, D. Phillips, E. York, D. Wrobleski, and D. Wroe provided invaluable field assistance. J. McPhee piloted aircraft used to obtain telemetry locations. We acknowledge T. Chapin for assistance with study design, and W. Halteman for statistical advice. The manuscript benefited from reviews by K. Aubry, K. Foresman, A. Magoun, G. Proulx, T. B. Wigley, and an anonymous reviewer. This is Scientific Contribution No. 2714 of the Maine Agricultural and Forest Experiment Station.

8. LITERATURE CITED

Bowman, J. C., and J.-F. Robitaille. 1997. Winter habitat use of American martens *Martes americana* within second-growth forest in Ontario, Canada. Wildlife Biology 3:97–104.

Brown, J. K. 1971. A planar intersect method for sampling fuel volume and surface area. Forest Science 17:96–102.

Bull, E. L., and T. W. Heater. 2000. Resting and denning sites of American martens in northeastern Oregon. Northwest Science 74:179–185.

Buskirk, S. W., S. C. Forrest, M. G. Raphael, and H. J. Harlow. 1989. Winter resting site ecology of marten in the central Rocky Mountains. Journal of Wildlife Management 53:191–196.

_____, and R. A. Powell. 1994. Habitat ecology of fishers and American martens. Pages 283–296 *in* S. W. Buskirk, A. S. Harestad, M. G. Raphael, and R. A. Powell, editors. Martens, sables and fishers: biology and conservation. Cornell University Press, Ithaca, New York, USA.

_____, and L. F. Ruggiero. 1994. American marten. Pages 7–37 *in* L. F. Ruggiero, K. B. Aubry, S. W. Buskirk, L. J. Lyon, and W. J. Zielinski, editors. The scientific basis for conserving forest carnivores: American marten, fisher, lynx, and wolverine in the western United States. U.S. Forest Service General Technical Report RM-254, Rocky Mountain Forest and Range Experiment Station, Fort Collins, Colorado, USA.

Chapin, T. G., D. J. Harrison, and D. M. Phillips. 1997a. Seasonal habitat selection by marten in an untrapped forest preserve. Journal of Wildlife Management 61:707–717.

_____, D. M. Phillips, D. J. Harrison, and E. C. York. 1997b. Seasonal selection of habitat by resting marten in Maine. Pages 166–181 *in* G. Proulx, H. N. Bryant, and P. M. Woodard, editors. *Martes*: taxonomy, ecology, techniques, and management. Provincial Museum of Alberta, Edmonton, Canada.

_____, D. J. Harrison, and D. D. Katnik. 1998. Influence of landscape pattern on habitat use by American marten in an industrial forest. Conservation Biology 12:1327–1337.

Corn, J. G., and M. G. Raphael. 1992. Habitat characteristics at marten subnivean access sites. Journal of Wildlife Management 56:442–448.

Cottam, G., and J. T. Curtis. 1956. The use of distance measures in phytosociological sampling. Ecology 37:451–460.

DeGraaf, R. M., D. P. Snyder, and B. J. Hill. 1991. Small mammal habitat associations in poletimber and sawtimber stands of four forest cover types. Forest Ecology and Management 46:227–242.

de Vries, P. G. 1986. Sampling theory for forest inventory. Springer-Verlag, Berlin, Germany.

Fretwell, S. D., and H. L. Lucas. 1970. On territorial behavior and other factors influencing habitat distribution in birds: part 1. Theoretical development. Acta Biotheoretica 19:16–36.

Fuller, A. K, D. J. Harrison, and H. J. Lachowski. 2004. Stand scale effects of partial harvesting and clearcutting on small mammals and forest structure. Forest Ecology and Management 191:373–386.

Griffith, B., and B. A. Youtie. 1988. Two devices for estimating foliage density and deer hiding cover. Wildlife Society Bulletin 16:206–210.

Hargis, C. D., and D. R. McCullough. 1984. Winter diet and habitat selection of marten in Yosemite National Park. Journal of Wildlife Management 48:140–146.

_____, J. A. Bissonette, and D. L. Turner. 1999. The influence of forest fragmentation and landscape pattern on American marten. Journal of Applied Ecology 36:157–172.

Hayes, J. P., and S. P. Cross. 1987. Characteristics of logs used by western red-backed voles, *Clethrionomys californicus*, and deer mice, *Peromyscus maniculatus*. Canadian Field-Naturalist 101:543–546.

Hensher, D., and L. W. Johnson. 1981. Applied discrete choice modeling. Croom Helm, London, England.

Hodgman, T. P., D. J. Harrison, D. M. Phillips, and K. D. Elowe. 1997. Survival of American marten in an untrapped forest preserve in Maine. Pages 86–99 *in* G. Proulx, H. N. Bryant, and P. M. Woodard, editors. *Martes*: taxonomy, ecology, techniques, and management. Provincial Museum of Alberta, Edmonton, Canada.

Irland, L. C., J. B. Dimond, J. L. Stone, J. Falk, and E. Baum. 1988. The spruce budworm outbreak in Maine in the 1970's-assessment and directions for the future. Maine Agricultural Experiment Station Bulletin 819, University of Maine, Orono, Maine, USA.

Katnik, D.D. 1992. Spatial use, territoriality, and summer–autumn selection of habitat in an intensively harvested population of martens on commercial forestland in Maine. Thesis, University of Maine, Orono, Maine, USA.

Kirkland, G. L., Jr. 1990. Patterns of initial small mammal community change after clearcutting of temperate North American forests. Oikos 59:313–320.

Koehler, G. M., and M. G. Hornocker. 1977. Fire effects on marten habitat in the Selway-Bitterroot Wilderness. Journal of Wildlife Management 41:500–505.

Koehler, G. M., J. A. Blakesley, and T. W. Koehler. 1990. Marten use of successional forest stages during winter in north-central Washington. Northwest Naturalist 71:1–4.

Lachowski, H. J. 1997. Relationships among prey abundance, habitat, and American marten in northern Maine. Thesis, University of Maine, Orono, Maine, USA.

Martin, S. K. 1994. Feeding ecology of American martens and fisher. Pages 297–315 *in* S. W. Buskirk, A. S. Harestad, M. G. Raphael, and R. A. Powell, editors. Martens, sables and fishers: biology and conservation. Cornell University Press, Ithaca, New York, USA.

McMahon, J. S. 1990. The biophysical regions of Maine: patterns in the landscape and vegetation. Thesis, University of Maine, Orono, Maine, USA.

Miller, D. H., and L. L. Getz. 1977. Factors influencing local distribution and species diversity of forest small mammals in New England. Canadian Journal of Zoology 55:806–814.

Monthey, R. W., and E. C. Soutiere. 1985. Responses of small mammals to forest harvesting in northern Maine. Canadian Field-Naturalist 99:13–18.

Nams, V. O. 1988. Effects of radiotelemetry error on sample size and bias when testing for habitat selection. Canadian Journal of Zoology 67:805–811.

Nordyke, K. A., and S. W. Buskirk. 1991. Southern red-backed vole, *Clethrionomys gapperi*, populations in relation to stand succession and old-growth character in the central Rocky Mountains. Canadian Field-Naturalist 105:330–334.

Payer, D. C. 1999. Influences of timber harvesting and trapping on habitat selection and demographic characteristics of American marten. Dissertation, University of Maine, Orono, Maine, USA.

_____, and D. J. Harrison. 2000. Structural differences between forests regenerating following spruce-budworm defoliation and clearcut harvesting: implications for marten. Canadian Journal of Forest Research 30:1965–1972.

_____, and _____. 2003. Influence of forest structure on habitat use by American marten in an industrial forest. Forest Ecology and Management 179:145–156.

_____, _____, and D. Phillips. 2004. Territoriality and home-range fidelity of American martens in relation to timber harvesting and trapping. Pages 99–114 *in* D. J. Harrison, A. K. Fuller, and G. Proulx, editors. Martens and fishers (*Martes*) in human-altered environments: An international perspective. Kluwer Academic Publishers, Boston, Massachusetts, USA.

Phillips, D. M. 1994. Social and spatial characteristics, and dispersal of marten in a forest preserve and industrial forest. Thesis, University of Maine, Orono, Maine, USA.

_____, D. J. Harrison, and D. C. Payer. 1998. Seasonal changes in home-range area and fidelity of martens. Journal of Mammalogy 79:180–190.

Potvin, F., L. Belanger, and K. Lowell. 2000. Marten habitat selection in a clearcut boreal landscape. Conservation Biology 14:844–857.

Ruggiero, L. F., D. E. Pearson, and S. E. Henry. 1998. Characteristics of American marten den sites in Wyoming. Journal of Wildlife Management 62:663–673.

Seymour, R. S., and M. L. Hunter, Jr. 1992. New forestry in eastern spruce-fir forests: principles and applications to Maine. Maine Agricultural and Forest Experiment Station Miscellaneous Publication 716, University of Maine, Orono, Maine, USA.

Sherburne, S. S., and J. A. Bissonette. 1994. Marten subnivean access point use: response to subnivean prey levels. Journal of Wildlife Management 58:400-405.

Spies, T. A., J. T. Franklin, and T. B. Thomas. 1988. Coarse woody debris in Douglas-fir forests of western Oregon and Washington. Ecology 69:1689–1702.

Thompson, I. D., and P. W. Colgan. 1987. Numerical responses of martens to a food shortage in northcentral Ontario. Journal of Wildlife Management 51:824–835.

Thompson, I. D., and W. J. Curran. 1995. Habitat suitability for marten of second-growth balsam fir forests in Newfoundland. Canadian Journal of Zoology 73:2059–2064.

U.S. Department of Agriculture. 1995. Field instructions for the fourth inventory of Maine. Version 1.4. Northeast Forest Experiment Station, Radnor, Pennsylvania, USA.

White, G. C., and R. A. Garrott. 1984. Portable computer system for field processing biotelemetry triangulation data. Colorado Division of Wildlife, Game Information Leaflet 110.

Chapter 9

EFFECT OF AMBIENT TEMPERATURE ON THE SELECTION OF REST STRUCTURES BY FISHERS

Richard Weir, Fraser Corbould, and Alton Harestad

Abstract: We examined the effect of ambient temperature on the selection of rest structures by 20 radio-tagged fishers in two areas of central British Columbia during 1991–1993 and 1996–2000. Fishers rested in tree cavities, on rust brooms or tree branches, under pieces of large coarse woody debris (CWD), and in burrows or rock crevices. We located fishers at 86 rest structures and recorded the local ambient temperature at nearby climate stations while these structures were occupied. The type of rest structure selected by fishers varied with local ambient temperature ($P = 0.005$). Temperatures were colder when fishers used CWD structures than when they used branch or cavity structures ($P < 0.05$). Large pieces of CWD may be important habitat elements for fishers during long periods of extremely low temperatures because they likely provide a more favorable thermal microenvironment than that found at other types of rest structures. Our results have implications for habitat management and conservation of old-forest structures for fishers in regions with cold climates.

1. INTRODUCTION

Fishers (*Martes pennanti*) use rest sites for a variety of purposes, including refuge from potential predators and thermoregulatory cover (Kilpatrick and Rego 1994). Fishers have been reported to use a wide variety of structures at rest sites: tree nests and cavities, logs (hollow or solid), root wads, willow (*Salix* spp.) thickets, ground burrows, and rock falls (Raine 1981, Arthur et al. 1989, Jones 1991, Powell 1993, Kilpatrick and Rego 1994, Gilbert et al. 1997).

Little is known about the factors that affect selection of rest structures by fishers. American martens (*M. americana*) are ecologically similar to fishers and utilize many of the same types of structures for resting (e.g., Buskirk et al. 1989, Martin and Barrett 1991, Gilbert et al. 1997, Raphael and Jones 1997).

Much of the selection for rest sites by martens has been attributed to ambient temperature; martens tend to select subnivean rest sites during periods of cold weather (e.g., Buskirk et al. 1989, Martin and Barrett 1991). The lower critical temperature (i.e., the lowest ambient temperature at which they can passively maintain normal body-core temperature, Harlow 1994) of martens when they rest is relatively high and therefore they select microenvironments that are energetically favorable for resting during periods of low temperature.

The effect of ambient temperature on the selection of rest sites by American martens is relatively well understood, but the relationship between temperature and selection of rest structures by fishers has not been examined thoroughly enough to determine if the same relationship holds true for fishers. Because fishers have a greater body mass than do martens, Buskirk and Powell (1994) hypothesized that thermal losses while resting are probably not as important to fishers. Indeed, Powell (1979) estimated that the lower critical temperature of resting fishers was -60°C for females and -120°C for males. According to this estimate, fishers are not exposed to temperatures that approach their lower critical temperature while resting throughout much of their range.

However, data suggests that fishers may select different structures for resting depending on ambient temperature. Raine (1981), Arthur et al. (1989), Jones (1991), and Kilpatrick and Rego (1994) all noted that fishers tend to use subnivean rest sites more frequently during winter and arboreal rest structures (i.e., tree nests and cavities) more frequently during spring, when temperatures are warmer. This evidence suggests that although ambient temperatures may be above their estimated lower critical temperature, fishers may make behavioral changes to reduce thermal losses above this critical limit.

In the northern portion of their range across Canada, arctic high-pressure weather systems can persist for several weeks during winter with ambient temperatures consistently below -25°C. The objective of our research was to perform an exploratory analysis to determine if a relationship exists between the local ambient temperature near the rest structure (i.e., coarse-scale ambient temperature) and the type of structures that fishers select. We hypothesized that fishers modify their selection of rest structures and use different structures during periods of cold temperature. This information may be useful for directing forest management practices that will aid in the conservation of fisher populations in areas with long periods of extreme cold.

2. STUDY AREAS

Our 1,800-km² northern study area (Williston) was centered 75 km northwest of Mackenzie, British Columbia (55° 30'N, 123° 02'W) and within the moist-cool and wet-cool subzones of the Sub-Boreal Spruce Biogeoclimatic (SBSmk and SBSwk) zone (Meidinger et al. 1991) to the west of the Williston Reservoir. Our 1,500-km² southern study area (Beaver Valley) was centered 65 km north-east of Williams Lake, British Columbia (52° 10'N, 122° 10'W) and entirely within the dry-warm subzone (SBSdw) of the SBS zone. Both study areas were ecologically similar; mean annual temperatures in Williston and Beaver Valley areas were from 1.2 to 3.6°C respectively, with mean annual precipitation between 585 (SBSdw) and 690 mm (SBSmk) (MacKinnon et al. 1990, Steen and Coupé 1997). The SBS zone is a heavily forested, coniferous, montane zone that dominates the landscape of the central interior of British Columbia and generally occurs from valley bottoms to about 1,200 m above sea level (Meidinger et al. 1991). The climate of the SBS zone is continental and characterized by severe, snowy winters and relatively warm, moist, and short summers. During the study, local ambient temperatures in the Beaver Valley study area ranged between -29 and 34°C, while in the Williston study area temperatures typically ranged between -32 and 33°C. Consistent snow cover persisted in both study areas from mid-November through until mid-April, with maximum accumulations reaching approximately 90 cm in the forest interior.

Forests in both study areas were dominated by lodgepole pine (*Pinus contorta* var. *latifolia*), hybrid white spruce (*Picea engelmannii* x *glauca*), and subalpine fir (*Abies lasiocarpa*), with minor deciduous components of trembling aspen (*Populus tremuloides*), paper birch (*Betula papyrifera*), and black cottonwood (*Populus balsamifera trichocarpa*). Douglas-fir (*Pseudotsuga menziesii* var. *glauca*) was a common mid- to late-successional species in the Beaver Valley study area. Common understory shrubs were prickly rose (*Rosa acicularis*), black huckleberry (*Vaccinium membranaceum*), black twinberry (*Lonicera involucrata*), kinnikinnick (*Arctostaphylos uva-ursi*), and black gooseberry (*Ribes lacustre*).

The dry and moist subzones of the SBS zone had a natural disturbance regime of frequent, large-scale fires on a cycle of about 150 years, with most stands burning every 100 years, while the wet subzone had typical fire return intervals of greater than 250 years (British Columbia Ministry of Forests and British Columbia Ministry of Environment, Lands and Parks 1995). Forest harvesting, using a variety of techniques, has occurred over the past 25 to 40 years and created a mosaic of seral stages and stand types throughout both of the study areas. Land clearing for cultivation and cattle grazing occurred along the valley bottom in the Beaver Valley study area.

3. METHODS

We captured and monitored 20 fishers (3 M, 17 F) as part of 2 larger studies on the ecology of fishers in British Columbia (Weir 1995, 2000). We live-trapped resident fishers in each area using baited wire cage traps (24.5 × 31 × 81 cm; Havahart Model 1081, Lititz, Pennsylvania, USA). We placed traps on beds of hybrid spruce or subalpine fir boughs and covered the top and sides of the traps with wax-coated cardboard boxes. Traps were also lined with hay and covered with more boughs so that snow or wind would not penetrate the trap. We baited each trap with approximately 500 g of meat from salmon (*Oncorhynchus* spp.), galliform or anseriform birds, or moose (*Alces alces*) and scented nearby trees with commercial trapping lure.

Upon capture, we immobilized each fisher with either a 10:1 mixture of ketamine:xylazine (Ketaset®, Ayerst Veterinary Laboratories, Guelph, Ontario, Canada; Rompun®, Bayer Inc., Toronto, Ontario, Canada) administered at 18 mg/kg or a 1:1 mixture of tiletamine:zolazepam (Telazol®; Fort Dodge Animal Health, Fort Dodge, Iowa, USA) administered at 8 mg/kg for radiotagging. We affixed radiocollars to healthy adult fishers captured in the Beaver Valley area. In the Williston area, we either affixed radiocollars or had intra-abdominal radiotransmitters surgically implanted by a wildlife veterinarian.

We monitored 9 fishers (1 M, 8 F) during 1991–1993 in the Beaver Valley study area and 11 fishers (2 M, 9 F) during 1996–2000 in the Williston study area. In both study areas, we identified resting structures used by radio-tagged fishers throughout the year by homing-in (White and Garrott 1990) on signals of stationary fishers to their rest sites (i.e., area surrounding a rest structure) and locating the structure with which the fisher was associated. Rest structures were identified throughout the year, but we did not include locations of female fishers with kits in natal or maternal dens in this analysis.

Within 0.5 hr of identifying the rest structure, local ambient temperature was recorded at either manual or automatic recording stations located within 20 km of each resting structure. In the Beaver Valley area, we recorded temperature at thermometers located in forest interior conditions located throughout the study area. In the Williston area, we recorded temperatures at either permanent thermometers or from remote temperature data loggers (Optic Stowaway®, Pocasset, Massachusetts) that recorded hourly temperatures. In both study areas, temperature stations were placed under normal forest canopy at least 10 m from the forest edge in well-ventilated, shaded locations approximately 2 m above the ground. These stations provided us with representative data on the local ambient temperature of the area near the rest structures.

We assessed the effect of local ambient temperature on the selection of rest structures by fishers using an Analysis of Variance. Because we occasionally recorded individual fishers using one type of resting structure more than once, we used each fisher as a replicate and performed the analysis of variance procedure on mean ambient temperatures for each fisher for each type of rest structure. We identified significant differences in mean ambient temperatures among the different types of rest structures using Tukey multiple comparison tests. We set the acceptable Type I error at 0.05.

4. RESULTS

We identified resting fishers associated with 4 distinct types of structures: branch, cavity, coarse woody debris (CWD), and ground. Branch rest structures were arboreal sites that typically involved abnormal growths (i.e., witches brooms) on live spruce trees (caused by spruce broom rust [*Chrysomyxa arctostaphyli*]) or on subalpine fir trees (caused by fir broom rust [*Melampsorella caryophyllacearum*]). We occasionally observed branch rest sites that were associated with exposed large limbs of black cottonwood and spruce trees. Cavity rest structures were chambers in decayed heartwood of the main bole of black cottonwood, aspen, or Douglas-fir trees that were declining in vigour, but still alive. Cavities were accessed by fishers through branch-hole entrances into heart-rot (black cottonwood, aspen, or Douglas-fir trees) or using excavations made by primary-cavity nesting birds (aspen trees only). Coarse woody debris rest structures were located inside, amongst, or under pieces of CWD. The source of the CWD was natural tree mortality, logging residue, or man-made piling. Coarse woody debris rest structures were usually comprised of a single large (>35 cm diameter) piece of debris, but occasionally involved several pieces of smaller diameter logging residue. Ground rest structures were those that involved large-diameter pieces of loosely arranged colluvium (e.g., rock piles) or burrows into the soil that were likely excavated by another animal.

We located 86 rest sites of 20 radio-tagged fishers over 6 years. We located fishers using structures for resting in both winter (59 locations) and non-winter seasons (27 locations). We recorded fishers using branch rest structures most frequently (57.0%), followed by cavity (19.8%), CWD (18.6%), and ground (4.6%) rest structures. We did not detect a significant difference in the frequency of selection of each type of rest structure by fishers between the Beaver Valley and Williston study areas ($\chi^2 = 3.32$, df = 3, $P = 0.34$). The simultaneous local ambient temperature near the rest sites ranged between -29.4 and 21.1°C. The frequency at which we located rest structures with respect to temperature did not appear to be different than that of local ambient temperatures in a typical year (Fig. 9.1), although we intentionally collected data at cold temperatures.

Fishers did not use each type of rest structure independently of local ambient temperature ($F_{3,37} = 7.45$, $P \leq 0.001$). Local ambient temperatures were significantly colder when fishers used CWD rest structures compared to when they used branch and cavity structures ($P < 0.05$; Table 9.1). We did not detect a significant difference in local ambient temperature among the use of branch, cavity, and ground rest structures, nor were they different between the use of ground and CWD rest structures ($P > 0.05$).

5. DISCUSSION

Fishers in our study used structures for resting in a pattern similar to that reported elsewhere. Fishers used arboreal branch and cavity sites most frequently, but used CWD sites when temperatures were colder. In Maine (Arthur

Figure 9.1. Sampling distribution of rest structures of radio-tagged fishers with respect to local ambient temperature in the Sub-Boreal Spruce Biogeoclimatic zone of British Columbia, 1991–1993 and 1996–2000. The bars represent the number of rest structures obtained in each temperature interval. The solid line represents the proportion of hourly temperature recordings that occurred in each temperature interval throughout a typical year (data from 1999, Williston area).

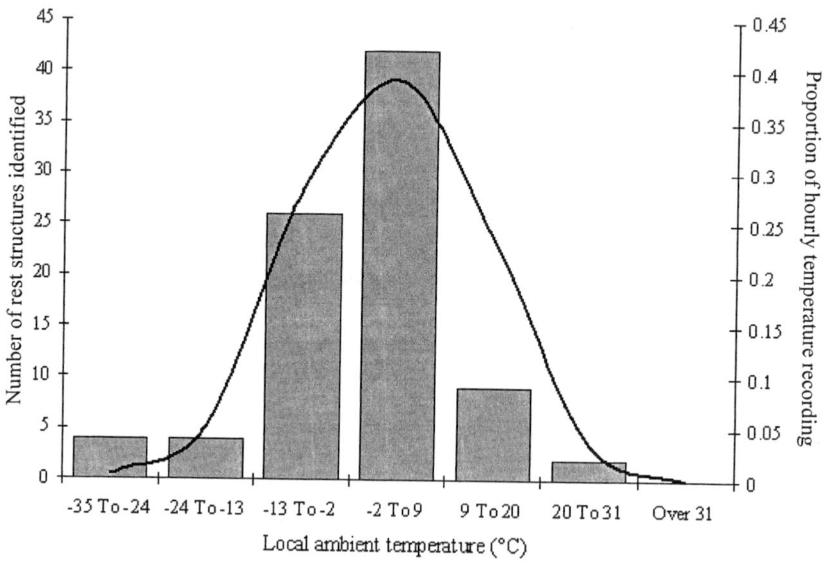

Table 9.1. Mean local ambient temperatures (°C) at which radio-tagged fishers used each type of rest structure in the Sub-Boreal Spruce Biogeoclimatic zone of British Columbia, 1991–1993 and 1996–2000. The local ambient temperatures for structures with the same letter were not significantly different. The local ambient temperature when fishers used CWD structures was significantly lower than when they used branch or cavity structures.

Type of structure	Mean (°C)	SE	Range (°C)	n	Contrast of means	
Branch	2.4	1.1	-13.1–20.9	49	A	
Cavity	1.3	2.0	-14.2–21.1	17	A	
Ground	-5.7	2.2	-10.9–0.3	4	A	B
CWD	-10.7	3.1	-29.4–3.2	16		B

et al. 1989), Idaho (Jones 1991), and Connecticut (Kilpatrick and Rego 1994), fishers also used arboreal sites most frequently and CWD or subnivean sites only on very cold days. Our data support the hypothesis that a relationship exists between local ambient temperature and the selection of rest structures by fishers.

Previous research has illustrated that each type of rest structure has differing thermal properties. Taylor and Buskirk (1994) measured and calculated the thermal attributes of branch, cavity, and CWD sites used by American martens in high-elevation forests of southern Wyoming. They found that CWD sites provided the warmest microenvironments only during periods of cold temperatures (<-5°C), deep snow pack (>15 cm), and high wind speed. Branch or cavity sites were warmer during all other combinations of ambient temperature, snow pack, and wind. In Manitoba, Raine (1981) measured the ambient temperature at a subnivean rest site of a fisher and found that, while the ambient temperature was -26°C, the temperature inside the subnivean cavity was -11°C.

The thermal attributes of the 4 types of rest structures may have affected the selection by fishers and helped to explain the patterns that we observed. Fishers probably used branch and cavity structures for resting during most of the year because these sites were relatively common and provided an adequate thermal environment for most combinations of ambient temperature, snow depth, and wind speed.

Although it is unlikely that fishers in our study areas encountered temperatures that were near their estimated lower critical temperature for resting, our data suggests that fishers modified their selection of rest structures during cold temperatures. Like American martens, fishers probably used CWD rest structures because they provided the warmest thermal environments during periods of cold temperatures. The mean temperature at which fishers used subnivean structures for resting (-10.7°C) was considerably lower than that reported for martens in Wyoming (-5.5°C, Buskirk et al. 1989). This finding is not unexpected because the lower critical temperature for fishers while resting is estimated to be substantially lower than that estimated for martens (fishers: female = -60°C, male = -120°C, Powell 1979; martens: 16°C, Buskirk et al. 1988).

In our study areas, we did not observe fishers using arboreal or ground sites for resting when temperatures were below -14.2°C; 6 of the 6 rest sites that we documented at temperatures <-14.2°C were in subnivean sites with CWD structures. The exclusive use of subnivean CWD structures at temperatures <-14.2°C suggests that fishers may select CWD structures for energetic benefits. It is unclear if this result represents a challenge to Powell's (1979) estimate of lower critical temperature for resting fishers, but this finding suggests that fishers modify their behavior during bouts of cold temperatures.

The reasons for selecting a specific rest structure probably change over time and thermoregulation is not the only factor that affects the selection of rest sites by fishers. Several authors have suggested that fishers rest close to food sources (de Vos 1952, Coulter 1966, Powell 1993). The resting sites in trees are generally more numerous than ground sites (Martin and Barrett 1991), hence, fishers may select tree sites opportunistically. Raphael and Jones (1997) speculated that arboreal structures offer greater protection from predators than do ground sites. Because of their elevated position, the detection of potential predators would probably be enhanced in tree sites that afford earlier olfactory or visual discovery of approaching predators. Similarly, elevated sites may provide greater detection of potential prey. In the absence of restrictive thermoregulatory demands (i.e., most of the year), we expect that fishers would select structures based upon factors other than temperature.

Local ambient temperature was not the best measure by which to assess the thermal environment facing fishers in our study. Our measurement methods likely ameliorated the fine-scale differences in ambient temperature among stands and patches of habitat. Site-specific abiotic factors, such as wind convection, solar radiation, and precipitation also affect the standard operative temperature that an animal experiences at any point in time (Taylor and Buskirk 1994) and examination of these factors would likely provide more insight into the effect of the thermal environment on rest structure selection by fishers. Also, the thermal attributes associated with each type of structure can change over time; Taylor and Buskirk (1994) showed that snow depth greatly affects the insulative capabilities of CWD sites. Habitat features, such as overhead cover surrounding the various rest structures undoubtedly affected the thermal properties of each respective structure. However, our research illustrated that a relationship existed between local ambient temperature and the selection of rest structures by fishers. Future research should be directed towards assessing these more detailed spatial- and temporal-specific attributes and determining their effects on the selection of rest structures by fishers.

6. MANAGEMENT IMPLICATIONS

Fishers use a variety of structures for resting, and most of these structures result from the natural processes of disease, death, and decay of trees. These structures accumulate over time and reach the greatest densities in mature- and late-successional forests (Cline et al. 1980). In regions where extended periods of extreme cold (<-15°C) occur, fishers, like American martens, probably rely upon the CWD component that is characteristic of later successional forests to provide thermal cover while resting. As mentioned by Raphael and Jones (1997), rest sites are important microhabitats that contribute to the overall fitness of an animal because the selection of rest structures affects an individual's thermoregulation and vulnerability to predators.

The boreal and sub-boreal forests of British Columbia are prominent timber-producing areas. The harvesting of mature and late-successional forests may have a detrimental effect on fisher populations by reducing the availability and recruitment of large CWD that is suitable for thermal cover. Silvicultural prescriptions that retain large CWD, and the ecological processes that create it, will be important components of forest management plans that encourage the persistence of fisher populations, especially in regions that experience extreme cold. However, before these silvicultural prescriptions can be implemented, we need to determine the appropriate size, density, and arrangement of the structures that fishers use for resting. This information could be used to identify targets for the retention and generation of resting structures in managed forests.

7. ACKNOWLEDGMENTS

This research was supported by different agencies in each study area. The Beaver Valley study was funded by the Fur Initiatives and Habitat Conservation Fund programs of the British Columbia Ministry of Environment, Lands and Parks; the Habitat Silviculture Protection Account of the British Columbia Ministry of Forests; the British Columbia Trappers Association; and the Science Council of British Columbia. The Williston study was funded by the Peace/Williston Fish and Wildlife Compensation Program, Forest Renewal British

Columbia, and the Slocan Group (Mackenzie Operations). We are indebted to A. Bowser, S. Bowsfield, H. Davis, J. McCormick, K. Webster, and R. Wright for their invaluable assistance. This manuscript was greatly improved by comments from W. A. Adair, H. Davis, G. Proulx, and 1 anonymous reviewer.

8. LITERATURE CITED

Arthur, S. M., W. B. Krohn, and J. R. Gilbert. 1989. Habitat use and diet of fishers. Journal of Wildlife Management 53:680–688.

British Columbia Ministry of Forests, and British Columbia Ministry of Environment, Lands and Parks. 1995. Biodiversity Guidebook. Province of British Columbia. Victoria, British Columbia, Canada.

Buskirk, S. W., H. J. Harlow, and S. C. Forrest. 1988. Temperature regulation in American marten (*Martes americana*) in winter. National Geographic Research 4:208–218.

_____, S. C. Forrest, M. G. Raphael, and H. J. Harlow. 1989. Winter resting site ecology of marten in the central Rocky Mountains. Journal of Wildlife Management 53:191–196.

_____ and R. A. Powell. 1994. Habitat ecology of fishers and American martens. Pages 283–296 *in* S. W. Buskirk, A. S. Harestad, M. G. Raphael, and R. A. Powell, editors. Martens, sables, and fishers: Biology and conservation. Cornell University Press, New York, USA.

Cline, S. P., A. B. Berg, and H. M. Wright. 1980. Snag characteristics and dynamics in Douglas-fir forests, western Oregon. Journal of Wildlife Management 44:773–786.

Coulter, M. W. 1966. Ecology and management of fishers in Maine. Dissertation, Syracuse University, Syracuse, New York, USA.

de Vos, A. 1952. Ecology and management of fisher and marten in Ontario. Ontario Department of Lands and Forests. Technical Bulletin, Wildlife Service Number 1. Toronto, Ontario, Canada.

Gilbert, J. H., J. L. Wright, D. J. Lauten, and J. R. Probst. 1997. Den and rest-site characteristics of American marten and fisher in northern Wisconsin. Pages 135–145 *in* G. Proulx, H. N. Bryant, and P. M. Woodward, editors. *Martes*: Taxonomy, ecology, techniques, and management. Provincial Museum of Alberta, Edmonton, Alberta, Canada.

Harlow, H. J. 1994. Trade-offs associated with the size and shape of American martens. Pages 391–403 *in* S. W. Buskirk, A. S. Harestad, M. G. Raphael, and R. A. Powell, editors. Martens, sables, and fishers: biology and conservation. Cornell University Press, New York, USA.

Jones, J. L. 1991. Habitat use of fisher in northcentral Idaho. Thesis, University of Idaho, Moscow, Idaho, USA.

Kilpatrick, H. J., and P. W. Rego. 1994. Influence of season, sex, and site availability on fisher (*Martes pennanti*) rest-site selection in the central hardwood forest. Canadian Journal of Zoology 72:1416–1419.

MacKinnon, A., C. Delong, and D. Meidinger. 1990. A field guide for identification and interpretation of ecosystems of the northwest portion of the Prince George Forest Region. British Columbia Ministry of Forests. Land Management Handbook number 21. Victoria, British Columbia, Canada.

Martin, S. K., and R. H. Barrett. 1991. Resting site selection by marten at Sagehen Creek, California. Northwestern Naturalist 72:37–42.

Meidinger, D. V., J. Pojar, and W. L. Harper. 1991. Chapter 14: Sub-boreal spruce zone. Pages 209–221 *in* D. V. Meidinger and J. Pojar, editors. Ecosystems of British Columbia. Volume Special Report Service Number 6. British Columbia Ministry of Forests, Research Branch,

Victoria, British Columbia, Canada.

Powell, R. A. 1979. Ecological energetics and foraging strategies of the fisher (*Martes pennanti*). Journal of Applied Ecology 48:195–212.

_____. 1993. The fisher: Life history, ecology, and behavior. Second edition. University of Minnesota Press, Minneapolis, Minnesota, USA.

Raine, R. M. 1981. Winter food habits, responses to snow cover and movements of fisher (*Martes pennanti*) and marten (*Martes americana*) in southeastern Manitoba. Thesis, University of Manitoba, Winnipeg, Manitoba, Canada.

Raphael, M. G., and L. L. C. Jones. 1997. Characteristics of resting and denning sites of American martens in central Oregon and western Washington. Pages 146–165 *in* G. Proulx, H. N. Bryant, and P. M. Woodward, editors. *Martes*: Taxonomy, ecology, techniques, and management. Provincial Museum of Alberta. Edmonton, Alberta, Canada.

Steen, O. A., and R. A. Coupé. 1997. A field guide to forest site identification and interpretation for the Cariboo Forest Region. British Columbia Ministry of Forests. Land Management Handbook 39. Victoria, British Columbia, Canada.

Taylor, S. L., and S. W. Buskirk. 1994. Forest microenvironments and resting energetics of the American marten *Martes americana*. Ecography 17:249–256.

Weir, R. D. 1995. Diet, spatial organization, and habitat relationships of fishers in south-central British Columbia. Thesis, Simon Fraser University, Burnaby, British Columbia, Canada.

_____. 2000. Ecology of fishers in the sub-boreal forests of north-central British Columbia: Year IV – Radiotelemetry monitoring and habitat sampling. Peace/Williston Fish and Wildlife Compensation Program Report Number 222. Prince George, British Columbia, Canada.

White, G. C., and R. A. Garrott. 1990. Analysis of wildlife radio-tracking data. Academic Press, San Diego, California, USA.

Part III

Research and Management Approaches

Chapter 10

ZOOGEOGRAPHY, SPACING PATTERNS, AND DISPERSAL IN FISHERS:

Insights Gained from Combining Field and Genetic Data

Keith Aubry, Samantha Wisely, Catherine Raley, and Steven Buskirk

Abstract: We demonstrate how research questions generated from radiotelemetry studies of fishers (*Martes pennanti*) can be further elucidated by combining field and genetic data. We genotyped a sample of 20 fishers at 9 polymorphic microsatellite loci and used these data to calculate observed and expected heterozygosity, estimate coefficients of relatedness, and exclude potential parent-offspring combinations. Previous research indicated that the population of fishers occurring in the southern Cascade Range in Oregon is reintroduced. Due to the presence of potentially strong ecological and anthropogenic barriers between that population and fishers occurring in the northern Siskiyou Mountains of Oregon, we hypothesized that they are geographically isolated from each other. Analyses of microsatellite genotypes supported this hypothesis by providing empirical evidence that genetic introgression of fishers from the northern Siskiyou Mountains into the southern Cascade Range has not occurred. Results from our field study indicated that male fishers may exhibit either of 2 distinct behavioral strategies during the breeding season. For 3 successive breeding seasons, 2 adult males in our study area remained resident on their non-breeding home ranges, whereas 1 or 2 other males (one died after year 1) abandoned their non-breeding home ranges and encroached on the home ranges of resident males. To determine which strategy resulted in greater reproductive success, we used field and genetic data to exclude potential parent-offspring relationships between these 4 males and 7 juveniles. In contrast to resident males, encroaching males could not have fathered any of the juveniles. These results suggest that maintaining intrasexual territoriality during the breeding season may provide a reproductive advantage to male fishers. Our research also provides the first empirical evidence of male-biased juvenile dispersal and female philopatry in fishers. In accordance with predictions from genetic theory, our analyses showed that adult females had significantly higher relatedness values than adult males, and that there were many more potential first-order relationships among adult females than among adult males. As these examples demonstrate, including genetic information in data analysis can substantially improve the heuristic value of field studies and enable researchers to study additional aspects of population biology.

1. INTRODUCTION

Researchers generally infer social structure and spacing patterns of mammals from behavioral observations, mark-recapture data, or radiotelemetry locations. Obtaining such data in the field is challenging, and resulting datasets often suffer from small sample sizes or lack of data for 1 or more study animals. Logistical constraints on the geographic extent of most radiotelemetry studies may prevent long-distance movements or dispersal events from being detected, and such movements can also be confused with transmitter failure. Furthermore, unsuccessful copulations, cuckoldry, or multipaternity litters can be impossible to detect with traditional research approaches, and familial relationships can be inferred incorrectly when adults who are not the biological parents are behaviorally or spatially associated with young (Avise 1994).

Recent developments in genetic techniques (Parker et al. 1998) and the widespread application of genetic data to wildlife conservation (Haig 1998) have led many field biologists, including those studying fishers (*Martes pennanti*) and American martens (*M. americana*), to routinely collect tissue samples from their study animals for potential use in genetic studies. Several researchers have measured genetic variation in fisher and marten populations (Mitton and Raphael 1990; Carr and Hicks 1997; McGowan et al. 1999; Williams et al. 1999, 2000; Kyle et al. 2000, 2001; Drew et al. 2003; Kyle and Strobeck 2003), but the use of genetic information to augment field studies of wild populations of fishers or martens has not been reported. Here, we demonstrate how genetic information can be used to test research hypotheses generated from field data and published literature, and provide insights into the biology of fishers both within and among populations that could not be obtained by analyses of field data alone.

1.1 The Study Population

From 1995 to 2001, we (KA and CR) conducted a radiotelemetry study of fishers on the west slope of the Cascade Range in southern Oregon (Fig. 10.1; Aubry and Raley 2002). Based on trapping results in our study area and extensive survey efforts by resource management agencies in Oregon using remote cameras and trackplate boxes (Aubry and Lewis 2003, K. Aubry, unpublished data), we concluded that the geographic distribution of our study population was restricted primarily to the shaded area shown in Fig. 10.1. Fisher populations in the Pacific states are unharvested; commercial trapping of fishers has been prohibited in Oregon and California since 1937 and 1946, respectively.

Figure 10.1. Distribution of fishers in southwestern Oregon and northwestern California. Open circles indicate capture sites of the 18 fishers from our study population in the southern Cascade Range in Oregon; the shaded polygon is the 100% minimum convex polygon (MCP) for all radio-marked animals in our study. Open triangles are the localities where 2 adult male fishers were trapped incidentally in the northern Siskiyou Mountains in Oregon. Solid circles are localities where fishers were detected during track-plate and remote-camera surveys in northwestern California (Zielinski et al. 1995) and in southwestern Oregon (Aubry and Lewis 2003, K. Aubry, unpublished data). Interstate Highway 5 is shown as a heavy banded line and county boundaries are shown as thin solid lines.

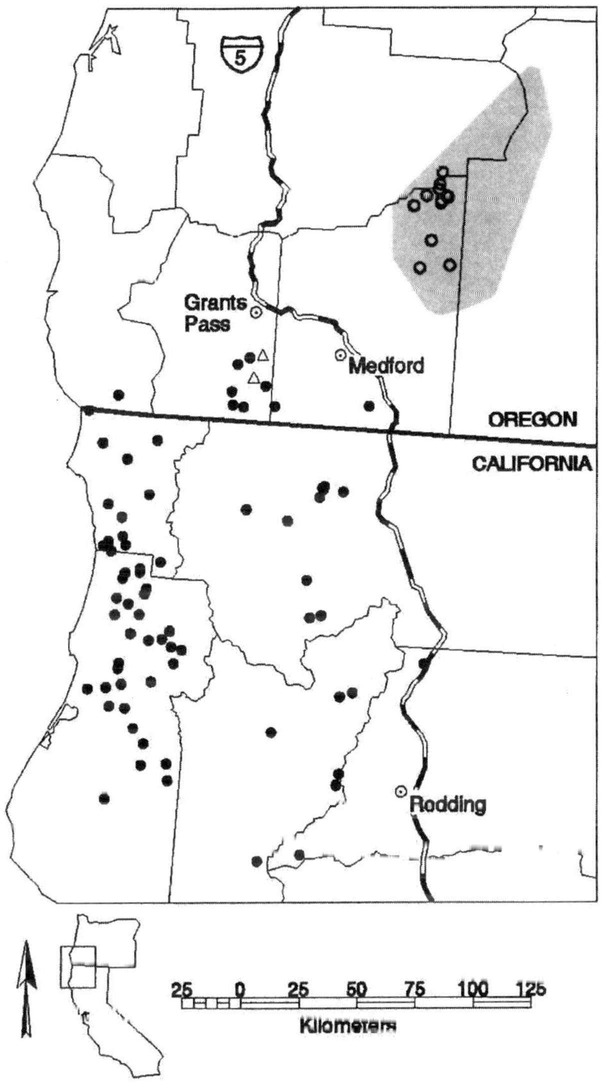

Several attempts have been made to reintroduce fishers to Oregon; between 1961 and 1981, 41 fishers were translocated to various locations in or adjacent to our study area (Kebbe 1961, Aubry and Lewis 2003). There is no evidence that fishers have been translocated into California. Extant populations of fishers in northwestern California (Fig. 10.1) are believed to be descended entirely from indigenous animals, which are significantly smaller in size (based on condylobasal length) than fishers from western and central Canada (Hagmeier 1959, Zielinski et al. 1995, Aubry and Lewis 2003). Based on historical records, and differences in body weights and mitochondrial DNA haplotypes between fishers in southern Oregon and northern California, Aubry and Lewis (2003) concluded that the extant population of fishers in the southern Cascade Range in Oregon was reintroduced to that area by a series of translocations from British Columbia and Minnesota in the late 1970s and early 1980s.

1.2 Research Hypotheses

1.2.1 Zoogeography

Habitat conditions in the area between our study population and fishers in the northern Siskiyou Mountains in Oregon (Fig. 10.1) are generally unsuitable for fishers. This area contains an interstate highway corridor (I-5), urban and agricultural development in and around the city of Medford, and extensive areas of open grassland and oak savannah in the interior Rogue River valley (Franklin and Dyrness 1973). We hypothesized that these ecological and anthropogenic barriers have resulted in the geographic isolation of our study population from fishers southwest of the I-5 corridor (Fig. 10.1). If so, then the 2 fishers we sampled from the northern Siskiyou Mountains of Oregon do not belong to the southern Cascade Range gene pool and will differ genetically from fishers in our study population.

1.2.2 Spacing Patterns

Fishers are polygynous and intrasexually territorial; home-range overlap is minimal within sexes, but extensive between sexes (Powell 1993). Male home ranges are larger than those of females; the large home ranges of males provide them with primary access to receptive females during the breeding season, whereas the smaller home ranges of females provide them with primary access to sufficient resources to survive and successfully raise kits (Leonard 1986; Powell 1993, 1994). Leonard (1986) hypothesized that male fishers could maximize their reproductive success by adopting either of 2 strategies during the breeding season: continuing to defend their non-breeding home range against other males and mating with as many resident females within their home range as possible (here referred to as "resident" males), or abandoning their territories and trying to mate with as many females as possible (here referred to as "encroaching" males). Based on field observations of extensive movements by adult males during the breeding season, Leonard (1986) predicted that the second strategy was more likely the typical breeding strategy for male fishers. Movements by adult males outside their home ranges during the breeding season have also been reported by other researchers (Buck 1982, Arthur et al. 1989), and are generally assumed to result in increased reproductive success.

Field observations revealed that adult males in our study population exhibited both of these breeding strategies (Fig. 10.2). We captured one adult male (05) within the home range of another male (02) during 3 successive breeding seasons (1996–1998); during this time, male 05 occupied a non-breeding home range about 30 km SE of the home range of male 02. Another male (06) also made a breeding-season movement into 02's home range in 1996, but died before the next breeding season. Both resident males (02 and 07) extended the boundaries of their home ranges slightly during the breeding season, but maintained stable and apparently exclusive home ranges throughout the year; neither made breeding-season movements similar to those exhibited by the encroaching males (05 and 06). Both resident males had home ranges that overlapped those of females, but we do not know if females were present within the non-breeding home ranges of the encroaching males. If Leonard's (1986) hypothesis is correct, then the reproductive success of the 2 encroaching males will be greater than that of the 2 resident males.

1.2.3 Dispersal

Polygyny and intrasexual territoriality are associated with an imbalance in parental investment between the sexes; male fishers do not contribute to raising young, whereas females make a substantial investment by gestating a litter of 2–5 kits, nursing them to weaning, and feeding them until they are independent (Powell 1993, 1994). Competition for mates is therefore much stronger among male fishers than among females. Accordingly, a territorial male would be expected to allow juvenile females to establish home ranges within or near his home range, but would not be expected to allow juvenile males to do so. Juvenile males must therefore disperse relatively far from their natal areas to find unoccupied home ranges, whereas females can remain close to their natal areas, where the risks associated with dispersal are low and where resource availability is known to be sufficient for survival and reproduction (Greenwood 1980, Dobson 1982). Thus, as in other polygynous and intrasexually

Figure 10.2. Breeding-season movements for 4 adult male fishers in the southern Cascade Range in Oregon. Thick solid lines indicate the home ranges of males 05 and 06 during the non-breeding season (NBRS = May–Jan), and the direction and distance of breeding-season (BRS = Feb–Apr) movements. Medium solid lines indicate NBRS home ranges for males 02 and 07; dashed lines indicate movements during BRS that extended beyond the NBRS home ranges. Thin dashed lines indicate home ranges of adult female fishers within the study area. Home range boundaries for adult males are 100% MCPs encompassing the following telemetry locations: male 02: n_{NBRS} = 104 (1995–1999), n_{BRS} = 51 (1996–2000); male 05: n_{NBRS} = 24 (1996–1997); male 06: n_{NBRS} = 18 (1996–1997); male 07: n_{NBRS} = 150 (1996–1999), n_{BRS} = 48 (1997–1999).

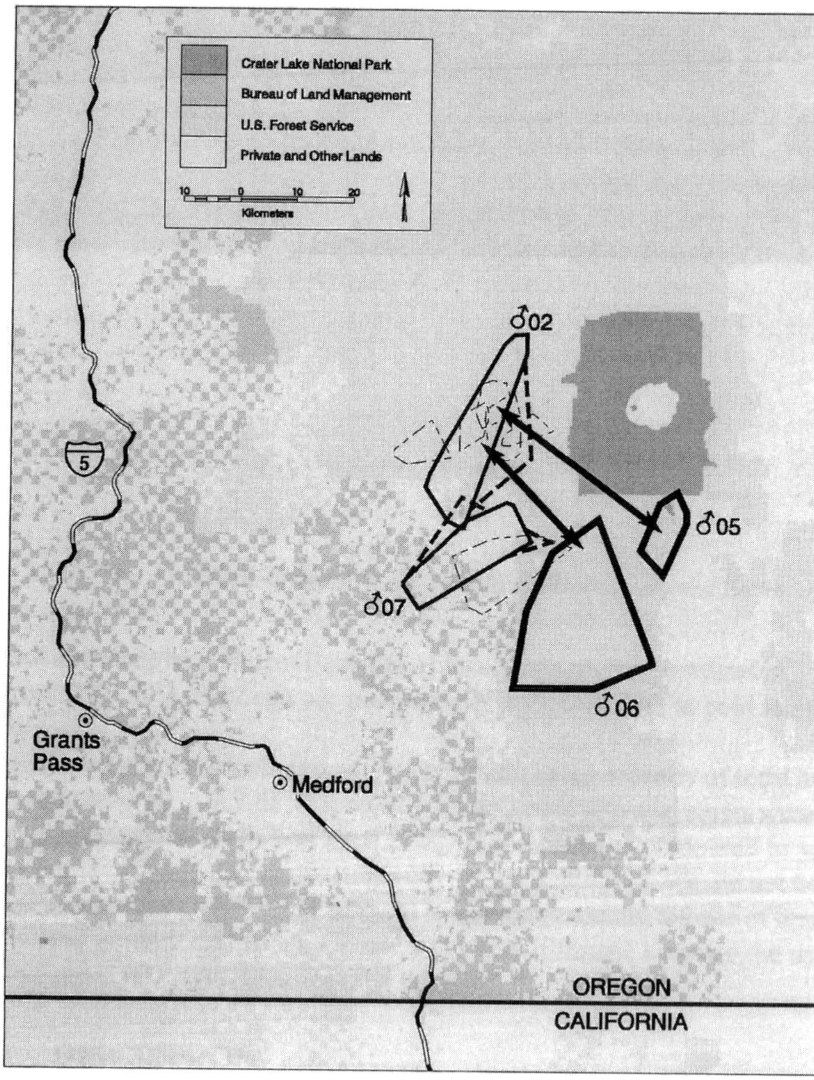

territorial species, we hypothesized that juvenile dispersal in fishers is male-biased, resulting in greater philopatry among females.

The minimum dispersal distances (*sensu* Arthur et al. 1993) we documented for 1 juvenile male and 1 juvenile female in our study population were consistent with these predictions. The female dispersed a relatively short distance (16.9 km) from her capture site and remained within our core study area, whereas the male dispersed a much greater distance (55.3 km), and settled well beyond the boundaries of our core study area (Fig. 10.3). Genetic theory predicts that for species with a polygynous mating system and female philopatry, adult females in a given area will be more closely related to one another than are adult males (Chesser 1991). This is because female offspring are more likely to disperse short distances and establish home ranges near their close relatives, whereas male offspring are more likely to disperse long distances and establish home ranges in different geographic areas than their close relatives. If our study population was socially structured due to male-biased juvenile dispersal and female philopatry, then adult females will have higher mean relatedness than adult males, and there will be more first-order (i.e., parent-offspring) relationships among adult females than among adult males.

2. METHODS

2.1 Molecular Markers

We obtained genotypic data on our study animals by measuring variation in microsatellite DNA extracted from ear or muscle tissue that we collected from 5 adult and 2 juvenile (<1 yr old) males, and 6 adult and 3 juvenile females in the southern Cascade Range in Oregon. We also collected tissue samples from 1 male and 1 female kit of known maternity that we found dead in a natal den. In addition, we obtained tissue samples from 2 adult male fishers that were trapped incidentally about 70 km SW of our study area in the Siskiyou Mountains of southwestern Oregon (Fig. 10.1).

Microsatellite DNA loci are repeated sequences of 2–5 nucleotides that occur in non-coding regions of nuclear DNA and have high mutation rates compared to mitochondrial DNA or genes that code for allozymes (McDonald and Potts 1997). With information on the unique genetic sequences complementary to the regions that flank a microsatellite locus (primers), researchers can use the polymerase chain reaction (PCR) to produce billions of copies of the target locus. Because fragments of different length (alleles) travel at different rates in an electrophoretic gel, unique alleles can be identified within and among individuals at each locus, providing a measure of genetic variation.

Figure 10.3. Dispersal of 2 juvenile fishers in the southern Cascade Range in Oregon.Thick solid lines (male) and thick dashed lines (female) show the direction and distance of dispersal, and post-dispersal home ranges. Thin solid lines (males) and thin dashed lines (females) indicate home ranges for adult fishers within the study area. Boundaries of post-dispersal home ranges are 100% MCPs encompassing the following telemetry locations: male 13, $n = 16$ (Jun–Dec 1999); female 04, $n = 117$ (Apr 1996–Jun 2000). Solid circles and open triangles are as described in Fig. 10.1.

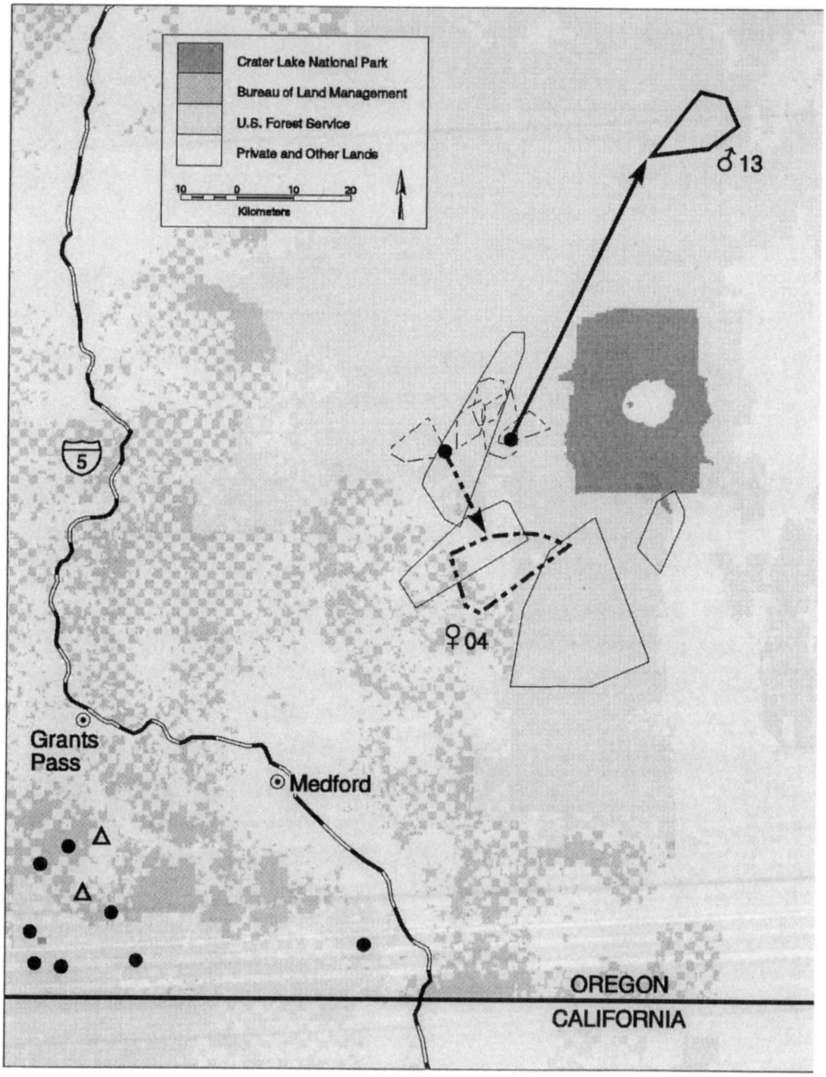

Microsatellite markers are useful for studying genetic differentiation over relatively short time frames both within and among populations (Avise 1994); thus, they are particularly appropriate for investigating the genetic affinities of radio-marked study animals.

2.2 Microsatellite Genotyping

We (SW and SB) used microsatellite DNA to describe the molecular genetic characteristics of fishers from southwestern Oregon. No microsatellite primers have been developed specifically for fishers, but primers developed for other carnivore species have been shown to amplify across species and genera (O'Connell et al. 1996). Consequently, we screened our sample of fishers for polymorphic loci using 19 primers developed for American marten, American mink (*Mustela vison*), ermine (*M. erminea*), American badger (*Taxidea taxus*), and black bear (*Ursus americanus*; Table 10.1).

We extracted DNA from 20 μg of ear plug or muscle tissue with 700 μl of lysis buffer that contained 20 μl of 1M DTT, 20 μl of 20 μg/μl proteinase k, and 10 μl of 10 μg/μl RNase. We agitated each sample in 55°C water for 40 hr; midway through the incubation, we added 10 μl of DTT, 10 μl of proteinase k, and 5 μl of RNase. We extracted DNA from the solution using 5 M ammonium acetate. To precipitate DNA from solution, we used a 70% ethanol wash. We resuspended DNA in a 1 X Tris EDTA solution and stored all samples at -20°C. We ran each reaction with a negative control to identify spurious results due to contamination. We used ear and muscle samples from the same individual as a positive control to verify that they produced similar results across loci. We resolved individual profiles electrophoretically with polyacrylamide gels using a LI-COR Model 4200 IR^2 Series DNA Sequencer (LI-COR Inc., Lincoln, NE). Resulting digital images characterized the size (number of nucleotide base-pairs) of alleles, and whether individuals were homozygous or heterozygous at each locus.

2.3 Genetic Analyses

2.3.1 Zoogeography

We used an assignment test (Paetkau et al. 1995) to determine whether the 2 fishers from the northern Siskiyou Mountains belonged to the southern Cascade Range gene pool. This test calculates the probability of an individual's genotype occurring in the population from which it was sampled and the probability of the genotype occurring in the other sampled population. We calculated probabilities with the software program GeneClass 1.0 (Cornuet et al.

1999) using a maximum-likelihood approach based on a Bayesian estimation of genotype frequencies to assign individuals to a population. We believe that our sample of 18 fishers from the southern Cascade Range represented a large proportion of the animals present in that population during our field study (Fig. 10.1); however, our sample of 2 fishers from the northern Siskiyou Mountains was clearly not a representative sample from that population. Accordingly, we only drew inferences from this analysis regarding the likelihood that the 2 Siskiyou fishers belonged to the Cascade Range gene pool and, consequently, whether or not they were to be included in subsequent genetic analyses.

We then tested for departures from Hardy-Weinberg (H-W) equilibrium in our resulting dataset. Departures from H-W equilibrium indicate an excess or deficit of heterozygosity in a population, and provide information on the potential influence of migration, genetic drift, or non-random mating on our analyses. In addition, many statistical analyses used in genetic studies are based on the assumption that populations are in H-W equilibrium. We tested for such departures by comparing observed (H_o) and expected heterozygosity (H_e) with a H-W exact test of P using the software program Arlequin 2.0 (Excoffier et al. 1996–2002). We adjusted for experiment-wise Type 1 error using a sequential Bonferroni adjustment (initial $\alpha = 0.05$, Rice 1989).

2.3.2 Spacing Patterns

To test the hypothesis that the 2 encroaching males had greater reproductive success than the 2 resident males, we used field and genotypic data to determine which of these adult males could be excluded as a potential father of the 7 juveniles we sampled during our field study. Due to delayed implantation of blastocysts in fishers, 11–12 months elapse between mating and parturition (Mead 1994); thus, we first excluded males as potential fathers if they had been dead ≥1 year prior to the birth of each offspring. We then excluded males as potential fathers if they did not share at least 1 allele at each locus with the offspring in question.

2.3.3 Dispersal

To test the hypothesis of female philopatry in fishers, we calculated genetic relatedness (R) among the entire sample of adult animals and within each sex. A positive R value indicates that a given group of animals are more closely related to one another than would be expected by random mating; a negative R value indicates the opposite. We estimated R and generated 95% confidence intervals around each estimate using a jackknife procedure in the software program Relatedness 5.0 (Queller and Goodnight 1989). To further investigate this hypothesis, we used genotypes to exclude potential first-order relationships among adult animals.

3. RESULTS

3.1 Zoogeography

Nine of the 19 microsatellite loci we screened produced clear polymorphisms, with an average of 3.0 ± 0.3 (± SE) alleles per polymorphic locus (Table 10.1). At 2 loci (Mvi 39 and Mvis 002), both fishers from the northern Siskiyou Mountains were homozygous for alleles that were not detected among fishers from the southern Cascade Range; at a third locus (Mer 041), they were homozygous for an allele that was rare among Cascade fishers (Table 10.2). Because of these strong genotypic differences, the assignment test resulted in probabilities <0.0001 that either fisher from the Siskiyou Mountains was a member of the southern Cascade Range gene pool. Consequently, all subsequent genetic analyses were conducted only on the sample of 18 fishers from

Table 10.1. Microsatellite loci screened for polymorphisms using DNA from fishers in southwestern Oregon. Allele lengths are expressed in numbers of base pairs.

Locus	Species for which primers were developed	No. of alleles	Allele length	Source
G1A	*Ursus americanus*	1	173	Paetkau et al. (1995)
Ma 1	*Martes americana*	5	218–226	Davis and Strobeck (1998)
Ma 14		1	219	
Ma 19		3	221–225	
Mer 030	*Mustela erminea*	1	278	Fleming et al. (1999)
Mer 041		4	175–181	
Mer 082		1	130	
Mer 095		1	178	
Mvi 39	*Mustela vison*	3	142–152	O'Connell et al. (1996)
Mvi 87		2	103–105	
Mvi 111		1	112	
Mvi 114		2	104–106	
Mvi 219		1	265	
Mvis 002	*Mustela vison*	3	220–228	Fleming et al. (1999)
Mvis 020		2	216–222	
Mvis 022		1	302	
Mvis 027		1	207	
Mvis 072		3	291–297	
Tt 4	*Taxidea taxus*	1	182	Davis and Strobeck (1998)

Table 10.2. Occurrence of microsatellite genotypes at selected loci in fishers from the southern Cascade Range and northern Siskiyou Mountains of Oregon.

Geographic location	Microsatellite loci		
	Mer 041[a]	Mvi 39[a]	Mvis 002[b]
Southern Cascade Range	175–177	142–144	220–220
	177–177	144–144	220–228
	177–179		228–228
	177–181		
Northern Siskiyou Mountains	181–181	152–152	224–224

[a]$n = 17$ for the southern Cascade Range, $n = 2$ for the northern Siskiyou Mountains.
[b]$n = 18$ for the southern Cascade Range, $n = 2$ for the northern Siskiyou Mountains.

our study population in the southern Cascade Range. We found 1 locus (Mvis 020) to have a heterozygote deficiency; after accounting for experiment-wise error, however, all loci were in Hardy-Weinberg equilibrium (Table 10.3).

3.2 Spacing Patterns

Parent-offspring exclusions for the 2 encroaching males, the 2 resident males, and the 7 juveniles we sampled during our field study included 5 by date and 19 by genotype (Table 10.4). Based on this analysis, neither of the encroaching males could have fathered any of these juveniles, whereas 1 of the resident males (02) was a potential father for 4 of the 7 juveniles.

Table 10.3. Observed heterozygosity (H_o), expected heterozygosity (H_e), and the exact probability for the test of Hardy-Weinberg equilibrium (HWE *P*) for 9 polymorphic loci among 18 fishers from the southern Cascade Range in Oregon.

Locus	n[a]	H_o	H_e	HWE P[b]
Ma 1	16	0.59	0.74	0.39
Ma 19	18	0.67	0.56	0.48
Mer 041	17	0.59	0.52	1.00
Mvi 39	17	0.23	0.27	1.00
Mvi 87	18	0.39	0.51	0.37
Mvi 114	15	0.41	0.40	1.00
Mvis 002	18	0.28	0.29	1.00
Mvis 020	18	0.11	0.34	0.04
Mvis 072	18	0.55	0.54	0.34

[a]Sample sizes vary among loci because some DNA samples failed to amplify.
[b]We adjusted for experiment-wise Type 1 error using a sequential Bonferroni adjustment (initial $\alpha = 0.05$).

Table 10.4. Inferred paternity of juvenile fishers among 2 resident (02, 07) and 2 encroaching (05, 06) males from our study population in the southern Cascade Range in Oregon. For each juvenile, we excluded adult males as potential fathers by date if they died ≥ 1 year prior to that juvenile's birth date, and by genotype if they did not share at least 1 allele at each locus with that juvenile.

Offspring	Potential fathers excluded by date	Potential fathers excluded by genotype	Potential fathers
04	none	02, 05, 06, 07	none
10	none	02, 05, 06, 07	none
12	06	02, 05, 07	none
13	06	05, 07	02
14	06	05, 07	02
16	06	05, 07	02
17	06	05, 07	02

3.3 Dispersal

Mean relatedness (R) among all adult fishers in our sample was –0.05 ± 0.11; mean R among males and females was –0.29 ± 0.15 and 0.19 ± 0.30, respectively. For R values, variance was represented by 95% confidence intervals; the lack of overlap between intervals for males and females indicated that adult females were significantly more related to each other than were adult males (Zar 1984). Separate genotypic comparisons among adult males and adult females showed that there were no possible first-order relationships among any of the 5 adult male fishers in our sample; in contrast, each of 6 females had at least 2 potential first-order relationships with another adult female (Table 10.5).

4. DISCUSSION

4.1 Zoogeography

The population of fishers we studied in the southern Oregon Cascade Range was reintroduced to that area approximately 25 years ago (Aubry and Lewis 2003). Since that time, there would have been numerous opportunities for juvenile dispersals or long-distance movements by adults from the Siskiyou Mountains into our study area (Fig. 10.1). Dispersals by juvenile fishers >50 km have been reported by several researchers (Leonard 1980, York 1996, this study), demonstrating that the southern Oregon Cascades are within the dispersal range of fishers in the northern Siskiyou Mountains (Figs. 10.1, 10.3). However, the

Table 10.5. Potential first-order (parent-offspring) relationships among consexuals for 11 adult fishers from the southern Cascade Range in Oregon. We excluded potential first-order relationships if they did not share at least 1 allele at each locus.

Adult fishers	Potential first-order relationships
Females	
01	03, 08, 18
03	01, 08, 11, 18
08	01, 03, 11, 18
09	11, 18
11	03, 08, 09, 18
18	01, 03, 08, 09, 11
Males	
02	none
05	none
06	none
07	none
15	none

large genotypic differences we found between fishers in our study population and those in the northern Siskiyou Mountains support our hypothesis that ecological or anthropogenic barriers in the intervening area have resulted in the geographic isolation of reintroduced fishers in the southern Cascade Range in Oregon. Thus, it appears that expanses of unsuitable habitat as narrow as 50 km might impede genetic exchange between fisher populations.

Our findings are consistent with recent genetic studies across Canada and in the northeastern U.S. which suggested that fishers may have relatively poor dispersal capabilities (Kyle et al. 2000, 2001; Kyle and Strobeck 2003). These studies revealed that fishers exhibited much more genetic structure than martens; i.e., there is less gene flow among fisher populations than among marten populations. These findings were unexpected because fishers are assumed to have better dispersal capabilities than martens due to their larger body size, which should result in less genetic structure among fisher populations. Kyle et al. (2001) speculated that this difference may be due to fisher populations being exposed to stronger anthropogenic influences (e.g., human development, transportation corridors, and habitat fragmentation) than marten populations. This hypothesis cannot be evaluated with field data, however, because information on juvenile dispersal is extremely limited for both fishers (Arthur et al. 1993, York 1996, this study) and American martens (Phillips 1994, Bull and Heater 2001, Fecske and Jenks 2002). This dearth of information underscores

Caughley's (1977) assertion that dispersal is the most challenging of all population processes to study in the field. Based on results of this study, we suggest that dispersal characteristics can be elucidated most effectively through a combination of field and genetic studies.

Recent genetic analyses of fishers from the region depicted in Fig. 10.1 indicated that fishers in the Siskiyou Mountains of Oregon represent the northern limit of the California gene pool. Ancillary to a phylogeographic study of fisher populations in the Pacific states (Wisely et al. 2004), an assignment test was used to evaluate the genetic affinities of 21 fishers from the southern Cascade Range in Oregon (including 16/18 analyzed in this study), the 2 fishers we genotyped from the northern Siskiyou Mountains in Oregon, and 23 additional animals from northwestern California. Both fishers from the Siskiyou Mountains in Oregon were assigned to the population in northwestern California (S. Wisely, unpublished data). These results, combined with a lack of strong ecological or anthropogenic barriers between these areas (Franklin and Dyrness 1973), indicate that fishers occurring southwest of I-5 in Oregon (Fig. 10.1) represent the only extant populations of indigenous fishers in Oregon.

4.2 Spacing Patterns

Contrary to Leonard's (1986) hypothesis that the typical breeding strategy for male fishers is to abandon their non-breeding home ranges and search for receptive females, our field observations indicated that adult males may exhibit either "resident" or "encroaching" behavioral strategies during the breeding season. Furthermore, our genetic results suggest that Sandell's (1986) findings for stoats (*Mustela erminea*), in which dominant, roaming males secured all of the mating opportunities, does not apply to fishers. Although several males encroached on the home range of a resident male whose home range encompassed those of several females (Fig. 10.2), these encroachments appear not to have resulted in successful matings (Table 10.4). Ours is the first study to explore the relationship between spacing patterns of male fishers or martens during the breeding season and subsequent reproductive success. However, results reported for martens in Maine appear to be consistent with our observations (Katnik et al. 1994). All of the 14 adult male martens that were monitored with radiotelemetry remained resident on their non-breeding home ranges during the breeding season. At least 11 of these males occupied home ranges that overlapped those of 1 or more females and 7 had ranges that overlapped with 2 or more, suggesting that males may have remained in established territories during the breeding season to maximize their reproductive potential (Katnik et al. 1994).

Our observations are consistent with Sandell's (1989) theory that territoriality in male solitary carnivores (including fishers) during the breeding season is flexible and influenced primarily by the availability of receptive females. Sandell (1989) predicted that the distribution of females will determine whether or not males maintain exclusive home ranges throughout the year, or adopt a roaming strategy during the breeding season. He reasoned that when females are densely and evenly distributed, a male will secure more matings by staying and mating with the females in his home range; in all other situations, a male will secure more matings by searching for receptive females beyond the boundaries of his home range.

The 2 males in our study population whose non-breeding home ranges overlapped those of 1 or more females maintained their home ranges throughout the year. We speculate that the 2 males who adopted an encroaching strategy during the breeding season did so because they were unable to establish a home range that encompassed 1 or more resident females. Although other researchers have reported similar roaming behavior by adult male fishers during the breeding season (Buck 1982, Leonard 1986, Arthur et al. 1989), they did not determine whether roaming behavior was typical of adult males in the populations they studied, nor if it resulted in successful matings. Additional research involving both field and genetic studies will be needed to further elucidate the causal mechanisms of these behavioral strategies.

4.3 Dispersal

Published information providing empirical support for the theory of male-biased juvenile dispersal and female philopatry in fishers is extremely limited. Heterozygote deficiencies found by Williams et al. (2000) in all 8 of the fisher populations they sampled from the north-central and northeastern portions of its range indicated the presence of fine-scale genetic structuring within populations. They speculated that such structuring may reflect gene correlations that have accrued over time due to male-biased juvenile dispersal and female philopatry.

Two studies have investigated juvenile dispersal in wild fishers, yet neither supported these predictions. In a heavily trapped population in south-central Maine, Arthur et al. (1993) measured straight-line distances from the sites where 13 fishers were captured as juveniles to the nearest locations in their adult home ranges, and found no significant difference between males (n = 8, mean = 10.8, range = 4.1–19.5 km) and females (n = 5, mean = 11.3, range = 5.0–18.9 km). Similarly, York (1996) found no significant difference in male and female dispersal distances among fishers in a population in central Massachusetts with lower trapping mortality and higher population density (males: $n = 10$, mean = 25, range = 10–60 km; females: $n = 19$, mean = 37, range = 12–107 km). However, in our study of an untrapped fisher population, the 2 juvenile dispersals we documented in the field, the significantly higher relatedness of adult females compared to adult males, and the strong disparity in potential first-order relationships among adult females compared to adult males provide empirical evidence of male-biased juvenile dispersal and female philopatry in fishers. As several authors have argued (Arthur et al. 1993, Powell 1994), trapping mortality is likely to disrupt many aspects of fisher population ecology, including spacing patterns and dispersal processes. We suspect that similarities in male and female juvenile dispersal distances in both Maine and Massachusetts were related to trapping mortality.

5. CONCLUSIONS

Combining information on the distribution, spatial organization, and movements of fishers obtained in the field with genotypic data enabled us to gain important new insights about the zoogeography and population ecology of fishers in the Pacific Northwest. These insights have potentially important implications for the conservation of fisher populations in the Pacific states, and for refining our current understanding of adult spacing patterns and juvenile dispersal in fishers. We have provided evidence that genetic introgression of fishers from the northern Siskiyou Mountains to the southern Cascade Range in Oregon has not occurred, suggesting that these populations are geographically isolated. These results bear directly on questions regarding connectivity and gene flow among extant fisher populations in the Pacific states. We have also provided the first empirical evidence that behavioral strategies of male fishers during the breeding season are flexible and appear to be influenced primarily by the density and distribution of receptive females, and that juvenile dispersal in fishers is male-biased, resulting in greater philopatry among females.

Genetic analyses can substantially improve the interpretive value of radiotelemetry data and, ultimately, the usefulness and applicability of wildlife field studies. To maximize the heuristic value of radiotelemetry studies, which involve significant investments of time and money, we urge researchers to consider incorporating genetic data into their field studies. As shown here, such data can be obtained through collaborative research efforts, which have the added benefit of bringing several disciplines together during the analytical and interpretive phases of the research process; alternatively, a number of commercial laboratories now provide genotyping services at relatively low cost.

6. ACKNOWLEDGMENTS

We thank Daniel Harrison, Angela Fuller, and several anonymous reviewers for many valuable suggestions that greatly improved the quality of this manuscript. Greg Russell and Natalie Schenker assisted with genotyping, David McDonald provided statistical consultation, and Kent van Wagtendonk prepared the figures. Funding for this research was provided by the Pacific Northwest Research Station, U.S. Forest Service.

7. LITERATURE CITED

Arthur, S. M., W. B. Krohn, and J. R. Gilbert. 1989. Home range characteristics of adult fishers. Journal of Wildlife Management 53:674–679.

_____, T. F. Paragi, and W. B. Krohn. 1993. Dispersal of juvenile fishers in Maine. Journal of Wildlife Management 57:868–874.

Aubry, K. B., and J. C. Lewis. 2003. Extirpation and reintroduction of fishers (*Martes pennanti*) in Oregon: implications for their conservation in the Pacific states. Biological Conservation 114:79–90.

_____, and C. M. Raley. 2002. Ecological characteristics of fishers in the southern Oregon Cascade Range. Final progress report on file at the Pacific Northwest Research Station, U.S. Forest Service, Olympia, Washington, USA.

Avise, J. C. 1994. Molecular markers, natural history and evolution. Chapman and Hall, New York, USA.

Buck, S. 1982. Habitat utilization by fisher (*Martes pennanti*) near Big Bar, California. Thesis, Humboldt State University, Arcata, California, USA.

Bull, E. L., and T. W. Heater. 2001. Home range and dispersal of the American marten in northeastern Oregon. Northwestern Naturalist 82:7–11.

Carr, S. M., and S. A. Hicks. 1997. Are there two species of marten in North America? Genetic and evolutionary relationships within *Martes*. Pages 15–28 *in* G. Proulx, H. N. Bryant, and P. M. Woodard, editors. *Martes*: taxonomy, ecology, techniques, and management. Provincial Museum of Canada, Edmonton, Alberta.

Caughley, G. 1977. Analysis of vertebrate populations. John Wiley and Sons, New York, USA.

Chesser, R. K. 1991. Gene diversity and female philopatry. Genetics 127:437–447.

Cornuet, J. M., S. Piry, G. Luikart, A. Estoup, and M. Solignac. 1999. New methods employing multilocus genotypes to select or exclude populations as origins of individuals. Genetics 153:1989–2000.

Davis, C. S., and C. Strobeck. 1998. Isolation, variability, and cross-species amplification of polymorphic microsatellite loci in the family Mustelidae. Molecular Ecology 7:1771–1788.

Dobson, F. S. 1982. Competition for mates and predominant juvenile male dispersal in mammals. Animal Behavior 30:1183–1192.

Drew, R. E., J. G. Hallett, K. B. Aubry, K. W. Cullings, S. M. Koepf, and W. J. Zielinski. 2003. Conservation genetics of the fisher (*Martes pennanti*) based on mitochondrial DNA sequencing. Molecular Ecology 12:51–62.

Excoffier, L., S. Schneider, and D. Roessli. 1996–2002. Arlequin ver. 2.0: a software for population genetic analysis. University of Geneva, Switzerland.

Fecske, D. M., and J. A. Jenks. 2002. Dispersal by a male American marten, *Martes americana*. Canadian Field-Naturalist 116:309–311.

Fleming, M. A., E. A. Ostrander, and J. A. Cook. 1999. Microsatellite markers for American mink (*Mustela vison*) and ermine (*Mustela erminea*). Molecular Ecology 8:1351–1362.

Franklin, J. F., and C. T. Dyrness. 1973. Natural vegetation of Oregon and Washington. U.S. Forest Service General Technical Report PNW-8.

Greenwood, P. J. 1980. Mating systems, philopatry and dispersal in birds and mammals. Animal Behaviour 28:1140–1162.

Hagmeier, E. M. 1959. A re-evaluation of the subspecies of fisher. Canadian Field-Naturalist 73:185–197.

Haig, S. M. 1998. Molecular contributions to conservation. Ecology 79:413–425.

Katnik, D. D., D. J. Harrison, and T. P. Hodgman. 1994. Spatial relations in a harvested population of marten in Maine. Journal of Wildlife Management 58:600–607.

Kebbe, C. E. 1961. Return of the fisher. Oregon State Game Commission Bulletin 16:3–7.

Kyle, C. J., C. S. Davis, and C. Strobeck. 2000. Microsatellite analysis of North American pine marten (*Martes americana*) populations from the Yukon and Northwest Territories. Canadian Journal of Zoology 78:1150–1157.

_____, J. F. Robitaille, and C. Strobeck. 2001. Genetic variation and structure of fisher (*Martes pennanti*) populations across North America. Molecular Ecology 10:2341–2347.

_____, and C. Strobeck. 2003. Genetic homogeneity of Canadian mainland marten populations underscores the distinctiveness of Newfoundland pine martens (*Martes americana atrata*). Canadian Journal of Zoology 81:57–66.

Leonard, R. D. 1980. The winter activity and movements, winter diet and breeding biology of the fisher (*Martes pennanti*) in southeastern Manitoba. Thesis, University of Manitoba, Winnepeg, Canada.

_____. 1986. Aspects of reproduction of the fisher, *Martes pennanti*, in Manitoba. Canadian Field-Naturalist 100:32–44.

McDonald, D. B., and W. K. Potts. 1997. Microsatellite DNA as a genetic marker at several scales. Pages 29–49 *in* D. Mindell, editor. Avian molecular evolution and systematics. Academic Press, New York, USA.

McGowan, C., L. A. Howes, and W. S. Davidson. 1999. Genetic analysis of an endangered pine marten (*Martes americana*) population from Newfoundland using randomly amplified polymorphic DNA markers. Canadian Journal of Zoology 77:661–666.

Mead, R. A. 1994. Reproduction in *Martes*. Pages 404–422 *in* S. W. Buskirk, A. S. Harestad, M. G. Raphael, and R. A. Powell, editors. Martens, sables, and fishers: biology and conservation. Cornell University Press, Ithaca, New York, USA.

Mitton, J. B., and M. G. Raphael. 1990. Genetic variation in the marten, *Martes americana*. Journal of Mammalogy 71:195–197.

O'Connell, M., J. M. Wright, and A. Farid. 1996. Development of PCR primers for nine polymorphic American mink (*Mustela vison*) microsatellite loci. Molecular Ecology 5:311–312.

Paetkau D., W. Calvert, I. Stirling, and C. Strobeck. 1995. Microsatellite analysis of population structure in Canadian polar bears. Molecular Ecology 4:347–354.

Parker, P. G., A. A. Snow, M. D. Schug, G. C. Booton, and P. A. Fuerst. 1998. What molecules can tell us about populations: choosing and using a molecular marker. Ecology 79:361–382.

Phillips, D. M. 1994. Social and spatial characteristics, and dispersal of marten in a forest preserve and industrial forest. Thesis, University of Maine, Orono, USA.

Powell, R. A. 1993. The fisher: life history, ecology, and behavior. Second edition. University of Minnesota Press, Minneapolis, USA.

_____. 1994. Structure and spacing of *Martes* populations. Pages 101–121 *in* S. W. Buskirk, A. S. Harestad, M. G. Raphael, and R. A. Powell, editors. Martens, sables, and fishers: biology and conservation. Cornell University Press, Ithaca, New York, USA.

Queller, D. C., and K. F. Goodnight. 1989. Estimating relatedness using genetic markers. Evolution 43:258–275.

Rice, W. R. 1989. Analyzing tables of statistical tests. Evolution 43:223–225.

Sandell, M. 1986. Movement patterns of male stoats *Mustela erminea* during the mating season: differences in relation to social status. Oikos 47:63–70.

_____. 1989. The mating tactics and spacing patterns of solitary carnivores. Pages 164–182 *in* J. L. Gittleman, editor. Carnivore behavior, ecology, and evolution. Volume 1. Cornell University Press, Ithaca, New York, USA.

Williams, R. N., L. K. Page, T. L. Serfass, and O. E. Rhodes, Jr. 1999. Genetic polymorphisms in fishers (*Martes pennanti*). American Midland Naturalist 141:406–410.

_____, O. E. Rhodes, Jr., and T. L. Serfass. 2000. Assessment of genetic variance among source and reintroduced fisher populations. Journal of Mammalogy 81:895–907.

Wisely, S. M., S. W. Buskirk, G. A. Russell, K. B. Aubry, and W. J. Zielinski. 2004. Phylogeography and genetic diversity of the fisher (*Martes pennanti*) in a peninsular and peripheral metapopulation. Journal of Mammalogy 85:in press.

York, E. C. 1996. Fisher population dynamics in north-central Massachusetts. Thesis, University of Massachusetts, Amherst, USA.

Zar, J. H. 1984. Biostatistical analysis. Second edition. Prentice-Hall, Inc., Englewood Cliffs, New Jersey, USA.

Zielinski, W. J., T. E. Kucera, and R. H. Barrett. 1995. Current distribution of fishers, *Martes pennanti*, in California. California Fish and Game 81:104–112.

Chapter 11

HARVEST STATUS, REPRODUCTION AND MORTALITY IN A POPULATION OF AMERICAN MARTENS IN QUÉBEC, CANADA

Clément Fortin and Michel Cantin

Abstract: Effects of trapping on a previously unexploited (since 1895) population of American martens (*Martes americana*) were investigated (1984–1994) in the Laurentides Wildlife Reserve, Québec. We used trapline-specific data on total harvest and trapping effort, and the age and sex distribution of 8,801 carcasses to evaluate a commonly applied age- and sex-ratio index for evaluating harvest sustainability. Percent of males in the harvest was linearly related to trapping success. However, juvenile:adult female and juvenile:ovulating female ratios were not significantly related to harvest or trapping success. A sex ratio in favor of males (1.3 to 1.6) male:female was positively associated with trapping success (1.3 to 1.5 captures/100TN) during the last 6 years of the study. Ovulation rate for 1-yr-old females was 44% during 1984–1985, and increased to 76% in 1990, a unique finding for American marten. Mean ovulation rate (4.11 corpora lutea/ovulating female) was exceptionally high for American martens and did not vary significantly among years. The number of ovulations per adult (>2 yrs) female was 3.21; this is among the highest values reported for marten in North America. Annual mortality rate for all ages classes (35%) was similar to the rate calculated for the >5-yr-old cohort. However, mortality rates for martens in the 0–3-yr cohort were 61% (1984–86) and 66% (1990–91). Although sex ratio seemed to be a reliable index of harvest intensity, other commonly used harvest indices (e.g., juveniles:adult females; juveniles:ovulating females) were not significantly related with trapping success. Use of traditional harvest indices for managing marten populations was not supported by our results.

1. INTRODUCTION

Because populations of American martens (*Martes americana*) are difficult to monitor and are sensitive to over-exploitation (Strickland 1994), indirect indices are often used to monitor harvest. Juvenile (0.5 years) /adult ratio (≥1.5 years) (J:A), the male to female sex ratio (M:F), the juvenile/adult female ratio (J:AF), and the juvenile/adult female ≥2.5 years old ratio (J:FM), have been used previously to monitor harvested populations of martens (Strickland and Douglas 1987). The advantage of these indices is that they are inexpensive and easy to obtain; however, they provide a *post facto* assessment of harvest levels, which hinders attempts to anticipate over-harvest (Thompson and Colgan 1987). Indices need to be validated across the distributional range of the species, but they may still provide inaccurate assessments of populations status during food shortages (Thompson and Colgan 1987, Graf 1993).

Since 1989, the Québec Government has implemented the use of a trapper's logbook to gather information about abundances, population trends, and trapping effort for martens and snowshoe hares (*Lepus americanus*). The analysis of 5 years of logbook data did not indicate a relationship between abundance and population characteristics (Garant et al. 1996), suggesting that population indices used to monitor harvest levels needed to be further assessed at a finer scale.

To better understand the effects of trapping on abundance and population trends of martens, Fortin and Cantin (1994) initiated the collection of data on trapping effort and population dynamics in the Laurentides Wildlife Reserve. This study is a continuation of that work; our objectives were to evaluate harvest level indices, and to determine reproductive and mortality rates for martens in the reserve.

2. STUDY AREA AND METHODS

The Laurentides Wildlife Reserve is located about 40 km north of Québec City and encompasses 7,934 km^2. This predominantly wilderness area is part of the Laurentides Highlands, the largest and highest section of the Laurentian Precambrian shield. Elevations range from 350 to 1,090 m. Temperatures in the central part of the study area averaged 15.5 °C in July, and −17 °C in January; average annual snowfall was 4.3 m. The study area is dominated by black spruce (*Picea mariana*) and balsam fir (*Abies balsamea*). Since 1984, most of the Wildlife Reserves in Québec have been open to regulated trapping. Prior to 1984, Laurentides Wildlife Reserve had not been trapped for over a century. Legal trapping began in the fall of 1984 without quotas and harvests of martens were monitored from fall 1984 until winter 1994, during 11 trapping seasons that extended from 18 October to 31 March in 1984–1987, 1 November to 15 December in 1988–1990, and 18 October to 15 December in 1991–1994 (Table 11.1). Trapping territories were allocated to 109 trappers through a computerized lottery system. Meetings were held with trappers before each trapping season to explain the purpose and methods of the study. All trappers were issued an information package, and before the beginning of the season, all trappers were contacted by mail to secure their cooperation.

Total harvests were estimated using trappers' mandatory reports and daily logbooks. Trapping effort (trap-nights [TN]) was based on the number of traps recorded daily in the logbooks. All trappers used rotating-jaw (Conibear 120 manufactured by Woodstream Oneida Victor, Cleveland, Ohio, USA) traps. Carcass deposit bins were placed on the 5 main highways that provided areas to the reserve. Trappers deposited carcasses in plastic bags with proper trapline identification, and individual dates of capture. Carcasses were collected weekly, anatomically sexed, and aged by radiographs of root canals (Dix and Strickland 1986, Fortin et al. 1988).

Microtome sections were taken from the ovaries of a sample of 183 adult females, (60, 53 and 70 in 1984, 1985 and 1990 respectively) and stained with a mixture of hematoyxline, phloxine, and saffron (Masson 1956). Corpora lutea were counted to estimate ovulation and fecundity rates (Strickland and Douglas 1987, Mead 1994, Aune and Schladweiler 1997). Data were analyzed using SAS/STAT (SAS 1988). In order to detect differences in sex and age ratio across years, Chi-square analyses were used to compare age and sex ratios among seasons, and to compare ovulation rate among age classes. After 7 years of trapping, changes in reproductive potential were verified using Chi-square analyses for comparing ovulation rates between females 1.5-yr-old and ≥2 yr-females. To evaluate whether regulatory changes in season length were effective in altering trapping effort and harvest of martens, we used ANOVA to test whether effort (trap-nights) and harvest density (captures/10 km^2) were different among 3 categories of trapping seasons of equal length within category (1984–87 = 135 days, 1988–90, 45 days, 1991–1994 = 59 days). ANOVA was also used to assess reproductive potential between 1.5-yr-old and ≥2 yr-old females, and to compare mortality rates between seasons and age classes. Linear regressions were used to evaluate if harvest was related to trapping effort and fur prices, and allowed us to adjust trapping effort for changes in fur prices. We assumed that trappers would continue to add more trapping effort if fur prices remained high. Linear regressions were also used to evaluate the relationships between harvest density and age ratios, and between trapping success and the percent males in the harvest.

Mortality and survival rates estimated from sex-age ratios of 454 martens captured during long (135 days) trapping seasons (1984–86) were compared to those of 379 martens harvested during short (45 days) trapping seasons. We use Ricker's (1980) capture regression, which allows an estimation of the survival rate using the natural logarithm of the number of individuals remaining in each year class, to estimate the degree of resiliency of martens to trapping.

3. RESULTS

3.1 Harvest

Harvests ranged from 610 to 2347 martens per trapping season (Table 11.1). Harvest size was linearly related with trapping effort ($r^2 = 0.48$; $n = 11$, $P = 0.02$) (Fig. 11.1), but not with fur price ($r^2 = 0.03$; $n = 11$, $P = 0.63$). However, there was a significant linear relationship between indexed prices (average price of a marten pelt based on consumer's price index for reference year 1991 = \$100) and trapping effort ($r^2 = 0.46$; $n = 11$, $P = 0.02$) (Fig. 11.2). On the basis of their length, trapping seasons of equal length (Table 11.1) were grouped into 3 classes and trapping effort varied significantly ($F_{2,8} = 6.31$; $P = 0.023$) between 1984–87 (156,737 TN), 1988–90 (71,621 TN), and 1991–94 (61,662 TN). Also, there was a significant difference between the number of captures/10 km^2 among trapping seasons of different length ($F_{2,8} = 5.31$; $P = 0.03$), which averaged 2.4 (1984–87), 1.3 (1988–90), and 1.2 (1991–94). There were no significant linear relationships between annual harvest density and the M:F ($r^2 = 0.06$; $n = 11$, $P = 0.48$), J:A ($r^2 = 0.04$; $n = 11$, $P = 0.54$) and J:AF ($r^2 = 0.003$; $n = 11$, $P = 0.88$) ratios (Tables 11.1 and 11.2).

Trapping success varied from 0.6 to 1.9 captures/100 TN, and remained very stable (1.3–1.5) during the last 6 years (Table 11.1). Trapping success did not differ among the 3 classes based on trapping season length ($F_{2,8} = 0.35$; $P = 0.720$). There was a significant linear relationship ($r^2 = 0.56$; $n = 11$, $P = 0.003$) between trapping success and the proportion of males in the harvest (Fig. 11.3). However, there were no significant ($P \geq 0.14$) linear relationship between trapping success and age and sex ratios. For similar trapping successes of 1.3–1.5 captures/100 TN, the J:A ratio varied from 0.7 and 3.9, and J:AF ratio ranged from 1.7 and 9.3. Similar J:A ratios, 1.9 for year 1985 and 2.0 for year 1988 were associated with very large (1985 = 2,347) and very small (1988 = 610) harvests (Tables 11.1 and 11.2), while trapping success was relatively stable for that period, ranging from 1.3 to 1.5 captures/100 TN.

3.2 Demography

The age and sex structure of the harvested population was based on the analysis of 8,801 carcasses, which represented 72% of the total harvest (Table 11.2). Adults represented nearly half of the harvest in 1984, but declined substantially during subsequent years (Table 11.2). There was a significant difference ($\chi^2 = 86.42$, df = 10, $P < 0.005$) in the proportion of adults of each sex in the harvest among years, but not among the 3 trapping season classes

Table 11.1. Characteristics of marten harvests in the Laurentides Wildlife Reserve, Québec, Canada, from 1984–1994.

Year	# days in trapping season	Harvest	Exploited Area (km²)	Harvest Density (captures/ 10km²)	Effort (trap-nights x 1000)	Success (captures/ 100 TN)
1984	135[a]	1,130	5,009	2.3	60.4	1.9
1985	135	2,347	6,529	3.6	173.2	1.4
1986	135	1,617	6,855	2.4	208.3	0.8
1987	135	1,037	6,875	1.5	185.0	0.6
1988	45[b]	610	6,342	1.0	74.8	0.8
1989	45	1,010	6,805	1.5	65.5	1.5
1990	45	1,059	7,056	1.5	74.6	1.4
1991	59[c]	1,063	7,083	1.5	75.0	1.4
1992	59	695	7,125	1.0	53.6	1.3
1993	59	763	6,791	1.1	52.6	1.5
1994	59	900	6,904	1.3	65.5	1.4

[a]18 October–31 March
[b]1 November 15 December
[c]18 October–15 December

Figure 11.1. Relationship between harvest and trapping effort of American martens harvested in the Laurentides Wildlife Reserve, Québec, Canada, 1984–1994.

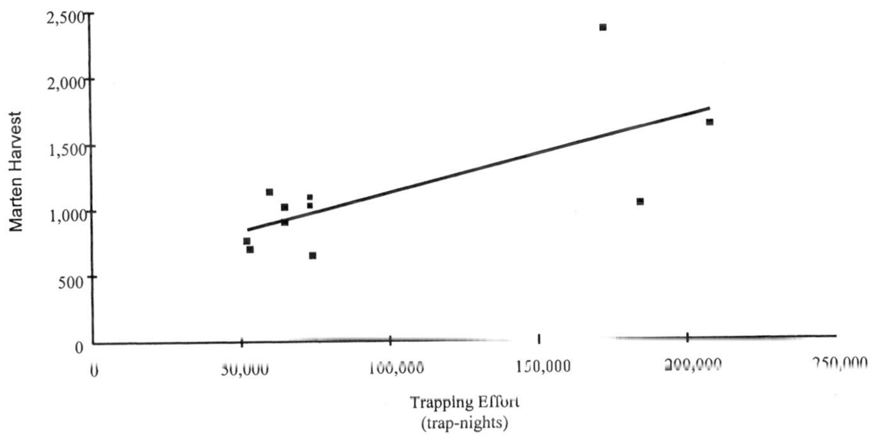

($\chi^2 = 3.88$; df = 2; $P > 0.10$). The M:F ratio differed significantly among years ($\chi^2 = 69.76$, df = 10; $P < 0.005$), but not among trapping season classes ($\chi^2 = 3.33$; df = 2; $P > 0.10$). Among juveniles, the sex ratio also differed significantly among years ($\chi^2 = 20.65$, df = 10, $P < 0.025$), but not among trapping

Figure 11.2. Relationship between trapping effort and the fur price for American martens harvested in Laurentides Wildlife Reserve, Québec, Canada, 1984–1994.

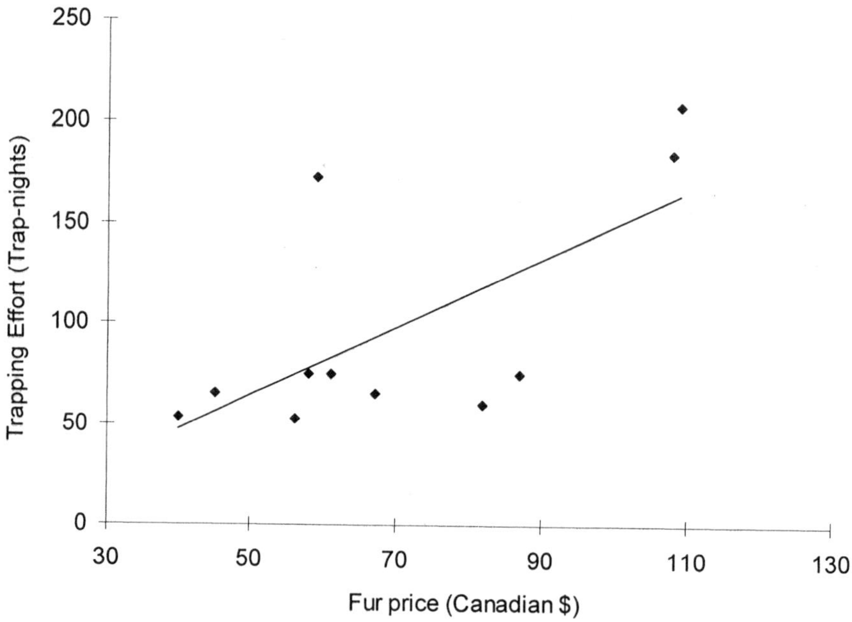

Table 11.2. Age and sex structure of American martens harvested in the Laurentides Wildlife Reserve, Québec, Canada, 1984–1994.

Year	Number of Males	% Males	Number of Females	% Females	M : F[a]	J : A[b]	J :AF[c]
1984	611	70.1	260	29.9	2.4	1.1	4.0
1985	1,126	60.3	742	39.7	1.5	1.9	5.8
1986	592	55.7	471	44.3	1.3	0.5	1.1
1987	418	52.2	383	47.8	1.1	1.2	2.3
1988	215	57.2	161	42.8	1.3	2.0	4.6
1989	426	57.7	312	42.3	1.4	3.9	9.3
1990	447	57.4	332	42.6	1.3	2.0	5.5
1991	387	57.8	282	42.2	1.4	0.7	1.7
1992	311	60.4	204	39.6	1.5	1.0	2.8
1993	309	61.1	197	38.9	1.6	2.4	6.3
1994	358	59.1	248	40.9	1.4	2.7	6.6
TOTAL	5,200	59.1	3,592	40.9	1.4	1.4	3.6

[a] Male:female ratio
[b] Female:adult ratio
[c] Juveniles:adult female ratio

Figure 11.3. Relationship between trapping success and the percent of male martens harvested in Laurentides Wildlife Reserve, Québec, Canada, 1984–1994.

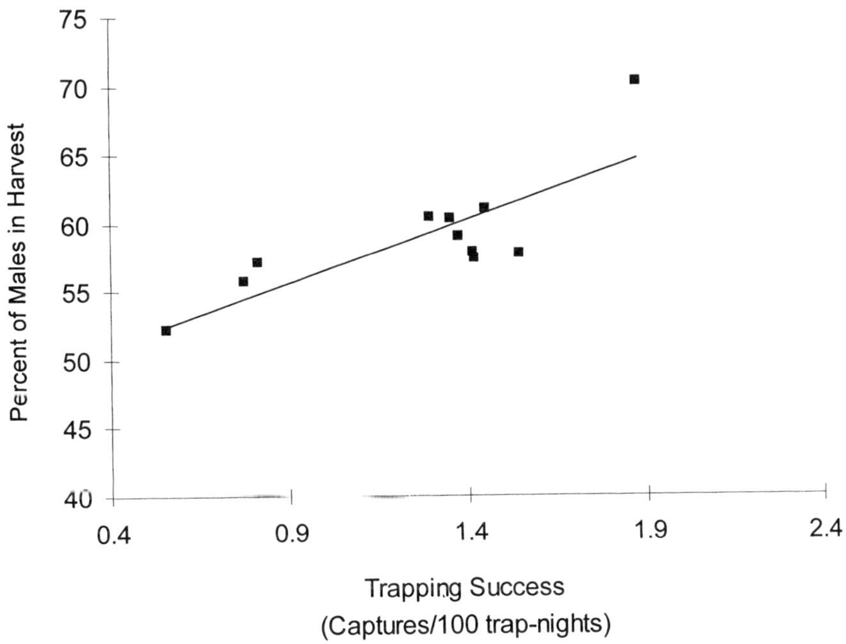

season classes ($\chi^2 = 1$, df = 2, $P > 0.50$). The J:A ratio varied significantly among years ($\chi^2 = 655.06$, df = 10, $P < 0.005$) and among trapping season classes ($\chi^2 = 200.25$, df = 2, $P < 0.005$). Similarly, the J:AF ratio was significantly different among years ($\chi^2 = 501.51$, df = 10, $P < 0.005$) and among season classes ($\chi^2 = 88.55$, df = 2, $P < 0.005$).

3.3 Mortality Factors

Estimated mortality rate was 35% for 0–12 yr-old martens and differed significantly ($F_{1,2} = 174.16$, $P = 0.0001$) between years. Similarly, estimated mortality rate was 33% for ≥1.5-yr-old martens, and also differed significantly between years ($F_{1,10} = 186.19$, $P = 0.0001$) (Fig. 11.4). Mortality rates for ≥5-yr-old martens were similar (35%) between 1984–1986 ($F_{1,6} = 56.43$, $P = 0.0003$) and 1990–1992 ($F_{1,4} = 8.96$, $P = 0.04$). The mortality rate of 0 to 3 yr-old martens was markedly higher than for older martens, and was 61% in 1984–1986 ($F_{1,2} = 23.17$, $P = 0.04$) and 66% in 1990–1992 ($F_{1,2} = 204.06$, $P = 0.005$).

Figure 11.4. Estimated survival (Ln) of martens harvested in the Laurentides Wildlife Reserve, Québec, Canada, 1984–1991.

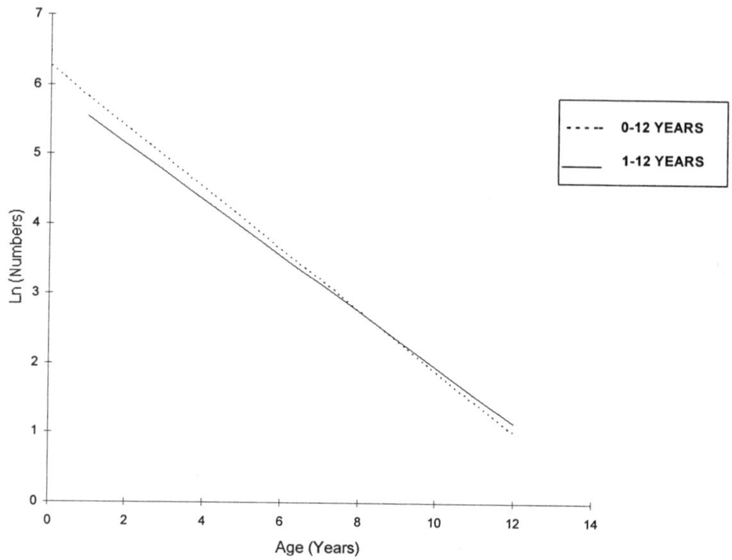

3.4 Reproduction

Reproductive potential was estimated using corpora lutea counts on 183 >1.5-yr-old adult females pooled from 1984 ($n = 60$), 1985 ($n = 53$), and 1990 ($n = 70$). Ovulation rate did not differ among the 3 years ($\chi^2 = 1.06$, df = 1, $P > 0.05$), and was 78% for the entire sample (Table 11.3). Ovulation rate of 1.5 yr-old females was significantly lower than for females ≥1.5 years ($\chi^2 = 26.61$, df

Table 11.3. Ovulation rate of American martens in Laurentides Wildlife Reserve, Québec, Canada, by age class during 1984–85 (1984), 1985–86 (1985) and 1990–91 (1991).

		Age (Years)						
Year		1	2	3	4	5	> 5	All
1984	%	40	100	100	100	80	67	73
	n	20	7	9	10	5	9	60
1985	%	48	100	100	100	86	100	76
	n	21	9	3	8	7	5	53
1990	%	76	100	100	50	100	88	83
	n	42	12	1	2	5	8	70
All	%	60	100	100	95	88	82	78
	n	83	28	13	20	17	22	183

$= 1$, $P < 0.005$) (Table 11.3). Also, the combined ovulation rate of females (44%) during 1984–85 was significantly lower than that for 1.5-yr-old females in 1990 (76%) ($\chi^2 = 7.8$, df $= 1$, $P < 0.01$) (Table 11.3). The mean number of corpora lutea per ovulating female was 4.11 (SE ± 0.7) (Table 11.4). The mean number (pooled) of corpora lutea for all adult females sampled in 1984, 1985, and 1990 was 3.21 (SE 0.14) (Table 11.5). There was a significant difference ($F_{5,177} = 11.09$, $P = 0.0001$) between the mean number of corpora lutea between 1.5-yr-old and ≥2.5-yr-old females (Table 11.5).

Table 11.4. Production of corpora lutea per ovulating females by age class of American martens in Laurentides Wildlife Reserve, Québec, Canada, 1984–85(1984), 1985–86 (1985), and 1990–91 (1991).

		Number of Corpora Lutea						
1984	Mean	3.87	4.00	4.56	4.90	4.75	4.33	4.41
	(SE)	(0.35)	(0.44)	(0.18)	(0.23)	(0.25)	(0.33)	(0.13)
	n	8	7	9	10	4	6	44
1985	Mean	3.27	4.25	4.00	4.25	4.17	4.60	4.00
	(SE)	(0.27)	(0.37)		(0.31)	(0.40)	(0.81)	(0.17)
	n	11	8	3	8	6	5	41
1990	Mean	3.72	4.25	4.00	5.00	4.2	4.29	3.97
	(SE)	(0.13)	(0.13)			(0.20)	(0.18)	(0.09)
	n	32	12	1	1	5	7	58
All	Mean	3.65	4.19	4.38	4.63	4.33	4.39	4.11
	(SE)	(0.11)	(0.16)	(0.14)	(0.19)	(0.19)	(0.24)	(0.07)
	n	51	27	13	19	15	18	143

Table 11.5. Numbers of corpora lutea observed in ovaries of adult females martens in Laurentides Wildlife Reserve, Québec, Canada, 1984–85 (1984), 1985–86 (1985), and 1990–91 (1991).

		Number of Corpora Lutea						
1984	Mean	1.55	4.00	4.56	4.90	3.80	2.89	3.23
	(SE)	(0.46)	(0.44)	(0.18)	(0.23)	(0.97)	(0.75)	(0.27)
	n	20	7	9	10	5	9	60
1985	Mean	1.71	3.78	4.00	4.25	3.57	4.60	3.09
	(SE)	(0.39)	(0.57)		(0.31)	(0.69)	(0.81)	(0.27)
	n	21	9	3	8	7	5	53
1990	Mean	2.83	4.25	4.00	2.50	4.20	3.75	3.29
	(SE)	(0.27)	(0.13)		(2.50)	(0.20)	(0.56)	(0.19)
	n	42	12	1	2	5	8	70
All	Mean	2.24	4.04	4.38	4.10	3.82	3.3.59	3.21
	(SE)	(0.21)	(0.22)	(0.14)	(0.29)	(0.39)	(0.42)	(0.14)
	n	83	28	13	20	17	22	183

4. DISCUSSION

4.1 Trapping and Population Indicators

Harvest density (captures/10 km^2) dropped by approximately half between the first and last 3 trapping seasons. However, yearly variations in the yield were not reflected in the J:A and J:AF ratios. This is in agreement with Garant et al. (1996) who did not find a significant difference between trapping success and the J:A ratio. However, our finding disagree with Strickland and Douglas (1987), who reported that J:AF ratios >3:1 indicated that harvests were within "acceptable levels".

Thompson and Colgan (1987) endorsed use of the J:AF and J:FM ratios, but recommended that food availability indices be developed to detect food shortages that could cause populations to decrease without affecting J:AF ratios. Aune and Schladweiler (1997) argued that J:AF ratios could be useful for assessing harvest status of martens if used "with caution".

Trapping success was linearly related with only the proportion of males in the harvest. Maintaining a sex ratio in favor of males (1.3 to 1.6) resulted in a relatively stable trapping success (1.3 to 1.5 captures/100TN) during the last 6 years of the study. Trapping success was comparable to levels reported by Potvin and Breton (1997) for elsewhere in Québec (1.4 captures/100 TN) and by Katnik et al. (1994) for Maine (1.6 captures/100 TN), but lower than values reported by Soutière (1979) for Maine (3.8 captures/100 TN), and Lofroth (1993) for British Columbia (2.7 to 3.9 captures/100 TN). Males are more vulnerable to trapping than females (Strickland and Douglas 1987, Fortin and Cantin 1994, Hodgman et al. 1994, Strickland 1994, Aune and Schladweiler 1997) because of their larger home range and greater movements (Banci and Proulx 1999). Thus, changes in sex ratio allow trappers and managers to retrospectively (1 yr delay) adjust harvest regulations, and to avoid removing an excess of females. Yeager (1950), Quick (1953), Soukkala (1983), and Archibald and Jessup (1984) noted that the harvest sex ratio was a function of trapping pressure; a harvest in which the sex ratio is nearly equal or in favor of females probably indicates overharvest. However, Strickland and Douglas (1987) claimed that sex ratios may not be reliable indicators of population status because the ratios in the population may vary with food abundance. On the basis of our findings, we suggest that the sex ratio may be a more useful index than J:A and J:AF ratios for monitoring population status of martens. This finding should be further evaluated in other regions where American martens are intensively harvested.

4.2 Fecundity

The demographic structure of this population of martens was considerably modified by 11 years of harvest. The population exhibited a change in age structure; the proportion of animals ≥5-yr-old declined from 13% in 1984–1986 to only 3.6% in 1990–1992 period. The overall ovulation rate of 83% (females >2.5 years) reported during this study was greater than observed in the Yukon (74%) (Archibald and Jessup 1984), but less than observed in southern Ontario (87%) (Strickland and Douglas 1987), southeastern Alaska (93%) (Flynn and Schumacher 1994), or Montana (87%) (Aune and Schladweiler 1997). After 7 years of harvest, the reproductive potential of 1-yr-old females increased significantly. This phenomenon, already known among canids (Gagnon and Fortin, unpublished report, 1987), is observed here for the first time in a population of American martens and probably results from greater availability of territories following the harvest of older, resident females.

The ovulation rate (all years pooled) for both females 1.5-yr-old (60%) and ≥2.5-yr-old (93%) was greater than observed in Ontario (Thompson and Colgan 1987). However, these rates are less than those reported by Strickland and Douglas (1987) for southern Ontario (80% and 92%), and by Aune and Schaldweiler (1997) for Montana (85.5% and 95.5%). The average number of corpora lutea per ovulating female (4.11, SE 0.14), that we observed was higher than has previously been reported for American martens by Lensink (1953: 2.8) and Flynn and Schumacher (1994: 3.7) in Alaska, by Archibald and Jessup (1984: 3.3), in the Yukon, Canada, by Thompson Colgan (1987: 3.2) and Strickland and Douglas (1987: 3.5) in Ontario, and by (Aune and Schaldweiler (1997: 2.6) in Montana. The fecundity rate of 3.21 corpora lutea per adult female that we observed in the Laurentides Wildlife Reserve was also among the highest reported in North America. Previously, fecundity rates of 1.26 to 3.25 have been reported (Strickland and Douglas 1987, Thompson and Colgan 1987, Archibald and Jessup 1984, Aune and Schaldweiler 1997, Flynn and Schumacher 1994).

4.3 Mortality Factors

The estimated trapping mortality rate calculated for all years was 35 ± 5%, and was similar to rates reported for Québec (35%) for martens ≥4 years old (Fortin and Cantin unpublished report, 1990) and for Ontario (38%) (Fryxell et al. 1999). Natural mortality rates reported for Quebec (Potvin and Breton 1997), Newfoundland (Bisonnette et al, unpublished report, 1988), Ontario (Thompson 1994), and Maine (Hodgman et al. 1997) were, in some instances, quite similar to the trapping mortality rate that we estimated. The mortality rate calculated among animals aged 0 to 3 yr-old was 61% in 1984–1986, and 66% in 1990–1992; these values are much higher than the pooled mortality rate (35%) that we estimated for martens aged 0–12 years old. This higher mortality among the 0–3 yr-old cohorts might be explained by the greater vulnerability of juveniles to trapping (Francis and Stephenson 1972, Strickland and Douglas 1987, Fortin and Cantin 1994).

Thompson and Colgan (1987) were the first to hypothesize that mortality by trapping could be additive to natural mortality. Hodgman et al. (1994) reported that trapping was the main cause of death in a forest with extensive road access, and that trapping mortality occurred above naturally sustainable levels. Payer (1999) concluded that high mortality by trapping became additive in the male segment of the population at high levels of access; however, natural mortality rates of males were lower in trapped populations.

5. MANAGEMENT RECOMMENDATIONS

The American marten population of the Laurentides Wildlife Reserve was heavily harvested, particularly during the first 4 years after trapping was resumed. Harvests altered the age structure and reduced the mean age of the population. Young animals may be more affected by natural mortality and human-induced mortality; therefore this population remains vulnerable to overharvest, despite that reproductive potential is high relative to other areas where American martens have been studied. To avoid overharvesting, we propose that an index to harvest level be implemented. Specifically, we recommend that the proportion of males in the harvest be monitored through time. We propose that maintaining a sex ratio of 1.5 M:F or of ≥60% males in the harvest may prevent overharvesting. This management tool should be used in conjunction harvest quotas and information on food availability during the kit-rearing season. We conclude that J:A and J:AF ratios must be used with caution since they can provide an inaccurate assessment of harvest sustainability. We recommend further studies on the relationship between trapping success and population density and on the cumulative effects of habitat loss (e.g., logging) on harvested populations.

6. ACKNOWLEDGMENTS

We thank the late M. A. Stickland from the Ontario Ministry of Natural Resources for her invaluable input. We also thank J. Beauchemin, C. Picard, C. Caron, J. G. Frenette, and J. L. Brisebois for their technical assistance. We thank Y. Garant, and I. Thompson for reviewing the original french manuscript, G. Proulx for his review in both languages, and M. Brown and an anonymous reviewer for comments on the English manuscript. We are particularly grateful to the trappers of Laurentides Wildlife Reserve.

7. LITERATURE CITED

Archibald, W. R., and R. H. Jessup. 1984. Population dynamics of the pine marten (*Martes americana*) in the Yukon Territory. Pages 81–96 *in* Olson, R., R. Hastings, and F. Geodes, editors. Northern Ecology and Resource Management. University of Alberta Press, Edmonton, Alberta, Canada.

Aune, K. E., and P. Schladweiler. 1997. Age, sex structure, and fecundity of the American marten in Montana. Pages 61–77 *in* G. Proulx, H. N. Bryant, and, P. M. Woodard, editors. Martes: taxonomy, ecology, techniques, and management. Provincial Museum of Alberta, Edmonton, Alberta, Canada.

Banci, V., and G. Proulx. 1999. Resiliency of furbearers to trapping in Canada. Pages 175–203 *in* G. Proulx, editor. Mammal trapping. Alpha Wildlife Research & Management Ltd., Sherwood Park, Alberta, Canada.

Dix, L. M., and M. A. Strickland. 1986. Use of tooth radiographs to classify martens by sex and age. Wildlife Society Bulletin 14:275–279.

Flynn, R. W., and T. Schumacher. 1994. Ecology of martens in southeast Alaska. Federal Aid in Wildlife Restoration Progress Report, Project W-24-2, Study 7.16. Alaska Department of Fish and Game, Juneau, Alaska, USA .

Fortin, C., M. Cantin and M. Fortin. 1988. Expérimentation d'une méthode radiographique pour la détermination du sexe et l'estimation de l'âge chez la martre d'Amérique. Ministère du Loisir, de la Chasse et de la Pêche, Service de l'aménagement et de l'exploitation de la faune, Région de Québec.

_____, and _____. 1994. The effects of trapping on a newly exploited American marten population. Pages 179–191 *in* S.W. Buskirk, A. G. Harestad, M. G. Raphael, and R. A. Powell, editors. Martens, sables, and fishers: biology and conservation. Cornell University Press, Ithaca, New York, USA.

Francis, G. R., and A. B. Stephenson. 1972. Marten ranges and food habits in Algonquin Provincial Park, Ontario. Report number 91, Ontario Ministry of Natural Resources, Toronto, Ontario, Canada.

Fryxell, J. M., J. M. Falls, E. A. Falls, R. J. Brooks, L. Dix and M. A. Strickland. 1999. Density dependence, prey dependence, and population dynamics of marten in Ontario. Ecology 80: 1311–1321

Garant, Y., R Lafond, and R. Courtois. 1996. Analyse du système de suivi de la martre d'Amérique (*Martes americana*) au Québec. Ministère de l'Environnement et de la Faune, Direction de la faune et des habitats, Québec, Québec.

Graf, R. P. 1993. Experimental overharvest of martens, (*Martes americana)*, in Northwest Territories, Canada. Pages 229–232 *in* I. D. Thompson, editor. Proceedings of the International Union of Game Biologists XXI Congress, Halifax, Nova Scotia, Canada.

Hodgman, T. P., D. J. Harrison, D. D. Katnik and K. D. Elowe. 1994. Survival in an intensively trapped marten population in Maine. Journal of Wildlife Management 49:593–600.

_____, _____, D. M. Phillips and K. D. Elowe. 1997. Survival of American marten in an untrapped forest preserve in Maine. Pages 86–99 *in* G. Proulx, H. N. Bryant, and P. M. Woodard, editors. *Martes*: taxonomy, ecology, techniques, and management. Provincial

Museum of Alberta, Edmonton, Alberta, Canada.
Katnik, D. D., D. J. Harrison and T. P. Hodgman. 1994. Spatial relations in a harvested population of marten in Maine. Journal of Wildlife Management 58:600–607.
Lensink, C. J. 1953. An investigation of the marten in interior Alaska. Thesis, University of Alaska, Fairbanks, Alaska, USA.
Lofroth, E. C. 1993. Scale dependent analyses of habitat selection by marten in the sub-boreal biogeoclimatic zone, British Columbia. Thesis, Simon Fraser University, Burnaby, British Columbia, Canada.
Masson, P. 1956. Tumeurs humaines, histologie, diagnostic et techniques. Second edition, Paris, Maloine.
Mead, R. A. 1994. Reproduction in *Martes.* Pages 404–422 *in* S. W. Buskirk, A. G. Harestad, M. G. Raphael, and R. A. Powell, editors. Martens, sables, and fishers: biology and conservation. Cornell University Press, Ithaca, New York, USA.
Payer, D.C. 1999. Influences of timber harvesting on habitat selection and demographic characteristics of marten. Dissertation, The University of Maine, Orono, Maine, USA.
Potvin, F. 1998. La martre d'Amérique (*Martes americana*) et la coupe à blanc en forêt boréale : une approche télémétrique et géomatique. Dissertation, Faculté de forestrie et de géomatique. Université Laval, Québec, Canada.
_____, and L. Breton. 1997. Short term effects of clearcutting on martens and their prey in the boreal forest of western Québec. Pages 452–474 *in* G. Proulx, H. N. Bryant, and, P. M. Woodard, editors. *Martes:* taxonomy, ecology, techniques, and management. Provincial Museum of Alberta, Edmonton, Alberta, Canada.
Quick, H. F. 1953. Wolverine, fisher, and marten studies in a wilderness region. Transactions of the North American Wildlife Conference 18:512–533.
Ricker, W. E. 1980. Calcul et interprétation des statistiques biologiques des populations de poissons. Bulletin de l'office des recherches sur les pêcheries du Canada. Ottawa. 191F.
SAS. 1988. SAS/STAT user guide, Sixth edition. SAS Institute, Cary, North Carolina, USA.
Soukkala, A.M. 1983. The effects of trapping on marten populations in Maine. Thesis, University of Maine, Orono, Maine, USA.
Soutière, E. C. 1979. Effects of timber harvesting on marten in Maine. Journal of Wildlife Management 43:850–860.
Strickland, M. A. 1994. Harvest management of fishers and American martens. Pages 149–164 *in* S. W. Buskirk, A. G. Harestad, M. G. Raphael and R. A. Powell, editors. Martens, sables, and fishers: biology and conservation. Cornell University Press, Ithaca, New York, USA.
_____, C. W. Douglas. 1987. Marten. Pages 599–612 *in* M. Novak, J. A. Baker, M. E. Obbard and B. Malloch, editors. Wild furbearer management and conservation in North America. Ontario Trappers Association, North Bay, Ontario, Canada.
Thompson, I. D. 1994. Marten populations in uncut and logged boreal forests in Ontario. Journal of Wildlife Management 58:272–280.
_____, and P. W. Colgan. 1987. Numerical responses of martens to a food shortage in northcentral Ontario. Journal of Wildlife Management 51:824–835.
Yeager, L. E. 1950. Implications of some harvest and habitat factors on pine marten management. Pages 319–334 *in* Transactions of the fifteenth North American wildlife conference.

Chapter 12

ARE SCAT SURVEYS A RELIABLE METHOD FOR ASSESSING DISTRIBUTION AND POPULATION STATUS OF PINE MARTENS?

Johnny Birks, John Messenger, Tony Braithwaite, Angus Davison, Rachael Brookes, and Chris Strachan

Abstract: Systematic searches for marten feces or 'scats' have been used since 1980 for assessing the status of protected populations of pine martens (*Martes martes*) in Britain. Previous surveys using scats have relied on unsubstantiated assumptions that martens typically defecate along roads and trails, that martens inhabit primarily woodland habitats, and that scats from martens can reliably be distinguished from those of other carnivores. Results of scat surveys have drawn conflicting conclusions about population status, which has lead to disagreement about conservation action, and doubts about the reliability and validity of assumptions associated with the technique. We reviewed the recent history of survey programs for pine marten populations in Great Britain. We examined the assumptions made in different surveys and considered these critically. The scat survey technique has several limitations, and is likely to be least reliable where populations of martens are low and where distribution is uneven. New DNA testing approaches revealed the inaccuracy of marten scat identification in the field. We recommend that scat surveys should be conducted only when genetic verification is available to confirm scat identity.

1. INTRODUCTION

Surveying wildlife populations is an important tool for management and conservation because distribution and abundance data derived from systematic surveys are needed to make policy decisions. In the UK, monitoring of wildlife populations is essential if the Government is to meet its obligations to maintain or restore the favorable conservation status of key species, under the European Commission's Habitats and Species Directive (e.g., Macdonald et al. 1998, Toms et al. 1999).

Carnivores present particular problems for those devising programs to survey or monitor populations (Gese 2001). For martens, data may be derived from trapping returns (e.g., Strickland 1994, Aune and Schladweiler 1997, Helldin 1998); however, martens are strictly protected in some states, so alternative approaches to detection and monitoring are necessary. Although marten sightings and carcasses obtained from road-kills may provide useful information on distribution and abundance, they are of limited value for monitoring because sampling effort cannot be controlled. Snow-tracking can be used as a source of winter data on marten in some countries (e.g., Lindström et al. 1995), but snowfall in most of Britain is limited and unpredictable. Track plates and camera traps have been used successfully on martens in some states (Zielinski and Kucera 1995), though not in Britain. Genetic analysis of hairs recovered from bait stations or at dens has been used to confirm the identity of other species (e.g., Woods et al. 1999, Sloane et al. 2000), and has potential for use on marten via hair snagging tubes (Messenger and Birks 2000).

Pine martens are the only *Martes* native to Britain. Outside its Scottish stronghold the species is scarce or absent as a consequence of habitat loss and persecution in previous centuries (Langley and Yalden 1977, Tapper 1992). During the decline of pine martens in the 19th and early 20th centuries, information on distribution and abundance was derived primarily from reports of animals observed or killed by hunters and gamekeepers (e.g., Langley and Yalden 1977, Strachan et al. 1996, Webster 2001). However, the species has been partially protected by law in Britain since 1982, and fully protected since 1988. Instances of deliberate or accidental killing are rarely reported, especially where the species is scarce (e.g., Jefferies and Critchley 1994, Birks et al. 1997, Messenger et al. 1997). Since 1980, assessments of marten status in Britain have been based on systematic searches for scats. Conclusions drawn from such surveys have been used to inform national conservation policies and recovery programs (e.g., Bright and Harris 1994, Bright et al. 1995*a,b*, Bright and Smithson 1997). However, there is concern about the reliability of scat surveys, especially where populations are sparse (Messenger and Birks 2000).

Ecologists derive information on diet, populations, habitat use, and genetics from feces (review by Putman 1984, Boyce 1988, Kohn and Wayne 1997). Many mammals use feces in olfactory communication by depositing them in prominent places throughout their ranges, or at territory boundaries (Gorman and Trowbridge 1989). This 'signing' behavior has enabled ecologists to survey elusive species whose feces and other field signs are easier to find and count than the animals that produce them. For example, Europe's vulnerable populations of otters (*Lutra lutra*) have been monitored since the 1970s by systematic searches for 'spraints' (Mason and Macdonald 1987); however, there has been debate about whether reliable information on distribution and habitat use can be derived from these data (Kruuk et al. 1986, Jefferies 1986, Mason and Macdonald 1987). Despite concerns about correct field identification of scats, and the uncertain relationship between scat abundance and animal population density, surveys have been applied to many carnivores to detect presence or absence and to document distribution (review by Gese 2001).

Recent surveys for pine martens (Table 12.1) have used a method adapted from otter surveys (Lenton et al. 1980). Searches for scats are conducted along linear features, such as forest trails and paths. This technique arose from the work of Lockie (1964), who first suggested a relationship between the numbers of scats and martens. This presumed but untested relationship has encouraged the development of an inexpensive approach to monitoring. A single field surveyor, searching a large number of pre-selected sites, can gather repeated sample data on marten presence over a wide geographical area. However, survey design and interpretation may involve assumptions about habitats utilized by martens, about territorial marking behavior, spatial and temporal patterns of scat-deposition, and field surveyors' identification skills.

We reviewed the application of scat searches as a survey tool, and we evaluated the implications of new DNA techniques used to assess the reliability of scat identification in the field. We assessed the use of scat surveys for inventory and monitoring by addressing three primary questions: (1) Are survey methods and objectives appropriate?, (2) Are scats correctly identified?, and (3) How does the pattern of scat abundance influence results?

2. REVIEW OF SURVEY OBJECTIVES AND METHODS

We evaluated the objectives for 8 previous scat surveys of martens in Britain and 1 in Spain (Table 12.1). We also reviewed the approaches to survey design, considering the selection of geographical areas and habitats for survey, and the sampling approaches adopted (e.g., distribution and density of sampling points, size and nature of specific features targeted for scat searches).

Survey objectives predominantly focused on inventory goals, such as determining the 'point in time' distribution and population status of pine martens at a state-wide or local scale (Table 12.1). Some researchers also pursued secondary objectives such as assessing habitat selection (Velander 1983, Strachan et al. 1996). Bright et al. (1995*a*) used scat surveys to determine the influence of woodland area and isolation of woodland patches on marten distribution. Some authors attempted to use variations in the abundance of marten scats to distinguish between established and non-breeding populations (e.g., Balharry

Table 12.1. A review of scat-based surveys of pine marten distribution, status, and abundance in Europe.

Author(s)	Geographical extent	Coverage	Objectives	Period	Duration	Number of transects	Transect length (km)	Density and arrangement of sampling points
Velander (1983)	292 10x10 km squares[a] in Scotland, England and Wales	National	Distribution and status	1980–1982 continuous	20 months	313	0.7–2.0	Mean of 1.0 transect per 10 x 10 km square in Scotland, and 1.9 per 10x10 km square in England and Wales.
Strachan et al. (1996)[b]	144 10 x 10 km squares in England and Wales	National	Distribution and status	1987–1988 continuous	19 months	896 (of which 212 were resurveyed)	2.0	Mean of 6.22 transects per 10x10 km square.
Clevenger (1993)	Minorca and Majorca Islands, Spain	Island-wide	Distribution and status	March–December 1990	10 months	39 (20 on Minorca; 19 on Majorca)	0.5–Minorca 2.5–Majorca	1 transect per 2x2 km square on Minorca; 1 transect per 5x5 km square on Majorca.
Bright and Harris (1994)	Parts of northern England	Local	Status	October & November 1993	< 8 weeks	91	2.0	Transect density notspecified, although similar to Strachan et al. (1996) above. Transects usually clumped 3 or 4 together, with each about 1 km apart.
McDonald et al. (1994)	39 10x10 km squares in mid and north Wales	Local	Distribution	June–October 1994	4 months	257	2.0	Mean of 6.1 transects per 10x10 km square. Transects not grouped closely together, but spaced 1–2 km apart.

Table 12.1. Continued.

Author(s)	Geographical extent	Coverage	Objectives	Period	Duration	Number of transects	Transect length (km)	Density and arrangement of sampling points
Bright et al. (1995*a*)	Ten 10x10 km squares in the Kielder Region of England-Scotland borders	Local	Determine status	May and June 1995	2 weeks	115	2.0	Mean of 11.5 transects per 10x10 km square. Transects not grouped closely together, but spaced 1–2 km apart.
Bright et al. (1995*b*)	Highland Region, north-west Scotland (mainland)	Local	Autecological study	October & November 1995	< 8 weeks	135 (in 63 woodlands)	2.0	Maximum of 3 transects per woodland. Transects grouped together "to avoid them falling between pine marten territories where scats may not be deposited".
Balharry et al. (1996)	82 4–5 km² 'search areas' covering parts of Scotland	Local	Map range expansion	May–October 1994	6 months	306	1.0	Four separate 1 km transects in each of 82 search areas of 4–5 km².
Bright and Smithson (1997)	South–west Scotland	Local	Distribution and population expansion	1995–1996	not specified	191	2.0	Six 2 km transects searched in each discrete forest block or larger forested area sub-divided on basis of rivers that form territory boundaries.

[a]A 10x10 km square is a square mapping unit with an area of 100 km².
[b]This survey was completed in 1988 but not published until 1996, after some subsequent surveys were published. Pre-publication drafts were made available to some authors of subsequent surveys; hence some references in the text may not appear chronological.

et al. 1996, Bright and Smithson 1997). True monitoring, involving repeated inventories to assess changes in population status and distribution (e.g., Strachan et al. 1990), was never an objective stated by survey authors. However, some authors inferred changes in range, status, or abundance by comparing the results of successive surveys organized by different authors (Bright and Harris 1994, McDonald et al. 1994).

Among the surveys considered in this review, only Clevenger (1993) attempted to achieve complete geographical coverage of a survey area. The other studies surveyed marten distribution within 10 × 10 km squares selected on the basis of locations of previous sightings or carcass collections (e.g., Velander 1983, Strachan et al. 1996). Because of the large extent of target areas, some authors delimited survey areas on the basis of concentrations of presumed suitable habitat, such as extensive woodland cover (e.g., McDonald et al. 1994, Balharry et al. 1996).

The selection of habitats chosen for survey reflects the predominant view that pine martens are animals of mature woodland and forest (Balharry 1993). However, some authors also searched non-wooded habitats (Table 12.2), which may be especially relevant in the British Isles, where marten populations have survived despite extensive deforestation that reduced woodland cover to only 4% of the land area by the early 20th century (currently 12%) (Anonymous 1998). Gradual deforestation in England created low and fragmented woodland cover that has existed for nearly 2,000 years (Rackham 1990). Under such conditions, martens probably faced strong pressure to exploit alternative three-dimensional habitats, enabling populations to survive in the absence of woodland and forest. Such adaptation may have left a legacy of habitat use by martens, persisting to the present day. There is abundant anecdotal evidence of martens occupying, or even favoring, open, rocky landscapes in Britain (e.g., Macpherson 1892, Corbet 1966, Hurrell 1968, Webster 2001). However, this possibility has not been reflected in the design of most scat surveys. The choice of habitats surveyed is not consistent across studies (Table 12.2). Some surveys encompassed a wide range of wooded and unwooded habitats, while others focused heavily on commercial conifer forests with transects concentrated in thicket stage plantations where “martens are likely to concentrate their activity” (Balharry 1993). Commercial conifer plantations in Britain are more extensive (Anonymous 1998) and are aggregated in larger blocks than other woodland types; thus, this habitat best satisfies the requirements of surveys that target areas of high forest cover, with the result that other woodland types may be less well represented in surveys. Most surveys involved sampling in a limited range of habitats, yet authors often drew wider inferences about presence or status of martens.

Table 12.2. Habitats sampled and specific features searched during scat surveys.

Author(s)	Habitats sampled	Transect characteristics
Velander (1983)	Predominantly mixed conifer plantations, but also other woodlands, open moorland, pasture, coastal, and scrub.	Trails and paths.
Strachan et al. (1996)	Mainly woodland, but also pasture, moorland, rock outcrops, crags, and scrub.	Trails, paths, and forest roads.
Clevenger (1993)	Pine forests, Holm oak woodlands, and forest or shrub cover.	Unpaved roads and trails that traversed the largest area with Mediterranean shrublands.
Bright and Harris (1994)	Woodland and forest.	Trails, paths, and woodland edges.
McDonald et al. (1994)	Predominantly woodland and forestry plantations.	Woodland trails, paths, woodland edges, streams within woodland, and crossing points between woodlands.
Bright et al. (1995*a*)	Commercial conifer forest and woodland.	5-m-wide woodland trails only, chosen to pass topographical intersections (e.g. a trail over a stream).
Bright et al. (1995*b*)	Woodland.	A range of trails and paths through woodland, and woodland edges. Transect routes selected to pass through habitats and past landscape nodes.
Balharry et al. (1996)	Woodland.	Woodland trails (ideally of dirt or stone, not grass-covered), chosen to include intersections with streams or other trails and to exclude areas with high vehicle or human use.
Bright and Smithson (1997)	Commercial conifer forest.	Forest trails (5 m wide).

The reliability of scat surveys depends upon sampling strategies that coincide with sites where scats of martens are deposited. Adult pine martens in captivity each produce an average of 5 scats per day (T.B. personal observation and M. Noble, personal communication). Since martens are believed to mark trails with their scats (Lockie 1964, Pulliainen 1982), transects are typically surveyed along such features (Table 12.2). Because martens may mark most heavily where their own trails cross man-made trails or other linear features such as streams, some surveyors also selected transects to include such intersections. Two studies searched transects along woodland edges, using the assumption that martens might not mark intensively along trails where woodlands are sparsely vegetated at ground level (e.g., Bright and Harris 1994). Bright et al. (1995*b*) suggested that scats were more likely to be found on wider trails within woodland; thus, 2 surveys limited the selection of transects to 5 m-wide trails in woodland/forest, and all narrower woodland/forest paths and trails, woodland edges, and non-wooded habitats were excluded. Thus, there has been some inconsistency resulting from *a priori* assumptions regarding habitat associations and behaviors of martens in many previous surveys.

Trail-marking behavior may be a particular feature of strongly territorial populations of martens (Balharry et al. 1996). However, survey protocols have not considered the possibility that martens may not defecate on trails and paths where populations are low and, consequently, the need for territorial marking is greatly reduced. Scat surveys involve searching the ground, despite that martens spend much of their time resting or active above ground (Birks 2002). An unknown proportion of scats may be deposited in ways that reflect this three-dimensional lifestyle. In the Netherlands, marten scats are concentrated on branches or tree bases beneath arboreal dens in the holes made by black woodpeckers (*Dryocopus martius*); therefore, surveyors concentrate their search for fresh scats beneath woodpecker holes (Kleef 1997). The untested assumption that scats of martens occur disproportionately on man-made trails is a weakness common to most surveys. Concerns about detection of scats by human surveyors searching only accessible features such as trails could be addressed by involving trained dogs, which use their scenting ability to search more representatively than humans (Smith et al. 2001).

Scat surveys for martens have used transect lengths of 0.5–2.0 km, with authors selecting transect length in response to local conditions and survey goals. Several authors justified their choice of transect length by estimating the probability of detecting scats over different lengths. For example, Velander (1983) reported that scats were detected within the first 500 m on 81.2% of positive transects, within the first 700 m on 94.1%, and within 1 km on 98.6%. On this basis she adopted 700 m as the minimum and 1 km as the preferred transect length in her study. However, most subsequent surveys have used the 2 km transect approach adopted by Strachan et al. (1996) on the basis that Velander's (1983) 1 km transects were too short to detect martens at low population densities. The method based on groups of 4 1-km transects adopted by Balharry et al. (1996) was tested in the core of the range of martens in Wester Ross, Scotland. The probability that at least one scat would be found was 85.3% if only 1 km was searched, and 97.8% if 2 km were searched. Bright and

Smithson (1997) reported that the probability of detecting scats reached an asymptote after 8 km of transect. Thus, they concluded that their choice of 6 2-km transects was adequate for detecting presence of martens. No validation of the effect of transect length on probability of scat detection has been attempted outside Scotland. Variations in transect length among surveys matter little where simple detection of marten occurrence is the goal. However, difficulties arise where authors seek to compare results or infer trends from independent surveys.

Some authors argued that the spacing of transects was important to ensure that they were not located between marten territories. For example, Bright and Harris (1994) suggested that Strachan et al. (1996) might have missed marten sites because most transects were spaced more than one territory diameter apart. To overcome this effect, some subsequent surveys have clumped or spaced transects only 1–2 km apart, which has the potential drawback of repeatedly sampling the same individual.

3. SCAT IDENTIFICATION

Confidence in the results of scat-based marten surveys is dependent on the correct identification of marten scats. Caution is needed because feces from foxes (*Vulpes vulpes*), polecats and polecat-ferrets (*Mustela putorius*), mink (*Mustela vison*), and stoats (*Mustela erminea*) may appear similar to those of martens (McDonald et al. 1994, Balharry et al. 1996). We reviewed the approaches adopted by different surveys to ensure accurate identification of marten scats. We also considered new genetic evidence for assessing the accuracy of scat identification in the field.

Several authors have sought to build confidence in their methodology by specifying the criteria applied when identifying scats, though the degree of rigor varies considerably (Table 12.3). Some surveyors also recorded additional evidence, such as clear footprints, to indicate marten presence (e.g., Strachan et al. 1996). Some studies refer to the distinctive sweet, musky odor as being critical to the correct identification of marten scats. As a result, some surveys specified that only fresh scats (a few days old) that had not lost their smell were taken as evidence of marten presence (e.g., Bright and Smithson 1997). McDonald et al. (1994) suggested that Strachan et al. (1996) may have misclassified scats from other carnivores as those of martens, leading to "an exaggerated estimate of marten abundance". Those 2 surveys, separated by a period of 6 years, offered different conclusions about the status of martens in Wales. Strachan et al. (1996) concluded that the population was extant and "static or showed a very moderate spread", and McDonald et al. (1994) con-

Table 12.3. Criteria applied to the identification of pine marten scats during surveys conducted in Europe.

Author(s)	Identification criteria
Velander (1983)	Scats were "examined according to size, texture and odour to ensure that they were those of martens".
Strachan et al. (1996)	No criteria specified, although variations in size, shape and color were described.
Clevenger (1993)	No criteria specified: "Faeces that might have been confused with those of weasel *Mustela nivalis* or feral cat *Felis catus* were discarded".
Bright and Harris (1994)	"Marten scats were recognised by their distinctive sweet odour (...like cranberry sauce) and frequent characteristic long, twisted shape; they were not recognised on the basis of shape alone"
McDonald et al. (1994)	As for Bright & Harris (1994) above. "Since single scats cannot be classified as those of martens with total certainty, it was considered that several scats needed to be found on a transect to confirm that martens were present".
Bright et al. (1995*a*)	"Marten scats were recognised on the basis of both their distinctive odour and morphology".
Bright et al. (1995*b*)	"Scats were identified on the basis of both their distinctive sweat (sic) musky odour and morphology, being 8 mm in diameter and 40–80 mm long. They are also frequently, but not always twisted". "Because marten scat odour and morphology vary to some extent with diet, several scats need to be found at one site to reliably indicate marten presence".
Balharry et al. (1996)	"A fresh scat has a unique smell of marten. Older marten scats can be identified by size and shape although they overlap in size with stoat at one extreme and with fox at the other and shape is highly dependent on diet. Confusion may also occur with ferret and mink scats. Only those scats which were obviously marten were collected".
Bright and Smithson (1997)	Marten scats "were recognised by their morphology and sweat-musky (sic) odour". "Only fresh scats (a few days old) with characteristic odour were classified as being from pine martens. This avoided confusion with those of fox *Vulpes vulpes* or stoat *Mustela erminea* which overlap in morphology, but not in odour".

cluded that no viable populations remained, and that martens in Wales were on the brink of extinction.

Because of the variation in scat odor and morphology, some studies only inferred marten presence if several fresh scats were found on a 2 km transect (McDonald et al. 1994), or if at least 3 scats were found within a woodland site (Bright et al. 1995*b*). In their survey of the Kielder Region (northern England), Bright et al. (1995*a*) recorded 27 scats that had similar morphology to marten scats, but all lacked odor. Moreover, some were found in characteristically marten-like groups, and appeared too fresh to have lost the pungent odor typical of scats produced by other carnivores. The authors concluded that they "might therefore have been produced by martens, but we cannot be certain". These scats might have been accepted as more certain evidence of martens by other studies with more inclusive criteria (e.g., Strachan et al. 1996). Clearly, differences in identification rigor between surveys preclude objective comparison of results.

The assumption that field identification of marten scats is accurate has only recently been tested by the application of DNA techniques. Such techniques are currently too expensive to be applied widely as an aid to surveys, but they can help to validate new or established field protocols (e.g., Hansen and Jacobsen 1999). A genetic study by Davison et al. (2002) revealed that 3 experienced surveyors misclassified 18% of fresh 'marten' scats ($n = 56$) collected in the field in Scotland. Based on DNA evidence, misclassified scats in this sample were from red foxes. DNA was successfully extracted and amplified from only 53% of fresh scats collected, and this has implications for the wider application of this approach to the verification of scat identity. Individual surveyor misclassification varied (9–29%) and this level of error is conservative because surveyors were both experienced and aware that their skills were being evaluated. Regardless, 2 surveyors misclassified scats that they had categorized as 'certain' marten on the basis of morphology and odor. The surveyor who performed most reliably (9% error) in Scotland misclassified all scats ($n = 12$) collected from the sparser populations of martens in England and Wales. This new genetic evidence of a significant error factor undermines the central assumption on which all scat surveys have been based.

4. VARIATION IN ABUNDANCE AND DETECTABILITY

We evaluated the use of scats for determining presence and population status of martens by reviewing patterns of abundance revealed by surveys. We also considered the role of seasonal factors in influencing scat abundance. We assessed attempts by some authors to relate scat abundance to marten residency status, and we examined the inferences drawn by authors where no scats were found.

Following Lockie's (1964) pioneering work, authors have noted temporal variations in the abundance of scats and have suggested possible explanations. Most have noted that scat numbers on transects are highest in summer, and suggest that surveying outside this period may be problematic (Bright et al.

1995*b*). Velander (1986) reported that scat density on a series of forest trails varied greatly from month to month, being more than 100 times greater in July (12 scats/km) than in January (0.1 scats/km). Clearly, seasonal variations may have a profound influence on the results generated by surveys. Where martens are scarce this may lead to conclusions that the species is absent at sites that would prove positive at other times of year. This effect was reported by Strachan et al. (1996) after an absence of scats was observed on transects in areas of sparse marten populations that yielded scats when resurveyed a few months later. Some surveyors have interpreted changes in scat abundance as evidence of seasonal range shifts (e.g., Velander 1983, Strachan et al. 1996). However, it is likely that seasonal changes in marten numbers, general activity levels, and the intensity of social marking behavior also contribute to the observed pattern (Helldin and Lindström 1995). Certainly, the observed pattern fits the prediction that marking should be most intense during the summer mating season (July/August), when adults socialize actively and the population is increased by the presence of young. Conversely, pine martens greatly reduce their activity during winter months (Zalewski 2000) when many scats are probably deposited at resting sites. However, most wide-scale and some local surveys have not concentrated on the ideal summer months (Table 12.1). As a consequence, a significant proportion of survey effort has occurred when the available scats were predictably scarce, which influences survey results, especially at low population densities.

A feature of all scat-based surveys has been the sizeable proportion of negative survey transects or search areas. These pose a problem of interpretation for authors who may be tempted to infer that martens are absent. Survey authors have conceded that it is impossible to prove that pine martens are absent from an area (e.g., Bright and Harris 1994), and some have taken other evidence (e.g., footprints, reported sightings, interviews with local naturalists) into account before drawing conclusions. The risks of inferring absence falsely from negative scat surveys are emphasised by the work of Velander (1983), who recorded 32 10 x 10 km squares in Scotland that were negative on the basis of scat surveys, yet they yielded carcasses or sightings of martens (these 'false negatives' comprised 21.3% of the total number of positive 10 x 10 km squares). In Bright and Smithson's (1997) survey in south-west Scotland, no scats were found at several locations where other recent evidence had indicated that martens were present. The very limited results in England and Wales (see Table 12.4) occurred when other evidence (footprints, reported sightings and carcasses) indicated that martens were present. Following interviews with local naturalists, Velander (1983) concluded that 4 main marten populations were still present in England and Wales, despite observing no scats during field

Table 12.4. Scat densities recorded during surveys of pine martens.

Author(s)	Geographical extent	Range of scat densities recorded on 'positive' transects	Proportion of transects or sampling areas with no field evidence of martens
Velander (1983)	Scotland	1–7 scats/500 m (mean 2.64 ± 0.21)	42.8% (n = 173)
Velander (1983)	England & Wales	No positive transects	100% (n = 36)
Strachan et al. (1996)	England & Wales	1–10 scats/2 km (mean 0.54 ± 0.04)	88.8% (n = 896)
Clevenger (1993)	Minorca & Majorca, Spain	No data (searches terminated when first scats found)	
Bright and Harris (1994)	England	1 scat/2 km (only two scats found in entire survey)	97.8% (n = 91)
McDonald et al. (1994)	Wales	1 scat/2 km (only three scats found in entire survey)	98.8% (n = 257)
Bright et al. (1995a)	England	No certain marten scats found, though possible marten scats found on 8 transects at rates of 2–4/500 m.	100% (n = 115)
Bright et al. (1995b)	Scotland	Mean distance walked to find three scats was 1.3 km±0.7km.	39.1% (n = 63)
Balharry et al. (1996)	Scotland	1–27 scats/1 km	58.8% of all transects (n = 306) 41.5% of search areas (n = 82)
Bright and Smithson (1997)	Scotland	Mean 5.4 ± 0.24 scats per 'breeding potential marten area'[a] (total 12 km transect searched per area); mean 1.8±0.37 scats per 'occupied potential marten area'[b].	Not stated

[a]These areas, with higher scat densities, generated sightings of female martens with young.
[b]These areas, with lower scat densities, generated less evidence of reproducing martens.

surveys. She explained the failure of scat surveys to detect these populations as "due presumably to the difficulties in finding evidence of martens when in low numbers" (Velander 1983). Other authors have placed greater confidence in scat data alone, even where the animals are scarce. McDonald et al. (1994) argue that the intensity of marking with scats is density dependent. They speculate that where martens are scarce, population densities would not be low in all areas, but would be high enough locally in some areas for scats to be observed during surveys. On this basis, and because few marten scats were found in surveys of both England and Wales, authors concluded that pine martens were on the brink of extinction with no viable populations remaining (Bright and Harris 1994, McDonald et al. 1994). This pessimistic assessment contrasts markedly with authors who have interpreted scat abundance data more cautiously, and have considered other evidence (e.g., Velander 1983, Strachan et al. 1996). Clearly, there are circumstances where it is misleading to base status assessments exclusively on the basis of scat occurrences.

Our own work in Wales has revealed a further influence that must reduce the detectability of marten scats where they are scarce. Foxes were observed to destroy, by aggressive scratching, several scats (from captive martens) that had been placed on forest trails to stimulate counter-marking by wild martens. Dor beetles (*Geotrupes* sp.) were observed to remove and bury scats of martens, and great black slugs (*Arion ater*) were observed to completely consume fresh scats within as few as 48 hrs (Braithwaite et al., The Vincent Wildlife Trust, Ledbury, UK. unpublished data).

Additional to determining presence of martens on the basis of scats, some authors have used variations in scat abundance to determine residency (Balharry et al. 1996, Bright and Smithson 1997). However, no empirical evidence supports the assumption that areas with fewer scats contain only dispersing or non-breeding marten. Nor did these attempts to define marten population status by reference to relative scat abundance account for seasonal influences on scat deposition rates (Velander 1986).

Some authors have tried to define thresholds of scat abundance as indicators of relative, but not absolute, absence of martens. For example, Bright et al. (1995b) considered that martens were absent from, or not regularly using, a woodland if fewer than 3 distinctive scats were found. However, the same survey team adopted a different criterion elsewhere in Scotland where areas with 1–3 scats (mean 1.8 over 12 km searched) were regarded as occupied by marten (Bright and Smithson 1997). Such arbitrary assumptions seem unwise in the absence of a clear understanding of the relationship between scat abundance and the numbers of martens.

5. CONCLUSIONS AND RECOMMENDATIONS

The revelation that experienced marten researchers misclassify fresh scats undermines all confidence in the scat survey method as it is currently applied. We recommend that the technique should not be used for any survey goals in the absence of genetic verification of scat identity, especially in areas where martens occur at low densities. It might be argued that scat surveys may have a role in determining presence of martens where independent and incontrovertible evidence indicates that they are common, but this circular argument would seem to render the technique irrelevant.

Survey methods and objectives were questionable because all were based on assumptions that surveyors could identify scats accurately. Regardless of this major flaw, methodologies have been based on assumptions that appear unreasonable in the absence of thorough field-testing. Scat survey protocols have not been validated across the full range of seasonal, habitat, and population conditions. Notably, protocols have never been adequately tested and shown to be reliable in areas where martens occur at low densities. We recommend that future application of scat surveys for inventory and monitoring goals should be preceded by a program of practical and statistical validation. We also recommend that inferences drawn from future surveys should be limited to the habitats sampled.

The field relationship between scat abundance on transects and marten numbers has not been established. Consequently, it is unsafe to use scat abundance data for inferring marten abundance, or for monitoring population trends. Particular problems of interpretation arise where scats are scarce or absent in areas known, from other evidence, to be occupied by martens. Few conclusions can safely be drawn where no marten scats are found, beyond the possibility that the animals are scarce in such areas. Where martens and their scats are apparently common, the influence of identification errors prevents the reliable use of scat abundance indices for assessing abundance and population trends. Thus, we recommend that genetic verification be included as an essential component of all scat surveys. Nevertheless, even with genetic verification, scat abundance indices could be meaningless if seasonal variation in scat deposition patterns is not controlled for. These issues can only be addressed through behavioral studies of martens across a range of season, habitat, and population conditions.

Even prior to the genetic confirmation of significant surveyor error (Davison et al. 2002), others have warned against the use of marten scat surveys, including those advising the UK Government on future mammal monitoring. Toms et al. (1999) warned that "In areas with low population densities or containing only transitory individuals, the degree of scatting is likely to be greatly reduced, making it difficult to apply a transect approach based on field signs. Territorial behavior in other mustelids has been shown to break down altogether at low population densities potentially making this method ineffective in some regions". Similarly, Macdonald et al. (1998) warn that "scat surveys may be unreliable at low population densities where they are less territorial".

Scat surveys are unreliable without genetic verification; therefore, conclusions drawn primarily from scat data by the authors of surveys reviewed in this paper are questionable. In particular, the use of scat abundance data to infer that no viable populations remain in England (Bright and Harris 1994) and Wales (McDonald et al. 1994) is unsupportable. There is clearly a need to develop and refine approaches to detecting and monitoring pine martens. This need is especially great where the species is scarce and difficult to detect. Under such circumstances, we recommend the systematic deployment of a range of methods such as sighting surveys (Messenger and Birks 2000), camera traps (Zielinski and Kucera 1995), tracker dogs (Smith et al. 2001), hair snagging stations (Messenger and Birks 2000), or track plates (Zielinski and Kucera 1995).

6. ACKNOWLEDGMENTS

This work is part of a project run by The Vincent Wildlife Trust. We are indebted to all who have helped to shape our thoughts on marten scat surveys, notably David and Liz Balharry, Paul Bright, Don Jefferies, Robbie McDonald, Colin Simms, Rob Strachan, Kathy Velander and John Webster. Mary Gough, Robbie McDonald, Bill Zielinski and an anonymous referee made constructive comments on an earlier draft of this paper.

7. LITERATURE CITED

Anonymous. 1998. The Forestry Industry Council of Great Britain Handbook, 1998. Forestry Industry Council of Great Britain, Stirling, UK.

Aune, K. E., and P. Schladweiler. 1997. Age, sex structure, and fecundity of the American marten in Montana. Pages 61 - 77 *in* G. Proulx, H. N. Bryant, and P. M. Woodard, editors. Martes: taxonomy, ecology, techniques and management. Provincial Museum of Alberta, Edmonton, Alberta, Canada.

Balharry, D. 1993. Factors affecting the distribution and population density of pine martens (*Martes martes* L.) in Scotland. Dissertation, University of Aberdeen, Aberdeen, UK.

Balharry, E. A., G. M. McGowan, H. Kruuk, and E. Halliwel 1996. Distribution of pine martens in Scotland as determined by field survey and questionnaire. SNH Survey and Monitoring Report No. 48. Scottish Natural Heritage, Edinburgh, UK.

Birks, J. D. S. 2002. The Pine Marten. The Mammal Society, London.

_____, J. E. Messenger, and A. Davison. 1997. A 1994 pine marten *Martes martes* (L.) record for Lancashire, including a preliminary genetic analysis. Naturalist 122:13–18.
Boyce, N. 1988. Bowels of the beasts. New Scientist 22 August:36–39.
Bright, P. W., and S. Harris. 1994. Reintroduction of the pine marten: feasibility study. English Nature Contract Report F72-11-10. University of Bristol, Bristol, UK.
_____, R. McDonald, and S. Harris. 1995*a*. Survey of pine martens in the Kielder Region. Pages 3–6 *in* Initiating a Recovery Programme for the Pine Marten in England and Wales. People's Trust for Endangered Species, London, UK.
_____, _____, and _____. 1995*b*. Determining the minimum woodland area requirements to sustain pine marten populations. Pages 7 - 15 *in* Initiating a Recovery Programme for the Pine Marten in England and Wales. People's Trust for Endangered Species, London, UK.
_____, and T. J. Smithson. 1997. Species Recovery Programme for the pine marten in England: 1995–96. English Nature Research Report No. 240. English Nature, Peterborough, UK.
Clevenger, A. P. 1993. The European Pine Marten *Martes martes* in the Balearic Islands, Spain. Mammal Review 23:65–72.
Corbet, G. B. 1966. The Terrestrial Mammals of Western Europe. G. T. Foulis, London, UK.
Davison, A., J. D. S. Birks, R. C. Brookes, A. C. Braithwaite, and J. E. Messenger. 2002. On the origin of faeces: morphological versus molecular methods for surveying rare carnivores from their scats. Journal of Zoology (London) 257:141–143.
Gese, E. M. 2001. Monitoring of terrestrial carnivore populations. Pages 372–396 *in* J. L. Gittleman, S. M. Funk, D. W. Macdonald, and R. K. Wayne, editors. Carnivore conservation. Cambridge University Press, Ithaca, New York, USA.
Gorman, M. L. and B. J. Trowbridge. 1989. The role of odour in the social lives of carnivores. Pages 57–88 *in* J. L. Gittleman, editor. Carnivore behavior, ecology and evolution. Cornell University Press, Ithaca, New York, USA.
Hansen, M. M., and L. Jacobsen. 1999. Identification of mustelid species: otter (*Lutra lutra*), American mink (*Mustela vison*) and polecat (*Mustela putorius*), by analysis of DNA from faecal samples. Journal of Zoology, London 247:177–181.
Helldin, J. O. 1998. Pine marten (*Martes martes*) population limitation: food, harvesting or predation? Acta Universitatis Agriculturae Sueciae, Silvestria 60.
____, and E. R. Lindström. 1995. Late winter social activity in pine marten (*Martes martes*) – false heat or dispersal? Annales Zoologica Fennici 32:145–149.
Hurrell, H. G. 1968. Pine Martens. Forest Record No. 64. HMSO, London, UK.
Jefferies, D. J. 1986. The value of otter *Lutra lutra* surveying using spraints: an analysis of its successes and problems in Britain. Journal of the Otter Trust 1:25–32.
_____, and C. H. Critchley. 1994. A new pine marten *Martes martes* (L.) record for the North Yorkshire Moors: skull dimensions and confirmation of species. Naturalist 119:145–150.
Kleef, H. L. 1997. Boommarterinventarisatie in Nederland: aanpak en resultaten, toegespitst op Noord-Nederland. Pages 11–22 *in* K. J. Canters and H. J. W Wijsman, editors. Wat Doen we met de boommarter. Werkgroep Boommarter Nederland, Utrecht, Netherlands.
Kohn, M. H., and R. K. Wayne. 1997. Facts from feces revisited. Trends in Ecology and Evolution 12:223–227.
Kruuk, H., J. W. H., Conroy, U. Glimmerveen, and E. J. Ouwerkerk. 1986. The use of spraints to survey populations of otters *Lutra lutra*. Biological Conservation 35:87–94.
Langley, P. J. W., and D. W. Yalden. 1977. The decline of the rarer carnivores in Great Britain during the nineteenth century. Mammal Review 7:95–116.
Lenton, E. J., P. R. F. Chanin, and D. J. Jefferies. 1980. Otter Survey of England 1977–79. Nature Conservancy Council, London, UK.
Lindström, E. R., S. M., Brainerd, J. O. Helldin, and K. Overskaug. 1995. Pine marten-red fox interactions: a case of intraguild predation? Annales Zoologica Fennici 32:123–30.

Lockie, J. D. 1964. Distribution and fluctuations of the pine marten, *Martes martes* (L.), in Scotland. Journal of Animal Ecology 33:349–356.

Macdonald, D. W., G. Mace, and S. Rushton. 1998. Proposals for future monitoring of British mammals. Department of the Environment, Transport and the Regions, London, UK.

Macpherson, H. A. 1892. A Vertebrate Fauna of Lakeland. Douglas, Edinburgh, UK.

Mason, C. F., and S. M. Macdonald. 1987. The use of spraints for surveying otter *Lutra lutra* populations: an evaluation. Biological Conservation 41:167–177.

McDonald, R., P. W. Bright, and S. Harris. 1994. Baseline Survey of Pine Martens in Wales. Report to the Countryside Council for Wales. Countryside Council for Wales, Bangor, UK.

Messenger, J. E., J. D. S. Birks, and D. J. Jefferies. 1997. What is the status of the pine marten in England and Wales? British Wildlife 8:273–279.

_____, and _____. 2000. Monitoring the very rare: pine marten populations in England and Wales. Pages 217–230 *in* H. I. Griffiths, editor. Mustelids in a modern world. Management and conservation aspects of small carnivore: human interactions. Backhuys, Leiden. Netherlands.

Pullianen, E. 1982. Scent marking in the pine marten (*Martes martes*) in Finnish forest Lapland in winter. Zeitschrift fur Saugetierkunde 47:91–99.

Putman, R. J. 1984. Facts from faeces. Mammal Review 14:79–97.

Rackham, O. 1990. Trees and woodland in the British Landscape. Phoenix, London, UK.

Sloane, M. A., P. Sunnucks, D. Alpers, L. B. Beheregaray, and A. C. Taylor. 2000. Highly reliable genetic identification of individual northern hairy-nosed wombats from single remotely collected hairs: a feasible censusing method. Molecular Ecology 9:1233–1240.

Smith, D., K. Ralls, B. Davenport, B. Adams, and J. E. Maldonado. 2001. Canine assistants for conservationists. Science 291:435.

Strachan, R., J. D. S. Birks, P. R. F. Chanin, and D. J. Jefferies. 1990. Otter survey of England 1984–86. Nature Conservancy Council, Peterborough, UK.

_____, D. J. Jefferies, and P. R. F. Chanin. 1996. Pine marten survey of England and Wales 1987 –1988. Joint Nature Conservation Committee, Peterborough, UK.

Strickland, M. 1994. Harvest management of fishers and American martens. Pages 149–164 *in* S. W. Buskirk, A. S. Harestad, M. G. Raphael and R. A. Powell, editors. Martens, sables and fishers: biology and conservation. Cornell University Press, Ithaca, New York, USA.

Tapper, S. 1992. Game Heritage. Game Conservancy Trust. Fordingbridge. UK.

Toms, M. P., G. M. Siriwardena, and J. J. D. Greenwood. 1999. Developing a mammal monitoring programme for the UK. BTO Research Report No. 223. British Trust for Ornithology, Thetford, UK.

Velander, K. A. 1983. Pine marten survey of Scotland, England and Wales 1980–1982. The Vincent Wildlife Trust, London, UK.

_____. 1986. A study of pine marten ecology in Inverness-shire. Nature Conservancy Council CSD Report 651. Nature Conservancy Council, Peterborough, UK.

Webster J. A. 2001. A review of the historical evidence of the habitat of the pine marten in Cumbria. Mammal Review 31:17–32.

Woods, J. G., D. Paetkau, D. Lewis, B. N. McLellan, M. Proctor, and C. Strobeck. 1999. Genetic tagging of free-ranging black and brown bears. Wildlife Society Bulletin 27:616–627.

Zalewski, A. 2000. Factors affecting the duration of activity by pine martens (*Martes martes*) in the Białowieża National Park, Poland. Journal of Zoology (London) 251:439–447.

Zielinski, W. J., and T. E. Kucera. 1995. American marten, fisher, lynx, and wolverine: survey methods for their detection. General Technical Report PSW-157. US Department of Agriculture Forest Service, Pacific Southwest Research Station, Berkeley, California, USA.

Chapter 13

POSTNATAL GROWTH AND DEVELOPMENT IN FISHERS

Herbert Frost and William Krohn

Abstract: Postnatal growth and development of fishers were quantified for 14 litters of kits born in captivity between 1991–93. Male (n = 22) and female (n = 16) body weight did not differ within 48 hrs of birth (P = 0.64); however, by 90 days of age the mean daily gain for males (48.1 g/day) was more than double that of females (21.6 g/day). Males grew approximately 1.49 times faster than females. Females attained mature body weight by 180 days and males by 200 days. Kits were altricial with few discernible morphological changes or behaviors observed during the first 30 days of life. Teeth could be palpated through the gums at 40 days. Eyelids and ear canals opened at approximately 48 (± 4) days. Kits began to eat solid food soon after their eyes opened and thereafter their weight increased significantly. Kits were not observed outside the nest box until an average of 70 days after parturition. Agonistic behaviors also became common after food was introduced at about 70 days.

1. INTRODUCTION

Patterns of postnatal growth and development are known for many species of carnivores (Gittleman 1986), but patterns of growth in fishers (*Martes pennanti*) have been based on small sample sizes (Coulter 1966, LaBarge 1991, Powell 1993). Early descriptions of growth in fishers were provided by naturalists (Seton 1937) and fur farmers (James 1934, Thomassen 1940), but were primarily anecdotal. Coulter (1966), Powell (1993) and LaBarge et al. (1991) each reported on growth and development of only 1 litter of fishers. However, both Powell (1993) and LaBarge et al. (1991) removed the young from their mothers and raised them by hand, which may have influenced growth and rate of morphological development.

The purpose of this study was to document and quantify postnatal growth, behavioral development, and morphological development of fishers. Our specific objectives were to compare growth patterns of male and female kits to determine when they become sexually dimorphic and to document first appearance of morphological developments and behaviors. When possible, we compared our results to observations made during other studies of fishers in the wild.

2. METHODS

Fishers were captured at 7 locations throughout central and eastern Maine. Thirty-seven female fishers were captured in live traps (Tru-Catch, Mechanicsburg, PA; use of manufacturer's name does not imply endorsement of commercial products) during the fall trapping seasons of 1990–92. All fishers were brought into captivity and housed at the University of Maine's Animal Research Facility in Orono, Maine (Frost and Krohn 1994). Fishers were housed individually in pens located under conifer trees and exposed to natural photoperiod. Each cage was 1.2 m wide by 1.2 m tall by 2.4 m long, and was wrapped with 14 gauge, 2.54 cm^2 wire mesh and elevated approximately 10–20 cm above the ground. Maternal nest boxes were equipped with plexiglass windows so observations could be made with minimal disturbance to the animals. Water was given *ad libitum*, and daily rations were composed of commercial mink feed (50%), meat (40%), and beef liver (10%). Fishers were weighed monthly throughout the year and weekly throughout the gestation period. If they became overweight (>15% over target weight), rations were decreased. Target body weights for adult fishers were 4.71 kg for males and 2.23 kg for females (Frost and Krohn 1994).

Ages of animals captured from the wild were initially estimated from examination of the sagittal crest and later from cementum annuli of the first premolar tooth (Arthur et al. 1992). In February of each year, pregnancy tests (ICG Canine Genetics, Inc, Malvern, Pennsylvania) and ultrasound examinations were conducted to asses progesterone levels for fishers determined to be ≥1 yr old. Maternal nest boxes were monitored daily beginning the first week of March until all pregnant females gave birth. In 1991, kits were not examined until they were 2½ weeks old. They were then weighed and measured at 10-day intervals through the end of June, and monthly throughout the rest of the year. In 1992 and 1993, kits were examined 24–48 hrs after birth. They were weighed and measured at weekly intervals through the end of June, and monthly throughout the rest of the year. All kits were raised by their mothers.

When kits were examined, the adult females were either anesthetized or moved to a separate nest box. Kits were measured with a measuring tape and were weighed on an electronic scale until 5 months of age. All adults and kits

>5 months were weighed on a triple beam scale. Kits were examined unrestrained until about 80 days. Thereafter, kits were anesthetized with 10:1 mixture of ketamine hydrochloride (100 mg/ml) and acepromazine maleate (10 mg/ml) delivered at a mean dose of 14 mg/kg of body weight (Frost and Krohn 1994). Total length was measured from the tip of the nose to the tip of the tail, and head circumference was measured at the widest part of the head. Kits were left in the same cage with their mother and siblings until September; and thereafter housed in individual cages. One litter, consisting of two males, was left with the female until late November of 1992 to observe behaviors.

We pooled data among years because sample sizes were small. We used non-parametric statistics for all of the analyses, but we present means and standard errors to make the data comparable to those in the literature. We used Chi-square analysis to test for differences between sex ratios at birth. Body masses between males and females were compared with a Mann-Whitney *U*-test. A Kruskal-Wallis test with pairwise comparisons (Zar 1984) was used to compare weights of males and females by month. Gompertz growth equations best fit the data (Ricklefs 1967), and were fitted to male and female body mass data, and the Gompertz growth coefficient (K) was used to compare growth rates between sexes.

First appearance of morphological developments and behaviors monitored were made ancillary to body measurements; definitions of features and behaviors that were monitored are presented in Table 13.1. To minimize handling times, not all ancillary variables were measured during each handling period.

Table 13.1. Behaviors and morphological features monitored in fisher kits born in captivity, University of Maine, Orono, USA, 1991–93.

Event	Definition
Fur darkens	Noticeable darkening of the fur.
First teeth	Teeth could be palpated.
Ears open	Both ear canals open
Eyes open	Both eyes open.
Adult vocalizations	Characteristic chuckling sound, different from the crying made at birth.
Crawling	Kits pulling themselves forward using both front and rear legs.
Out of nest box	Earliest date a kit observed out of the nest box.
Agonistic encounter	First agonistic interaction involving body contact.
Foot glands	First time glands on rear feet pads observed.
Permanent teeth	Permanent teeth visibly erupting.

3. RESULTS

Ages of female fishers captured ranged from 6 months to 3 years with 14 of 37 females estimated to be ≥1 yr old (i.e., adults). All 14 adults were determined to be pregnant and subsequently gave birth. Five litters totaling 11 kits were born in 1991, 7 litters totaling 23 kits were born in 1992, and 2 litters totaling 4 kits were born in 1993. Birth dates ranged from 4 March to 1 April with a mean (± SE) birth date of 22 March (±3 days) (Table 13.2). Litter size ranged from 1–4 ($\bar{x}$ = 2.7 ± 0.22 SE). The observed sex ratio at birth (22 males: 16 females) did not differ significantly from 1:1 ($\chi^2 = 0.95$, $P = 0.33$) (Table 13.2). Kits in 6 litters were the same sex.

For pregnancies that appeared to go full term, mean (± SE) body mass of kits ≤ 48 hrs old ranged from 36.50 g (±0.50) to 58.00 g (±0.58 g) for 9 litters. Body mass of male ($\bar{x}$ = 58 g ± 0.02) and female ($\bar{x}$ = 51 ± 0.02) kits weighed within 48 hrs of birth did not differ significantly ($U = 117.5$, $P = 0.644$).

Growth of male and female kits were similar through the first 2 months of life. There was no difference in body mass between males ($\bar{x}$ = 173 ± 0.13)

Table 13.2. Birth dates, litter size, and sex ratios for 14 litters of fishers born in captivity, University of Maine, Orono, USA, 1991–93.

Birth Year	Female Age	Birth Date	Sex	
			Male	Female
1991	3[a]	28 Mar	3	0
	2	22 Mar	2	1
	2	4 Mar	2	1
	3	23 Mar	0	2
	2	26 Mar	1	1
1992	3	21 Mar	2	0
	2	25 Mar	1	2
	3	26 Mar	0	3
	2	31 Mar	2	1
	3	27 Mar	3	1
	2	18 Mar	0	2
	4	28 Mar	3	1
1993	3	1 Apr	2	1
	2	24 Mar	1	0
Total			22[b]	16
Mean		23 March		
SE		± 3 days		

[a]Estimated age of female at time she gave birth based on cementum annuli.
[b]Sex ratio not different from 1:1 ($P = 0.33$).

and females ($\bar{x}$ = 158 ± 0.11) at 30 days (U = 1748.5, P = 0.609). At 90 days the mean daily weight gain for males (48.1 g/day) was more than double that of females (21.6 g/day) (Table 13.3). Males became larger than females by 2 months of age (Fig. 13.1, Table 13.3). Based on Gompertz growth coefficients (K), males grew approximately 1.49 times faster than females. Females reached adult body mass (2,400 g) by 180 days, whereas males attained adult body mass (3,800 g) by 200 days. Adult body mass was determined when the weights of fishers reached the asymptotic value derived from the Gompertz growth equation (Fig. 13.1).

Few morphological developments were observed through the first 30 days. Fishers were born altricial and barely moved. They had pink skin covered by a fine coat of grayish-white hair and made a high-pitched crying sound when disturbed. Both eyelids and ear canals were closed at birth. When 10 days old, dark dorsal hairs were interspersed with the fine grayish-white hairs, whiskers were present, and kits could pull themselves forward with their front legs. Fur thickened with age, and at 20 days, they were noticeably darker (Fig. 13.2). At

Figure 13.1. Change in body mass for male and female fishers during their first year of life, University of Maine, Orono, USA, 1991–93. Growth curves approximating Gompertz growth equations were fitted for all data (A = asymptotic value; K = Gompertz growth coefficient). The asymptotic value is equal to the final weight or mature body weight.

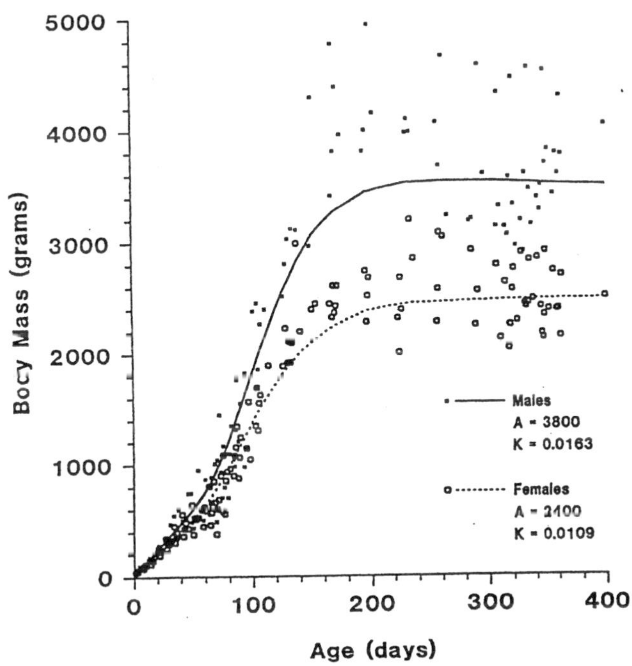

Figure 13.2. Means (center lines), standard deviations (edge of boxes), and ranges (bars) for time of first appearance of selected behaviors and morphological features in captive fishers, University of Maine, Orono, USA, 1991–93. Number of kits observed in parentheses. See Table 13.1 for descriptions of behaviors.

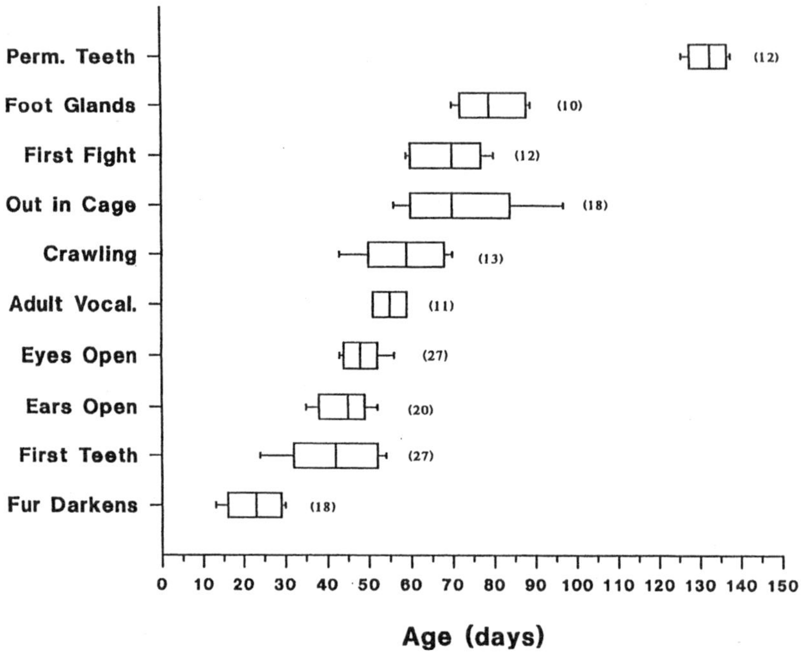

23 days, 1 kit was observed using it legs to push its body off the ground. Teeth could be palpated through the gums by 40 days. Deciduous premolars appeared first and were followed by the canines and incisors. All deciduous teeth were present at 64 days and permanent teeth had erupted by 133 days (Fig. 13.2). Eyes and ears canals opened about the same time (Fig. 13.2), approximately 48 days (±4 days) after birth. Within 2–3 days after kits opened their eyes, their mothers began provisioning solid food and growth of the kits accelerated (Fig. 13.1, Table 13.3).

4. DISCUSSION

Reported parturition dates for fishers range from mid-February to May (Powell 1993). This wide range for birth dates may be related to latitudinal differences among study areas (Powell 1993). Records from fur farms in British Columbia reported that most births occurred during March and April (Hodgson 1937, Douglas 1943). Similarly, in southcentral Maine, 12 litters

Table 13.3. Mean (± SE) values of growth parameters, by 30-day periods, for kits born in captivity, University of Maine, Orono, USA, 1991–93.

Age (days)	(n)	Sex	Mean Total Length (cm)	Mean Head Circumference (cm)	Mean Body Mass (g)	Mean Daily Gain (g)
0–30	(15)	M	22.57 ± 0.66	10.92 ± 0.32	173 ± 13[a]	12.5
	(12)	F	22.00 ± 0.64	10.51 ± 0.27	158 ± 11[b]	9.5
31–60	(14)	M	37.79 ± 0.71	15.99 ± 0.18	530 ± 24[a]	20.0
	(14)	F	36.32 ± 0.64	15.18 ± 0.16	470 ± 16[b]	14.3
61–90	(14)	M	50.01 ± 1.22	17.67 ± 0.23	903 ± 52[a]	42.5
	(13)	F	47.24 ± 0.83	16.77 ± 0.13	724 ± 29[b]	25.0
91–120	(13)	M	69.72 ± 2.62	19.96 ± 0.42	1780 ± 152[a]	48.1
	(13)	F	64.19 ± 1.33	18.23 ± 0.15	1330 ± 58[b]	21.6
121–150	(12)	M	84.02 ± 1.54	21.55 ± 0.30	2931 ± 113[a]	42.1
	(13)	F	79.46 ± 0.86	19.14 ± 0.15	2111 ± 78[b]	23.3
151–180	(9)	M	94.31 ± 2.01	23.05 ± 0.27	3894 ± 180[a]	34.3
	(13)	F	84.17 ± 0.72	20.29 ± 0.18	2452 ± 29[b]	7.7
181–210	(8)	M	96.61 ± 1.72	23.61 ± 0.26	4101 ± 161[a]	25.6
	(10)	F	85.13 ± 0.58	20.29 ± 0.16	2605 ± 81	NA[c]
211–240	(7)	M	98.64 ± 1.81	23.61 ± 0.46	4227 ± 205	NA
	(9)	F	85.90 ± 0.69	20.00 ± 0.14	2579 ± 121	NA
241–270	(7)	M	98.14 ± 1.27	23.03 ± 0.26	3862 ± 219	NA
	(5)	F	85.04 ± 1.14	19.28 ± 0.33	2813 ± 209	NA
271–300	(7)	M	97.94 ± 1.31	22.27 ± 0.54	3510 ± 225	NA
	(5)	F	86.22 ± 0.96	19.62 ± 0.30	2510 ± 144	NA
301–330	(8)	M	97.78 ± 1.01	22.60 ± 0.61	3362 ± 115	NA
	(5)	F	84.70 ± 1.54	19.62 ± 0.46	2504 ± 67	NA
331–360	(8)	M	99.73 ± 0.67	22.73 ± 0.47	3673 ± 84	NA
	(5)	F	84.54 ± 1.82	19.54 ± 0.48	2448 ± 71	NA

[a]Male mean body mass significantly different from surrounding 30-day periods ($P \leq 0.017$).
[b]Female mean body mass significantly different from surrounding 30-day periods ($P \leq 0.013$).
[c]Fishers had reached adult body mass. Differences in body mass was an artifact of the feeding regime.

from the wild were born between 3 March and 1 April (Paragi et al. 1996), and parturition dates of litters in Massachusetts ($n = 21$) ranged from March 4 to March 27 (York 1996). Because photoperiod may synchronize implantation within a population (Mead 1989) and day length varies with latitude, implantation and parturition also may vary with latitude (Powell 1993).

Mean birth dates were similar between wild fishers in Maine (Paragi et al. 1996) and captive fishers. Paragi et al. (1996) assumed that birth had occurred when they obtained 3 consecutive radiotelemetry locations of the mother at the same location, indirectly indicating denning behavior and birth. All litters during our study also were born between 3 March and 1 April, which is consistent with other reported parturition dates.

Mean litter sizes observed across the fisher's geographic range vary between 2.7 and 3.9 (Powell 1993). We observed a mean litter size at birth of 2.7; however, by 7 days post-partum the number of surviving kits had decreased to an average of 2.0 per litter. Many researchers do not know the actual age of a litter when first observed, which could be one reason litter size reported in the literature varies. Paragi et al. (1994) reported a mean litter size of 2.2 juveniles per female that whelped ($n = 4$); however, kits were first observed in the den. Fur farms provide the most information of fisher litter sizes at birth. Hall (1942) reported litter sizes ranging from 1–4 ($\bar{x} = 2.7$, $n = 26$) and Hodgson (1937) reported a single litter with 6 kits. Litters that we observed ranged between 1 and 4 kits per litter; however, the mode was 3. Sex ratios of kits are difficult to interpret from field data because sex ratios are usually determined after birth. Strickland et al. (1982) summarized data on sex ratios from 163 kits born in captivity and concluded that sex ratios did not differ from 1:1. We also observed a sex ratio that did not differ significantly ($P = 0.33$) from 1:1 for 38 kits born in captivity (22 males:16 females).

Sexual dimorphism in body mass is pronounced in adult fishers, with mass of males often twice that of females (Douglas and Strickland 1987). Powell (1993) reported that the male he raised was heavier than the female when they were first weighed at 18 days. However, Coulter (1966) found that a 44 day old male he had in captivity weighed less than a female from the same litter. Based on larger sample sizes, we observed no significant difference in body mass between males and females at birth or at 30 days of age. Powell (1993) hypothesized that growth in male fishers persists longer than for females and evidence from epiphyses fusion in long bones also support a longer period of growth in males (Wright and Coulter 1967, Dagg et al. 1975). Epiphyses in femurs from fishers collected in November were completely fused in females, whereas those in males were only partially fused. Our data for body mass also support the hypothesis that the period of growth in male fishers is of longer duration than for females. In captivity, both sexes would have continued to increase in mass if food were fed *ad libitum* (Frost, unpublished data). Therefore, when fishers reached similar mass to those captured from the wild (males 4.7 kg, $n = 7$; females 2.3 kg, $n = 31$), we reduced their food intake to avoid obesity. Mean (± SE) weights (kg) of juvenile (<1 yr) male ($n = 72$) and female ($n = 39$) fishers trapped during the winter in Ontario were 3.94 ± 0.59 and 2.17 ± 0.29, respectively (Douglas and Strickland 1987). This was similar to weights we observed for captive male (3.51 ± 0.23) and female (2.51 ± 0.14) fishers during winter.

Fishers were altricial from birth to 6 weeks. When first born they can cry, suckle, and defecate, but rely on maternal provisioning for other needs. The first noticeable change in young fishers is when their fur begins to change from a grayish-white to brown at about 23 days. Powell (1993) reported that by 18 days, kits could grasp with their claws, and by 21 days they could crawl. We also noticed kits trying to pull themselves forward 10 days after birth.

Powell (1993) reported that teeth began to break through the gum line at 40 days. We observed the first appearance of teeth at 42 days, with the deciduous premolars erupting first, followed by canines, upper incisors, and lower incisors. Mean ages for eyelids and ear canals to open were 45 and 48 days, respectively, which is within the range of other reported dates for these developments (Coulter 1966, LaBarge et al. 1991, Powell 1993).

Growth in young fishers increased rapidly soon after their eyelids and ear canals opened. Powell (1993) stated that kits were completely dependent on milk for 8–10 weeks (56–70 days), and that he started to feed kits a venison mash at 8 weeks. We could not determine when kits began to eat solid food because their mothers would take food into the nest box, regardless of whether they had kits. Coulter (1966) observed a captive adult female taking meat to kits at 62 days. Coulter (1966) reported difficulty in determining when fishers were weaned; however, he estimated weaning occurred at approximately 4 months, although kits continued to nurse through 114 days. During our study, 1 female escaped and abandoned her 2 female kits when they were 68 days old. Both were subsequently fed a regular diet for adult fishers with no apparent differences in growth from other litters born that year. One litter of 2 males was kept with their mother until 13 November (230 days old) for behavioral observations. After we removed the mother from the kits, the height of her nipples decreased from a mean of 11.6 mm on 9 November to 6.5 mm on 8 December, suggesting that kits may have tried to continue to nurse through 230 days.

Fisher kits in our study began to resemble adult fishers, both in appearance and behavior, soon after their eyelids and ear canals opened. They became more mobile and started moving around the nest box and into the cage. They also started to make the characteristic adult vocalization or chuckling sound. The mean date we observed kits leaving the nest box and moving around in the cage was 70 days. Kits also became more aggressive when food was introduced and agonistic encounters became common after 70 days. Paragi et al. (1996) reported that the mean interval that 25 wild fishers used natal dens was 71 days (range 58–90 days). This was consistent with the patterns of increased mobility and increased aggression we observed in captive fishers at this age, suggesting that fisher kits that we raised developed at a rate similar to kits in the wild. Our observations indicate that much of the physical development of fisher kits occurs in the natal den.

5. ACKNOWLEDGMENTS

This research was funded by the Maine Department of Inland Fisheries and Wildlife (MDIFW) through Federal Aid to Fish and Wildlife Restoration Project W-82-R. Additional support was provided by the Maine Cooperative Fish and Wildlife Research Unit (Department of the Interior, United States Geological Survey, University of Maine, and MDIFW, cooperating), and the Department of Ecology, University of Maine. This is publication number 2648 of the Maine Agricultural and Forest Experiment Station. We thank D. J. Harrison, C. R. Wallace, K. D. Elowe, and H. C. Gibbs for reviewing earlier versions of this manuscript and for all the students who helped feed and maintain the fishers in captivity. This research was done in accordance with an approved animal welfare protocol at the University of Maine.

6. LITERATURE CITED

Arthur, S. M., R. A. Cross, T. A. Paragi, and W. B. Krohn. 1992. Precision and utility of cementum annuli for estimating ages of fishers. Wildlife Society Bulletin 20:402–405.

Coulter, M. W. 1966. Ecology and management of fishers in Maine. Dissertation, State University College of Forestry, Syracuse University, Syracuse, New York, USA.

Dagg, A. I., D. Leach, and G. Sumner-Smith. 1975. Fusion of the distal femoral epiphyses in male and female marten and fisher. Canadian Journal of Zoology 53:1514–1518.

Douglas, W. O. 1943. Fisher farming has arrived. American Fur Breeder 16:18–2.

Douglas, C. W., and M. A. Strickland. 1987. Fisher. Pages 512–529 *in* M. Novak, M. E. Obbard, and B. Malloch, editors. Wild furbearer management and conservation in North America. Ontario Trappers Association, Ontario Ministry of Natural Resources.

Frost, H. C., and W. B. Krohn. 1994. Capture, care, and handling of fishers (*Martes pennanti*). Maine Agricultural and Forest Experiment Station, Technical Bulletin No. 157. University of Maine, Orono, Maine, USA.

Gittleman, J. L. 1986. Carnivore life history patterns: allometric, phylogenetic, and ecological associations. American Naturalist 127:744–771.

Hall, E. R. 1942. Gestation period in the fisher with recommendations for the animal's protection in California. California Game and Fish 28:143–147.

Hodgson, R. G. 1937. Fisher farming. Fur Trade Journal. Toronto, Canada. 104pp.

James, H. 1934. Fisher for the fur farmer. American Fur Breeder 5:6–7.

LaBarge, T., A. Baker, and D. Moore. 1991. Fisher (*Martes pennanti*): birth, growth, and development in captivity. Mustelid and Viverrid Conservation Newsletter. IUCN/SCC. Mustelid and Viverrid Specialist Group. Belgium 2:1–3.

Mead, R. A. 1989. The physiology and evolution of delayed implantation in carnivores. Pages 437–464 *in* J. L. Gittleman, editor. Carnivore behavior, ecology, and evolution. Cornell University Press, Ithaca, New York, USA.

Paragi, T. F., W. B. Krohn, and S. M. Arthur. 1994. Using estimates of fisher recruitment and survival to evaluate population trend. Northeast Wildlife 51:1–11.

____, ____, and S. M. Arthur. 1996. Importance of tree cavities as natal dens for fishers. Northern Journal of Applied Forestry 13:79–83.

Powell, R. A. 1993. The fisher: life history, ecology, and behavior. Second edition. University of Minnesota Press, Minneapolis, Minnesota, USA.

Ricklefs, R. E. 1967. A graphical method of fitting equations to growth curves. Ecology 48: 978–983.

Seton, E. T. 1937. The fisher. Pages 451–479 *in* E. T. Seton, editor. Lives of game animals. Volume 2. The Literacy Guild of America, New York, New York, USA.

Strickland, M. A., C. W. Douglas, M. Novak, and N. P. Hunziger. 1982. Fisher. Pages 586–598 *in* J. A. Chapman and G. A. Feldhamer, editors. Wild mammals of North America: biology, management, and economics. John Hopkins University Press, Baltimore, Maryland, USA.

Thomassen, O. K. 1940. Fisher in captivity. Canadian Silver Fox and Fur (February issue).

Wright, P. L., and M. W. Coulter. 1967. Reproduction and growth in Maine fishers. Journal of Mammalogy 31:70–87.

York, E. C. 1996. Fisher population dynamics in north-central Massachusetts. Thesis, University of Massachusetts, Amherst, Massachusetts, USA.

Zar, J. H. 1984. Biostatistical analysis. Second edition. Prentice Hall, Englewood Cliffs, New Jersey, USA.

Chapter 14

FIELD ANESTHESIA OF AMERICAN MARTENS USING ISOFLURANE

François Potvin, Laûrier Breton, and Robert Patenaude

***Abstract*:** Isoflurane was used in the field for the immobilization and attachment of radio collars on American martens (*Martes americana*) (n = 108 trials on 91 animals). Martens were captured in Tomahawk live traps and the traps were placed in a plexiglass box for induction with isoflurane. Martens were given 3 ml (78 trials) or 4 ml (30 trials) of isoflurane, a dose that produced induction in 93 trials. A second dose (2 ml) was needed to reach anesthesia in 15 trials. Induction time after the first dose did not differ between doses, sexes, and age classes (99 ± 5 sec [SE], range = 33–288 sec, n = 98). Recovery time for animals receiving a single dose was similar for 3 or 4 ml doses, males or females, and juveniles or adults (215 ± 17 sec, range = 30–580 sec, n = 62). Only 1 animal died while being handled. Based on telemetry data, ≥82% of the collared animals that we released survived >30 days after anesthesia. During this period, most mortalities that may have been associated with handling and collaring (6 cases) involved smaller animals and did not appear to be related to chemical anesthesia. We conclude that isoflurane is a safe and efficient drug for immobilizing marten with simple equipment when a short handling time (2–3 min) is required.

1. INTRODUCTION

Safe and efficient techniques are needed for immobilizing American marten (*Martes americana*) in the field because this medium size carnivore is very fast and has powerful teeth and claws. For ear tagging, a wire handling cone can be used to physically restrain the animal (Day et al. 1980, Archibald and Jessup 1984, Bull et al. 1996), while putting the tag in place through the wires. However, attaching radio collars to martens requires chemical anesthesia. Anesthetics for mustelids include injectable drugs, such as ketamine hydrochloride (often combined with xylazine) and volatile anesthetics, such as chloroform, ether, halothane, isoflurane, and methoxyflurane (Herman et al. 1982,

Davis 1983, Archibald and Jessup 1984, Seal and Kreeger 1987, Arthur 1988, Belant 1992, Arnemo et al. 1994, Bull et al. 1996, Mitcheltree et al. 1999).

Volatile anesthetics provide a short handling time that may be sufficient for ear tagging and collaring. Because of a faster recovery time, anesthetics may have an advantage over injectable drugs by enabling handling more animals during a given time. Volatile anesthetics might also be safer for the animal because the faster recovery decreases the risk of predation during this critical period. Halothane has been used to anethesize muskrats (*Odontra zibethicus*), striped skunks (*Mephitis mephitis*), and minks (*Mustela vison*) (Blanchette 1989, Larivière and Messier 1996, Larivière et al. 2000), and isoflurane has been used to anesthesize black-footed ferrets (*Mustela nigripes*) (Kreeger et al. 1998). Halothane, isoflurane, and methoxyflurane were also tested to restrain Arctic ground squirrels (*Spermophilus parryii*) (McColl and Boonstra 1999).

During a study on the response of martens to large clearcuts in western Québec (Potvin and Breton 1997, Potvin et al. 2000), we used isoflurane to anesthesize animals in the field to attach radiocollars. This paper presents the results of our work with isoflurane.

2. MATERIAL AND METHODS

Isoflurane (Bimeda MTC Animal Health Inc., Cambridge, Ontario) is a nonflammable, nonexplosive general anesthetic agent used for induction and maintenance of general anesthesia. In Canada, this product is a prescription drug and is normally obtained with a prescription from a licensed practitioner. The major advantages of this product are that is produces rapid induction, and recovery from anesthesia is rapid. Like other volatile anesthetics, it lacks analgesic properties. If used during painful procedures, the use of supplemental analgesic products should be considered. This product has a large safety margin for the circulatory system, but causes respiratory depression, making necessary the monitoring of breathing. The ideal situation for the utilization of this product is to use an anesthesia machine modified for field conditions to reduce the potential risks involved when using an open drop method, which can produce toxic concentrations, and possibly hypoxic conditions in the anesthesized animal.

Martens were captured in the fall (end of August to early December) from 1990 to 1993 with Tomahawk live traps 202 (15 × 15 × 48 cm) (Tomahawk Live Trap Co., Tomahawk, WI). Traps were attached with nails on logs inclined 30–45°, between 1.0–1.5 m from the ground, and were covered with moss and coniferous branches to minimize heat loss in rain or cold weather.

Traps were checked daily. To anesthesize the animal, the trap was placed in a plexiglass box (18 × 18 × 50 cm) and isoflurane (3 or 4 ml) was injected in liquid form with a syringe through a small hole (1 mm diameter) drilled at a 45° angle on the back of the box. A heavy gauge clear plastic bag (46 × 81 cm) was used in a few cases when the trap was distant from a road or a water course. The transparent box and bag allowed observation until martens became recumbent. A second dose (2 ml) was administered if the animal was not anesthesized after 2–3 min.

A 2-person crew was required to handle the animal, with one person physically restraining the marten, while the second person did the measurements and attachments. Anesthesized martens were removed from the cage, weighed, ear-tagged and radiocollared. Two types of radio collars were used, Lotek SMRC-5 (Lotek Engineering Inc., Newmarket, Ontario) weighing 40–42 g and Holohil MI-2 (Holohil Systems Ltd., Woodlawn, Ontario) weighing 31 g for females and 38–40 g for males. Collars had a mortality option that doubled the pulse rate after 4 hr of inactivity. When time permitted, a first upper premolar was extracted to facilitate aging. If the animal recovered before handling was completed, a supplemental dose was given in some trials by applying a can (5 cm diameter) containing a piece of absorbant cotton ball containing a small amount of gas (about 1 ml) over the muzzle of the animal. Induction time was defined as the interval between injection and lateral or sternal recumbency. Recovery time was defined as the interval between recumbency and the moment when the animal stood up when released. Induction times and recovery times between doses, age groups, and sexes were tested with a 1-way analysis of variance (ANOVA).

We used telemetry data to determine survivorship of released martens. Animals were located from an aircraft twice or more during the first month and then at least on a monthly basis. Collars that transmitted pulse rates indicating potential mortality were checked on the ground as soon as possible to determine the status and cause of death of the martens. Carcasses were brought to the lab for autopsy.

3. RESULTS

We performed 108 anesthesia trials on 91 martens (54 males, 37 females). All trials successfully immobilized the animal. Martens were given a first dose by injecting 3 ml of isoflurane in the box in 78 trials and 4 ml in 30 trials. This single dose was sufficient in 93 trials (64 with 3 ml, 29 with 4 ml) and a second injection (2 ml) was administered in 15 trials. To maintain anesthesia, we gave supplemental isoflurane (1 ml on a cotton ball in the can) on 11 trials (10 single injection trials, 1 double injection trial). Only one animal, a large male, died while being handled. The autopsy of this animal indicated that it had broken its right humerus in the cage before being handled and had probably died of acute shock.

Males were almost 50% heavier than females (Table 14.1). On average, adult males were 8% heavier than juvenile males (969 vs. 889 g) but adult females were no heavier than juvenile females (607 vs. 638 g). Data on induction and recovery times are complete for 79 trials and partial for 29 trials (Table 14.2). Induction time after the first dose did not differ between doses (100 ± 6 sec for 3 ml, 92 ± 6 sec for 4 ml, $P = 0.45$. Induction time was not significantly different between juveniles and adults within each sex ($P \geq 0.52$), or between males and females when all ages were combined ($P = 0.47$) (Fig. 14.1). Mean induction time was 99 ± 5 sec (SE) ($n = 98$) after the first dose and 99 ± 25 sec ($n = 8$) after the second dose.

Recovery time for animals receiving a single dose was similar for both doses (208 ± 22 sec for 3 ml, 231 ± 28 sec for 4 ml, $P = 0.57$), juveniles or adults within each sex ($P \geq 0.51$), and between males or females ($P = 0.91$) (Fig. 14.2, Table 14.2). Mean recovery time for these animals was 215 ± 17 sec ($n = 62$) after induction. Administering a supplemental dose with the can extended recovery time by approximately 100 sec. The recovery time of animals that received 2 injections was on average 272 sec. When the animal awoke, it slowly began moving its legs and head. This movement gradually increased and it took about 1 min for martens to achieve complete recovery. During that phase, the animal could still be physically restrained to complete tagging and collaring, but tooth removal was not possible. We were able to collect a first premolar on 35 of the 91 handled martens. On average, anesthesized martens could be handled for 2–3 min.

Table 14.1. Weights of martens anesthetized with isoflurane.

Sex	Age class	n [a]	Weight (g) $\bar{x}$	SE	Range
Males	Juveniles	17	889	32	600–1,100
	Adults	19	969	22	800–1,100
	Juveniles, adults, and unknown	64	937	18	550–1,550
Females	Juveniles	19	638	22	490–850
	Adults	5	607	20	550–675
	Juveniles, adults, and unknown	40	632	12	550–810

[a] Total number (104) exceeds the number of animals in the study because some animals were captured and weighed twice, over different years.

Figure 14.1. Induction time (± SE) after the first injection of martens anesthetized with isoflurane (3–4 ml), by sex and age group.

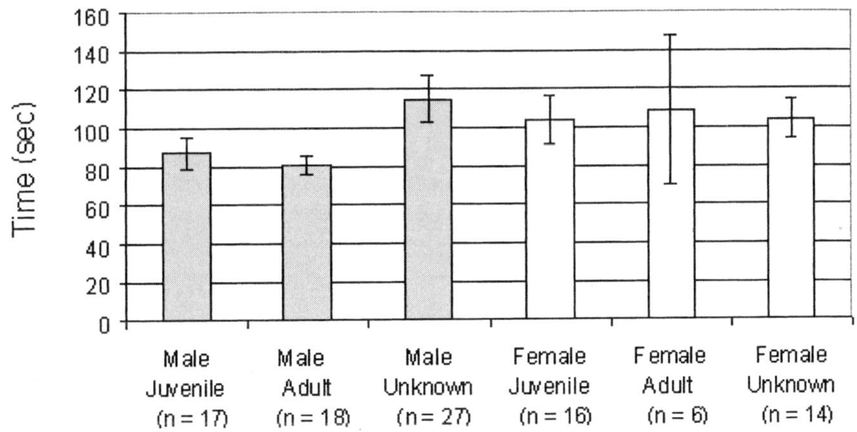

Figure 14.2. Recovery time (± SE) after induction of martens anesthetized with a single injection of isoflurane (3–4 ml, with no supplemental dose), by sex and age group.

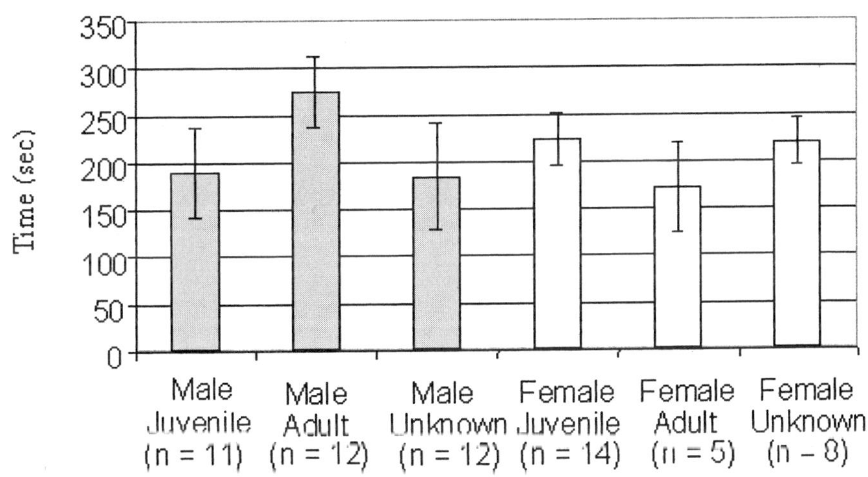

Two small females (≤550 g) were not radiocollared. We kept telemetry contact with 85 of the 88 collared animals. Based on telemetry data, ≥70 martens survived over 30 days after anesthesia. The fate of 4 animals is unknown because only the collar was found (3 cases) or the collar was underground and could not be reached (1). These animals either lost their collar or had died. The

Table 14.2. Induction times and recovery times for male and female martens anesthetized using isoflurane.

	Males				Females			
	n^a	$\bar{x}$	SE	Range	n^a	$\bar{x}$	SE	Range
Induction time (sec)								
First injection (3–4 ml)	62	97	6	40–288	36	102	8	33–270
Second injection (2 ml)	5	104	40	20–235	3	89	16	60–117
Recovery time (sec)								
One injection								
No supplemental dose	35	217	28	30–580	27	213	18	41–465
Supplemental dose	9	320	64	106–740	1	330		
Two injections	6	276	82	115–660	1	250		

[a]Total count for induction time and recovery time is lower than the number of trials (108) because of partial data for 29 trials.

11 documented mortality cases (5 males, 6 females) were caused by avian predation (1), distemper (1), trapping (1), hypothermia (1), unknown cause (1), and, possibly, stress related to handling or collaring (6). The marten that died from hypothermia was wet and shivering in the trap before handling. The 6 mortalities attributed to stress generally involved animals weighing <600 g (5 cases) and carried heavier collars (40 g) (4 cases). The sex and age class of those 6 mortalities is as follows: 2 males (1 juvenile, 1 unknown age) and 4 females (3 juveniles, 1 unknown).

4. DISCUSSION

Induction time (<2 min) with isoflurane was very short and similar to induction times reported for marten using ketamine-xylazine mixtures (Archibald and Jessup 1984, Belant 1992). Using medetomidine-ketamine, Arnemo et al. (1994) had slightly longer induction times (3.8–4.5 min). Typical recovery time with isoflurane was 3–4 min, as compared to 10–100 min with ketamine-xylazine (Archibald and Jessup 1984, Belant 1992, Bull et al. 1996). Recovery time can take 1–4 hr with medetomidine-ketamine, but may be reduced if atipamezole is provided as an antagonist (Arnemo et al. 1994). Recovery times reported by Herman et al. (1982) using halothane administered in a portable chamber were slightly shorter (3 min) than the recovery time that we observed using isoflurane.

Most animals immediately ran or jumped when released following anesthesia with isoflurane, suggesting that they had recovered completely. This behavior is quite different from the traumatic recovery of American minks anesthesized with halothane reported by Larivière et al. (2000). We lost only 1 marten during handling (mortality rate <1%). Further, based on telemetry data, ≥82% of the 88 collared animals survived over 30 days following their release. Six fatalities during this period may have been associated with handling and collaring. Most of these involved smaller animals carrying heavier collars. During the first 2 years of our study, mortality was abnormaly high for females carrying 40 g collars; with lighter collars (31 g), this rate became comparable to that of males in the 2 following years (Potvin and Breton 1997). Therefore, we suggest that these 6 mortalities were probably related to collaring rather than chemical anesthesia. Handling fatalities can be minimized under cold and rainy weather conditions by using procedures described by Bull et al. (1996), such as adding a protective wood box to the trap and covering it with a heavy gauge black plastic sheet.

The relative cost of each product for 1 anesthesia is $0.72 for isoflurane (4 ml) and $0.55 for ketamine-xylazine (60 mg/kg ketamine + 12 mg/kg xylazine [Kreeger 1996:186]) for a 900 g animal. Two persons are required to handle martens with isoflurane, but only 1 person is needed for handling when using a cone and injectable drugs. In spite of that, more animals can be handled during a given time with isoflurane because recovery is faster than with injectable drugs. We conclude that isoflurane is a safe and efficient drug for immobilizing marten with simple equipment when a short handling time is required (2–3 min). This period is long enough for attaching a radio collar, but may be too short if removing a tooth or taking blood samples. Larivière and Messier (1996) recommended halothane for short manipulations on skunks (about 1 min). When a longer handling time is needed, a portable anesthesia system (Kreeger et al. 1998) or injectable drugs should be used. For Arctic ground squirrels, McColl and Boonstra (1999) prefered methoxyflurane over isoflurane or halothane when a portable anesthesia system was not available, because of a lower potential of overdose. Methoxyflurane has not been tested and reported for use on martens under field situations.

5. LITERATURE CITED

Archibald, W. R., and R. H. Jessup. 1984. Population dynamics of the pine marten (*Martes americana*) in the Yukon Territory. Pages 81–97 *in* R. Olson, R. Hastings, and F. Geddes, editors. Northern ecology and resource management. University of Alberta Press, Edmonton, Alberta, Canada.

Arnemo, J. M., R. O. Moe, and N. E. Soli. 1994. Immobilization of captive pine marten (*Martes martes*) with medetomidine-ketamine and reversal with atipamezole. Journal of Zoological Wildlife Medicine 25:548–534.

Arthur, S. M. 1988. An evaluation of techniques for capturing and radiocollaring fishers. Wildlife Society Bulletin 16:417–421.

Belant, J. L. 1992. Field immobilization of American martens (*Martes americana*) and short-tailed weasels (*Mustela erminea*). Journal of Wildlife Disease 28:662–665.

Blanchette, P. 1989. Use of halothane to anesthetize muskrats in the field. Journal of Wildlife Management 53:172–174.

Bull, E. L., T. W. Heather, and F. G. Culver. 1996. Live-trapping and immobilizing American martens. Wildlife Society Bulletin 24:555–558.

Davis, M. H. 1983. Post-release movements of introduced marten. Journal of Wildlife Management 47:59–66.

Day, G. I., S. D. Shemnitz, and R. D. Taber. 1980. Capturing and marking wild animals. Pages 61–98 *in* S. D. Shemnitz, editor. Wildlife management techniques manual. Fourth edition. The Wildlife Society, Washington.

Herman, M. F., J. F. Pepper, and L. A. Herman. 1982. Field and laboratory techniques for anesthezing marten with halothane gas. Wildlife Society Bulletin 10:275–277.

Kreeger, T. J. 1996. Handbook of wildlife chemical immobilization. Wildlife Pharmaceuticals, Inc., Fort Collins, Colorado, USA.

Kreeger, T. J., A. Vergas, G. E. Plumb, and T. E. Thorne. 1998. Ketamine-metedomidine or isoflurane immobilization of black-footed ferrets. Journal of Wildlife Management 62:654–662.

Larivière, S., and F. Messier. 1996. Field anesthesia of striped skunks, *Mephitis mephitis*, using halothane. Canadian Field-Naturalist 110:703–705.

_____, L. R. Walton, and J. A. Virgl. 2000. Field anesthesia of American mink, *Mustela vison*, using halothane. Canadian-Field Naturalist 114:142–144.

McColl, C. J., and R. Boonstra. 1999. Physiological effects of three inhalant anesthetics on Arctic ground squirrels. Wildlife Society Bulletin 27:946–951.

Mitcheltree, D. H., T. L. Serfass, W. M. Tzilkowski, R. L. Peper, M. T. Whary, and R. P. Brooks. 1999. Physiological responses of fishers to immobilization with ketamine, ketamine-xylazine, or Telazol. Wildlife Society Bulletin 27:582–591.

Potvin, F., and L. Breton. 1997. Short-term effects of clearcutting on martens and their prey in the boreal forest of western Quebec. Pages 452–474 *in* G. Proulx, H. N. Bryant, and P. M. Woodward, editors. *Martes*: taxonomy, ecology, techniques, and management. Proceedings of the Second International *Martes* Symposium. The Provincial Museum of Alberta, Edmonton, Alberta, Canada.

Potvin, F., L. Bélanger, and K. Lowell. 2000. Marten habitat selection in a clearcut boreal landscape. Conservation Biology 14:844–857.

Seal, U. S., and T. J. Kreeger. 1987. Chemical immobilization of furbearers. Pages 191–215 *in* M. Novak, J. A. Baker, M. E. Obbard, and B. Mallock, editors. Wild furbearer management and conservation in North America. Ontario Ministry of Natural Resources, Ontario Trappers Association, North Bay, Ontario, Canada.

INDEX

A

B

C

D

L

M

N